Yamaha
YZ & WR 4-Stroke
Motocross & Off-road Bikes
Service and Repair Manual

by Alan Ahlstrand

Models covered
YZ250F, 2001 thru 2008
YZ400F, 1998 and 1999
YZ426F, 2000 thru 2002
YZ450F, 2003 thru 2008
WR250F, 2001 thru 2008
WR400F, 1998 thru 2000
WR426F, 2001 and 2002
WR450F, 2003 thru 2008

2689-9T2

ISBN-13: 978-1-62092-215-6
ISBN-10: 1-62092-215-0

Library of Congress Control Number 2008920336

Haynes Group Limited
Haynes North America, Inc.
www.haynes.com

08-256

Contents

LIVING WITH YOUR YAMAHA

Introduction

Daily (pre-ride) checks

MAINTENANCE

Routine maintenance and servicing

Contents

REPAIRS AND OVERHAUL

Engine, transmission and associated systems

Chassis and bodywork components

Wiring diagrams

REFERENCE

Index

Yamaha
Musical instruments to Motorcycles

The Yamaha Motor Company

The Yamaha name can be traced back to 1889, when Torakusu Yamaha founded the Yamaha Organ Manufacturing Company. Such was the success of the company, that in 1897 it became Nippon Gakki Limited and manufactured a wide range of reed organs and pianos.

During World War II, Nippon Gakki's manufacturing base was utilised by the Japanese authorities to produce propellers and fuel tanks for their aviation industry. The end of the war brought about a huge public demand for low cost transport and many firms decided to utilise their obsolete aircraft tooling for the production of motorcycles. Nippon Gakki's first motorcycle went on sale in February 1955 and was named the 125 YA-1 Red Dragonfly. This machine was a copy of the German DKW RT125 motorcycle, featuring a single cylinder two-stroke engine with a four-speed gearbox. Due to the outstanding success of this model the motorcycle operation was separated from Nippon Gakki in July 1955 and the Yamaha Motor Company was formed.

The YA-1 also received acclaim by winning two of Japan's biggest road races, the Mt. Fuji Climbing race and the Asama Volcano race. The high level of public demand for the YA-1 led to the development of a whole series of two-stroke singles and twins.

Having made a large impact on their home market, Yamahas were exported to the USA in 1958 and to the UK in 1962. In the UK the signing of an Anglo-Japanese trade agreement during 1962 enabled the sale of Japanese lightweight motorcycles and scooters in Britain. At that time, competition between the many motorcycle producers in Japan had reduced numbers significantly and by the end of the sixties, only the big-four which are familiar with today remained.

Yamaha Europe was founded in 1968 and based in Holland. Although originally set up to market marine products, the Dutch base is now the official European Headquarters and distribution centre. Yamaha motorcycles are built at factories in Holland, Denmark, Norway, Italy, France, Spain and Portugal. Yamahas are imported into the UK by Yamaha Motor UK Ltd, formerly Mitsui Machinery Sales (UK) Ltd. Mitsui and Co. were originally a trading house, handling the shipping, distribution and marketing of Japanese products into western countries. Ultimately Mitsui Machinery Sales was formed to handle Yamaha motorcycles and outboard motors.

Based on the technology derived from its motorcycle operation, Yamaha have produced many other products, such as automobile and lightweight aircraft engines, marine engines and boats, generators, pumps, ATVs, snowmobiles, golf cars, industrial robots, lawnmowers, swimming pools and archery equipment.

Two-strokes first

Part of Yamaha's success was a whole string of innovations in the two-stroke world. Autolube engine lubrication, pressed steel monocoque frame, electric starting, torque induction, multi-ported engines, reed valves and power valves kept their two-strokes at the forefront of technology.

In the 1960s and 70s the two-stroke engined YAS3 125, YDS1 to YDS7 250 and YR5 350 formed the core of Yamaha's range. By the mid-70s they had been superseded by the RD (Race-Developed) 125, 250, and 350 range of two-stroke twins, featuring improved 7 port engines with reed valve induction. Braking was improved by the use of an hydraulic brake on the front wheel of DX models, instead of the drum arrangement used previously, and cast alloy wheels were available as an option on later RD models. The RD350 was replaced by the RD400 in 1976.

Running parallel with the RD twins was a range of single-cylinder two-strokes. Used

The FS1-E - first bike of many sixteen year olds in the UK

in a variety of chassis types, the engine was used in the popular 50 cc FS1-E moped, the V50 to 90 step-thrus, RS100 and 125, YB100 and the DT trail range.

The air-cooled single and twin cylinder RD models were eventually replaced by the LC series in 1980, featuring liquid-cooled engines, radical new styling, spiral pattern cast wheels and cantilever rear suspension (Yamaha's Monoshock). Of all the LC models, the RD350LC, or RD350R as it was later known, has made the most impact in the market. Later models had YPVS (Yamaha Power Valve System) engines, another first for Yamaha - this was essentially a valve located in the exhaust ports which was electronically operated to alter port timing to achieve maximum power output. The RD500LC was the largest two-stroke made by Yamaha and differed from the other LCs by the use of its vee-four cylinder engine.

With the exception of the RD350R, now manufactured in Brazil, the LC range has been discontinued. Two-stroke engined models have given way to environmental pressure, and thus with a few exceptions, such as the TZR125 and TZR250, are used only in scooters and small capacity bikes.

The Four-strokes

Yamaha concentrated solely on two-stroke models until 1970 when the XS1 was produced, their first four-stroke motorcycle. It was perhaps Yamaha's success with two-strokes that postponed an earlier move into the four-stroke motorcycle market, although their work with Toyota during the 1960s had given them a sound base in four-stroke technology.

The distinctive paintwork and trim of the RD models

The XS1 had a 650 cc twin-cylinder SOHC engine and was later to become known as the XS650, appearing also in the popular SE custom form. Yamaha introduced a three cylinder 750 cc engine in 1976, fitted in a sport tourer frame and called the XS750, TX750 in the USA. The XS750 established itself well in the sport tourer class and remained in production with very few changes until uprated to 850 cc in 1980.

Other four-strokes followed in 1976, with the introduction of the XS250/360/400 series twins. The XS range was strengthened in 1978 by the four-cylinder XS1100.

The 1980s saw a new family of four-strokes, the XJ550, 650, 750 and 900 Fours. Improvements over the XS range amounted to a slimmer DOHC engine unit due to the relocation of the alternator behind the cylinders, electronic ignition and uprated braking and suspension systems. Models were available mainly in standard trim, although custom-styled Maxims were produced especially for the US market. The XJ650T was the first model from Yamaha to have a turbo-charged engine. Although these early XJ models have now been discontinued, their roots live on in the XJ600S and XJ900S Diversion (Seca II) models.

The FZR prefix encompasses the pure sports Yamaha models. With the exception of the 16-valve FZR400 and FZR600 models, the FZ/FZR750 and FZR1000 used 20-valve engines, two exhaust valves and three inlet valves per cylinder. This concept was called Genesis and gave improved gas flow to the combustion chambers. Other features of the new engine were the use of down-draught carburetors and the engine's inclined angle in the frame, plus the change to liquid-cooling. Lightweight Deltabox design aluminium frames and uprated suspension improved the bikes's handling. The Genesis engine lives on in the YZF750 and 1000 models.

The vee-twin engine has been the mainstay of the XV Virago range. Since 1981 XVs have been produced in 535, 700, 750, 920, 1000 and 1100

The XS650 led the way for Yamaha's four-stroke range

Yamaha's XS750 was produced from 1976 to 1982 and then uprated to 850 cc

engine sizes, all using the same basic air-cooled sohc vee-twin engine. Other uses of vee engines have been in the XZ550 of the early 1980s, the XVZ12 Venture and the mighty VMX-12 V-Max.

Anti-lock braking, engine management and catalytic converters are all features found on present-day models, ensuring that Yamaha remains at the forefront of technology.

Yamaha's entry into the field of four-stroke off-road competition came in 1998 with the introduction of the YZ400F motocrosser and its companion, the WR400F enduro bike. The WR was basically the same as the YZ, but with lights and wide-ratio transmission gears (hence the WR designation). Since the introduction, the four-strokes have enjoyed

growing success, to the point that they will probably replace two-stroke motocross and enduro bikes entirely within a few years.

The engine uses Yamaha's unique five-valve cylinder head (three intake valves and two exhausts). A flat-slide carburetor with throttle position sensor delivers the fuel. Power transmission is handled by a five-speed transmission in all models.

The YZ's displacement was increased from 399 cc to 426 cc with the 2000 model year (YZ426FM). The displacement increase was achieved by enlarging the cylinder bore from 92 to 95 millimeters. The WR received the same displacement increase a year later. In 2003, the YZ and WR were both increased to 450 cc with a stroke increase from 60.1 to 63.4 millimeters.

The 2000 model year saw the introduction of the YZ250 and WR250. The new, smaller-displacement engine used the same five-valve configuration as its bigger siblings.

The engine in all models uses a dry-sump lubrication system, where the engine oil is stored in a separate tank from the crankcase. On 2005 and earlier YZ400/450 models, as well as 2006 and earlier WR400/450 models, the oil tank is in the bike's frame. On 2006 and later WR450 models, as well as 2007 and later WR450 models, the oil tank is within the front of the crankcase, but the oil is kept separate from the spinning crankshaft by an interior crankcase wall.

The lubrication system on 2005 and earlier YZ250 models uses a frame oil tank. For 2006, the frame oil tank in the YZ250 was replaced by a separate oil tank mounted at the lower front of the bike between the frame members. The WR250 received this design change in the 2007 model year.

The front suspension consists of forks with an advanced design; cartridge forks are used in all models, with a change to twin-chamber forks in the 2006 YZ models and 2007 WR models. The rear suspension is a progressive rising rate design, which stiffens as suspension travel increases. This allows a fast response to minor unevenness in the terrain, along with good control over large bumps and jumps. As could be expected in a race bike, the front and rear suspensions include a full complement of adjustments for stiffness and damping.

All models are equipped with a kick starter. 2003 and later WR models also use an electric starter.

WR models beginning with the 2006 model year are equipped with an enduro meter, which Yamaha refers to as a multi-function display. This highly sophisticated meter has two settings, basic and race mode. In basic mode, it indicates the bike's speed, the time of day, distance traveled (on two tripmeters) and changes in tire diameter. In race mode, it indicates elapsed time, distance traveled, average speed and changes in tire diameter.

A new family of four-strokes was released in 1980 with the introduction of the XJ range

Acknowledgements

Our thanks are due to Grand Prix Sports of Santa Clara, California, for supplying the motorcycles used in the photographs throughout this manual; Tony Correa, service director, for arranging the facilities and fitting the project into his shop's busy schedule; and Craig Wardner, service technician, for doing the mechanical work and providing valuable technical information.

About this manual

The aim of this manual is to help you get the best value from your motorcycle. It can do so in several ways. It can help you decide what work must be done, even if you choose to have it done by a dealer; it provides information and procedures for routine maintenance and servicing; and it offers diagnostic and repair procedures to follow when trouble occurs.

We hope you use the manual to tackle the work yourself. For many simpler jobs, doing it yourself may be quicker than arranging an appointment to get the motorcycle into a dealer and making the trips to leave it and pick it up. More importantly, a lot of money can be saved by avoiding the expense the shop must pass on to you to cover its labor and overhead costs. An added benefit is the sense of satisfaction and accomplishment that you feel after doing the job yourself.

References to the left or right side of the motorcycle assume you are sitting on the seat, facing forward.

We take great pride in the accuracy of information given in this manual, but motorcycle manufacturers make alterations and design changes during the production run of a particular motorcycle of which they do not inform us. No liability can be accepted by the authors or publishers for loss, damage or injury caused by any errors in, or omissions from, the information given.

Engine and frame numbers

The frame serial number is stamped into the right side of the steering head. The engine number is stamped into the top rear of the crankcase and is visible from the right side of the machine. Both of these numbers should be recorded and kept in a safe place so they can be given to law enforcement officials in the event of a theft.

The frame serial number and engine serial number should also be kept in a handy place (such as with your driver's license) so they are always available when purchasing or ordering parts for your machine.

Buying spare parts

Once you have found all the identification numbers, record them for reference when buying parts. Since the manufacturers change specifications, parts and vendors (companies that manufacture various components on the machine), providing the ID numbers is the only way to be reasonably sure that you are buying the correct parts.

Whenever possible, take the worn part to the dealer so direct comparison with the new component can be made. Along the trail from the manufacturer to the parts shelf, there are numerous places that the part can end up with the wrong number or be listed incorrectly.

The two places to purchase new parts for your motorcycle – the accessory store and the franchised dealer – differ in the type of parts they carry. While dealers can obtain virtually every part for your motorcycle, the accessory dealer is usually limited to normal high wear items such as shock absorbers, tune-up parts, various engine gaskets, cables, chains, brake parts, etc. Rarely will an accessory outlet have major suspension components, cylinders, transmission gears, or cases.

Used parts can be obtained for considerably less than new ones, but you can't always be sure of what you're getting. Once again, take your worn part to the salvage yard for direct comparison.

Whether buying new, used or rebuilt parts, the best course is to deal directly with someone who specializes in parts for your particular make.

The engine number is stamped into the back of the crankcase

The frame serial number is stamped into the right side of the steering head

Professional mechanics are trained in safe working procedures. However enthusiastic you may be about getting on with the job at hand, take the time to ensure that your safety is not put at risk. A moment's lack of attention can result in an accident, as can failure to observe simple precautions.

There will always be new ways of having accidents, and the following is not a comprehensive list of all dangers; it is intended rather to make you aware of the risks and to encourage a safe approach to all work you carry out on your bike.

Asbestos

● Certain friction, insulating, sealing and other products - such as brake pads, clutch linings, gaskets, etc. - contain asbestos. Extreme care must be taken to avoid inhalation of dust from such products since it is hazardous to health. If in doubt, assume that they do contain asbestos.

Fire

● Remember at all times that gasoline is highly flammable. Never smoke or have any kind of naked flame around, when working on the vehicle. But the risk does not end there - a spark caused by an electrical short-circuit, by two metal surfaces contacting each other, by careless use of tools, or even by static electricity built up in your body under certain conditions, can ignite gasoline vapor, which in a confined space is highly explosive. Never use gasoline as a cleaning solvent. Use an approved safety solvent.

● Always disconnect the battery ground terminal before working on any part of the fuel or electrical system, and never risk spilling fuel on to a hot engine or exhaust.

● It is recommended that a fire extinguisher of a type suitable for fuel and electrical fires is kept handy in the garage or workplace at all times. Never try to extinguish a fuel or electrical fire with water.

Fumes

● Certain fumes are highly toxic and can quickly cause unconsciousness and even death if inhaled to any extent. Gasoline vapor comes into this category, as do the vapors from certain solvents such as trichloro-ethylene. Any draining or pouring of such volatile fluids should be done in a well ventilated area.

● When using cleaning fluids and solvents, read the instructions carefully. Never use materials from unmarked containers - they may give off poisonous vapors.

● Never run the engine of a motor vehicle in an enclosed space such as a garage. Exhaust fumes contain carbon monoxide which is extremely poisonous; if you need to run the engine, always do so in the open air or at least have the rear of the vehicle outside the workplace.

The battery

● Never cause a spark, or allow a naked light near the vehicle's battery. It will normally be giving off a certain amount of hydrogen gas, which is highly explosive.

● Always disconnect the battery ground terminal before working on the fuel or electrical systems (except where noted).

● If possible, loosen the filler plugs or cover when charging the battery from an external source. Do not charge at an excessive rate or the battery may burst.

● Take care when topping up, cleaning or carrying the battery. The acid electrolyte, even when diluted, is very corrosive and should not be allowed to contact the eyes or skin. Always wear rubber gloves and goggles or a face shield. If you ever need to prepare electrolyte yourself, always add the acid slowly to the water; never add the water to the acid.

Electricity

● When using an electric power tool, inspection light etc., always ensure that the appliance is correctly connected to its plug and that, where necessary, it is properly grounded. Do not use such appliances in damp conditions and, again, beware of creating a spark or applying excessive heat in the vicinity of fuel or fuel vapor. Also ensure that the appliances meet national safety standards.

● A severe electric shock can result from touching certain parts of the electrical system, such as the spark plug wires (HT leads), when the engine is running or being cranked, particularly if components are damp or the insulation is defective. Where an electronic ignition system is used, the secondary (HT) voltage is much higher and could prove fatal.

Remember...

✗ **Don't** start the engine without first ascertaining that the transmission is in neutral.

✗ **Don't** suddenly remove the pressure cap from a hot cooling system - cover it with a cloth and release the pressure gradually first, or you may get scalded by escaping coolant.

✗ **Don't** attempt to drain oil until you are sure it has cooled sufficiently to avoid scalding you.

✗ **Don't** grasp any part of the engine or exhaust system without first ascertaining that it is cool enough not to burn you.

✗ **Don't** allow brake fluid or antifreeze to contact the machine's paintwork or plastic components.

✗ **Don't** siphon toxic liquids such as fuel, hydraulic fluid or antifreeze by mouth, or allow them to remain on your skin.

✗ **Don't** inhale dust - it may be injurious to health (see Asbestos heading).

✗ **Don't** allow any spilled oil or grease to remain on the floor - wipe it up right away, before someone slips on it.

✗ **Don't** use ill-fitting wrenches or other tools which may slip and cause injury.

✗ **Don't** lift a heavy component which may be beyond your capability - get assistance.

✗ **Don't** rush to finish a job or take unverified short cuts.

✗ **Don't** allow children or animals in or around an unattended vehicle.

✗ **Don't** inflate a tire above the recommended pressure. Apart from overstressing the carcass, in extreme cases the tire may blow off forcibly.

✔ **Do** ensure that the machine is supported securely at all times. This is especially important when the machine is blocked up to aid wheel or fork removal.

✔ **Do** take care when attempting to loosen a stubborn nut or bolt. It is generally better to pull on a wrench, rather than push, so that if you slip, you fall away from the machine rather than onto it.

✔ **Do** wear eye protection when using power tools such as drill, sander, bench grinder etc.

✔ **Do** use a barrier cream on your hands prior to undertaking dirty jobs - it will protect your skin from infection as well as making the dirt easier to remove afterwards; but make sure your hands aren't left slippery. Note that long-term contact with used engine oil can be a health hazard.

✔ **Do** keep loose clothing (cuffs, ties etc. and long hair) well out of the way of moving mechanical parts.

✔ **Do** remove rings, wristwatch etc., before working on the vehicle - especially the electrical system.

✔ **Do** keep your work area tidy - it is only too easy to fall over articles left lying around.

✔ **Do** exercise caution when compressing springs for removal or installation. Ensure that the tension is applied and released in a controlled manner, using suitable tools which preclude the possibility of the spring escaping violently.

✔ **Do** ensure that any lifting tackle used has a safe working load rating adequate for the job.

✔ **Do** get someone to check periodically that all is well, when working alone on the vehicle.

✔ **Do** carry out work in a logical sequence and check that everything is correctly assembled and tightened afterwards.

✔ **Do** remember that your vehicle's safety affects that of yourself and others. If in doubt on any point, get professional advice.

● If in spite of following these precautions, you are unfortunate enough to injure yourself, seek medical attention as soon as possible.

1 Engine/transmission oil level check

⚠️ **Warning: On models with an oil level dipstick, never remove the dipstick to check oil level immediately after hard or high-speed riding. Hot oil could spurt out and cause burns. Let the engine cool to approximately 70-degrees C (150-degrees F) before removing the dipstick.**

Before you start:
✔ Start the engine and idle it for three minutes, then shut it off. This is necessary to move oil from the crankcase to the oil tank.
✔ Support the bike in an upright position with a motocross workstand or kickstand.
✔ On 2006 and later 250 models, the oil level is viewed through a window in the crankcase cover on the left-hand side of the machine. Wipe the glass clean before inspection to make the check easier.

Bike care:
● If you have to add oil frequently, you should check whether you have any oil leaks. If there is no sign of oil leakage from the joints and gaskets the engine could be burning oil (see *Troubleshooting*).

The correct oil
● Modern, high-revving engines place great demands on their oil. It is very important that the correct oil for your bike is used.
● Always top up with a good quality oil of the specified type and viscosity and do not overfill the engine.

Oil type
API grade SG or higher meeting JASO standard MA - the MA standard is required to prevent clutch slippage. See Chapter 1 for viscosity ratings.

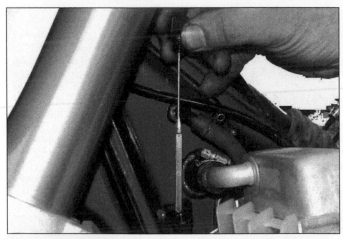

1 2005 and earlier models have a dipstick in the frame oil tank (on the top or left side of the frame, depending on model) to check the level. Unscrew the dipstick from the oil tank. Lift the dipstick out, wipe it clean and put it back (just let it rest on the threads, don't screw it in). Pull the dipstick back out and check the oil level on the dipstick scale. If it's low, add oil through the dipstick hole (2002 and earlier) or the oil filler plug on the left front corner of the crankcase (2003 and later).

2 2006 and later 250 models have a viewing window in the left side of the crankcase to check the oil level (right arrow). If the level is low, remove the oil filler cap (left arrow) and add oil as needed.

3 On 2006 and later 450 models, check oil level using the dipstick in the integral oil tank (left arrow). If the level is low, remove the filler cap (right arrow) and add oil as needed.

4 As a final check on the oil level of 2006 and later 450 models, unscrew the oil check bolt (arrow). If oil runs from the hole, let it continue running until it stops, then reinstall the check bolt.

2 Coolant level check

> ⚠️ *Warning: The radiator cap must be removed to check the coolant level on YZ models. DO NOT remove the cap while the engine is warm or scalding coolant will spray out. Let the engine cool down before removing the cap.*

> ⚠️ *Warning: Antifreeze is poisonous. Do not leave open containers of coolant about.*

> ⚠️ *Warning: Do not run the engine in an enclosed space such as a garage or workshop.*

Before you start:

✔ Make sure you have a supply of coolant available (a mixture of 50% distilled water and 50% corrosion inhibited ethylene glycol anti-freeze is needed).

✔ Always check the coolant level when the engine is cold.

✔ Ensure the motorcycle is held vertical while checking the coolant level. Make sure the motorcycle is on level ground.

Bike care:

● Use only the specified coolant mixture. It is important that anti-freeze is used in the system all year round, and not just in the winter. Do not top the system up using only water, as the system will become too diluted.

● Do not overfill the reservoir tank (WR models). If the coolant is significantly above the F (full) level line at any time, the surplus should be siphoned or drained off to prevent the possibility of it being expelled out of the overflow hose.

● If the coolant level falls steadily, check the system for leaks (see Chapter 1). If no leaks are found and the level continues to fall, it is recommended that the machine is taken to a Yamaha dealer for a pressure test.

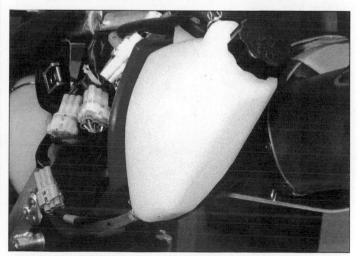

1 With the engine cold, unscrew the radiator cap on YZ models to check coolant level. It should be up to the bottom of the filler neck.

2 On WR models, check the coolant level in the reservoir or catch tank located inside the side cover. If it's below the lower mark with the engine cold, add the specified coolant.

3 Brake fluid level check

> ⚠ Warning: Brake fluid can harm your eyes and damage painted or some plastic surfaces, so use extreme caution when handling and pouring it and cover surrounding surfaces with rag. Do not use fluid that has been standing open for some time, as it absorbs moisture from the air which can cause a dangerous loss of braking effectiveness.

Before you start:

✔ Ensure the motorcycle is held vertical while checking the levels. Make sure the motorcycle is on level ground.

✔ Make sure you have the correct brake fluid. DOT 4 is recommended. Never reuse old fluid.

✔ Wrap a rag around the reservoir being worked on to ensure that any spillage does not come into contact with painted surfaces.

Bike care:

● The fluid in the front and rear brake master cylinder reservoirs will drop slightly as the brake pads wear down.

● If any fluid reservoir requires repeated topping-up this is an indication of a hydraulic leak somewhere in the system, which should be investigated immediately.

● Check for signs of fluid leakage from the hydraulic hoses and components – if found, rectify immediately.

● Check the operation of both brakes before taking the machine on the road; if there is evidence of air in the system (spongy feel to lever or pedal), it must be bled as described in Chapter 7.

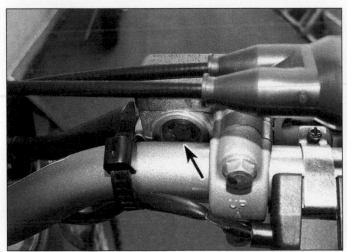

1 With the front brake fluid reservoir as level as possible, check that the fluid level is above the LOWER level line on the inspection window. If the level is below the LOWER level line, remove the cover screws (arrows) and lift off the cover and diaphragm. Top up fluid to the upper level line; don't overfill the reservoir, and take care to avoid spills (see WARNING above). Compress the diaphragm, reinstall the diaphragm and cover and tighten the screws securely.

2 The level in the rear master cylinder can be seen through the window on the rear side - the procedure is the same as for the front master cylinder.

4 Tire checks

The correct pressures:
● The tires must be checked when **cold**, not immediately after riding. Note that low tire pressures may cause the tire to slip on the rim or come off. High tire pressures will cause abnormal tread wear and unsafe handling.
● Use an accurate pressure gauge.
● Proper air pressure will increase tire life and provide maximum stability and ride comfort.

Tire care:
● Check the tires carefully for cuts, tears, embedded nails or other sharp objects and excessive wear. Operation of the motorcycle with excessively worn tires is extremely hazardous, as traction and handling are directly affected.

● Check the condition of the tire valve and ensure the dust cap is in place.
● Pick out any stones or nails which may have become embedded in the tire tread. If left, they will eventually penetrate through the casing and cause a puncture.
● If tire damage is apparent, or unexplained loss of pressure is experienced, seek the advice of a tire fitting specialist without delay.

Tire tread depth:
● Yamaha doesn't specify a minimum tread depth for motocross or off-road tires, but to maintain good performance and handling, check the profile of the knobs on the tires. When the knobs begin to get excessively rounded on their edges, the tire should be replaced (this usually happens far before the height of the knobs wears down very far).

Final drive checks:
● Make sure the drive chain slack isn't excessive and adjust it if necessary (see Chapter 1).
● Lubricate the chain if it looks dry (see Chapter 1).

Tire pressures	
1998	14 psi (98 kPa)
1999 and later	15 psi (100 kPa)

1 Check the tire pressures when the tires are **cold** and keep them properly inflated.

2 Yamaha doesn't specify a minimum tread depth for these bikes. A typical new tire looks like this. Bear in mind that a worn tire's poor traction can be dangerous, not just annoying - a sudden loss of traction can lead to loss of control and a fall.

Notes

Chapter 1
Tune-up and routine maintenance

Contents

Degrees of difficulty

Easy, suitable for novice with little experience	**Fairly easy,** suitable for beginner with some experience	**Fairly difficult,** suitable for competent DIY mechanic	**Difficult,** suitable for experienced DIY mechanic	**Very difficult,** suitable for expert DIY professional

Specifications

Engine

Spark plug type	NGK CR8E, Nippondenso U24ESR-N
Spark plug gap	0.7 to 0.8 mm (0.028 to 0.031 inch)
Valve clearances (engine cold)	
YZ250F, WR250F	
Intake..	0.10 to 0.15 mm (0.004 to 0.006 inch)
Exhaust..	0.17 to 0.22 mm (0.007 to 0.009 inch)
YZ400, WR400	
Intake..	0.15 to 0.20 mm (0.006 to 0.008 inch)
Exhaust..	0.25 to 0.30 mm (0.010 to 0.012 inch)
YZ426	
2000	
Intake..	0.15 to 0.20 mm (0.006 to 0.008 inch)
Exhaust	0.25 to 0.30 mm (0.010 to 0.012 inch)
2001 and later	
Intake..	0.10 to 0.15 mm (0.004 to 0.006 inch)
Exhaust	0.20 to 0.25 mm (0.008 to 0.010 inch)
YZ450	
Intake..	0.10 to 0.15 mm (0.004 to 0.006 inch)
Exhaust..	0.20 to 0.25 mm (0.008 to 0.010 inch)
WR426	
Intake..	0.10 to 0.15 mm (0.004 to 0.006 inch)
Exhaust..	0.20 to 0.25 mm (0.008 to 0.010 inch)
WR450	
Intake..	0.10 to 0.15 mm (0.004 to 0.006 inch)
Exhaust	0.20 to 0.25 mm (0.008 to 0.010 inch)
Engine idle speed	1700 to 1900 rpm

Engine (continued)

Pilot screw opening

YZ250F, WR250F	1-3/4 turns out
YZ400F	1-3/4 turns out

WR400F

1998	1-3/8 turns out
1999	1-1/8 turns out

2000

US	1-5/8 turns out
All others	1-1/2 turns out

YZ426F

2000

US	1-3/4 turns out
All others	1-1/2 turns out
2001	1-1/4 turns out

WR426F

2001

US	1-5/8 turns out
Except US	7/8 turn out

2002

US	1-5/8 turns out
Except US	1-1/8 turns out

YZ450F

2003 and 2004	2 turns out
2005	1-3/4 turns out
2006	2-1/8 turns out
2007	1-1/4 turns out
2008	1-1/2 turns out

WR450F

2003 and 2004

US	1-3/4 turns out
Except US	1-1/2 turns out
2005 and later	Not specified

Miscellaneous

Brake pad lining thickness limit (front and rear)	1.0 mm (0.04 inch)

Front brake lever distance from handlebar (1997 and later)

1998 through 2000	82.5 mm (3.25 inches)
2001 and later	95 mm (3.74 inches)

Rear brake pedal height

1998 through 2006	5 mm (0.20 inch) below top of footpeg

2007 and later

YZ models	5 mm (0.20 inch) below top of footpeg
WR models	10 mm (0.39 inch) above top of footpeg

Clutch lever freeplay

1998 through 2000	2 to 4 mm (0.08 to 0.016) at lever gap
2001 and later	8 to 13 mm (031 to 0.51 inch) at lever tip
Throttle grip freeplay	3 to 5 mm (1/8 to 1/4 inch)
Minimum tire tread depth	Not specified

Tire pressures (cold)

1998	14 psi (98 kPa)
1999 and later	15 psi (100 kPa)

Tire sizes

1998 and 1999

YZ400 front	80/100-21 51M
YZ400 rear	110/90-19 57M

WR400 front

US, Canada, New Zealand	80/100-21 51M
All others	90/90-21 54R

WR400 rear

US, Canada, New Zealand	110/100-18 64M
All others	120/90-18 65R

2000

YZ426 front

US, Canada, Australia, New Zealand, South Africa and France	80/100-21 51M
Europe (except France)	80/100-21 51R

YZ426 rear

US, Canada, Australia, New Zealand, South Africa and France	110/90-19 62M
Europe (except France)	110/90-19 NHS

WR400 front
    US, Canada, South Africa 80/100-21 51M
    Europe, Australia, New Zealand 90/90-21 54R
WR400 rear
    US, Canada, South Africa 110/100-18 64M
    Europe, Australia, New Zealand 120/90-18 65R
2001
  YZ250 front
    US, Canada, Australia, New Zealand, South Africa and France..... 80/100-21 51M
    Europe (except France) 80/100-21 MT300
  YZ250 rear
    US, Canada, Australia, New Zealand, South Africa and France..... 110/90-19 57M
    Europe (except France) 110/90-19 MT320
  WR250 front
    US, Canada, Australia, New Zealand, South Africa and France..... 80/100-21 51M
    Europe (except France) 80/100-21 MT320
  WR250 rear
    US, Canada, Australia, New Zealand, South Africa and France..... 110/90-19 57M
    Europe (except France) 110/90-19 MT320
  YZ426 front
    US, Canada, Australia, New Zealand, South Africa and France..... 80/100-21 51M
    Europe (except France) 80/100-21 51R
  YZ426 rear
    US, Canada, Australia, New Zealand, South Africa and France..... 110/90-19 62M
    Europe (except France) .... 110/90-19 NHS
  WR426 front
    US, Canada, South Africa 80/100-21 51M
    Europe, Australia, New Zealand 90/90-21 54R
  WR426 rear
    US, Canada, South Africa 110/100-18 64M
    Europe, Australia, New Zealand 120/90-18 69R
2002
  YZ250 front
    US, Canada, Australia, New Zealand, South Africa and France..... 80/100-21 51M
    Europe (except France) 80/100-21 MT320
  YZ250 rear
    US, Canada, Australia, New Zealand, South Africa and France..... 110/90-19 57M
    Europe (except France) 110/90-19 MT320
  WR250 front
    US, Canada and South Africa 80/100-21 51M
    Europe, Australia and New Zealand 90/90-21 54R
  WR250 rear
    US, Canada and South Africa 100/100-18 59M
    Europe, Australia, New Zealand 130/90-18 69R
  YZ426 front
    US, Canada and South Africa 80/100-21 51M
    Europe, Australia, New Zealand 90/90-21 54R
  YZ426 rear
    US, Canada, Australia, New Zealand, South Africa and France..... 110/90-19 62M
    Europe (except France) .... 110/90-19 NHS
  WR426 front
    US, Canada, South Africa 80/100-21 51M
    Europe, Australia, New Zealand 90/90-21 54R
  WR426 rear
    US, Canada, South Africa 110/100-18 64M
    Europe, Australia, New Zealand 130/90-18 69R
2003
  YZ250 front
    US, Canada, Australia, New Zealand, South Africa and France..... 80/100-21 51M
    Europe (except France) 80/100-21 51R
  YZ250 rear
    US, Canada, Australia, New Zealand, South Africa and France..... 110/90-19 57M
    Europe (except France) 110/90-19 MT320
  WR250 front
    US, Canada and South Africa 80/100-21 51M
    Europe, Australia and New Zealand 90/90-21 54R
  WR250 rear
    US, Canada and South Africa 100/100-18 59M
    Europe, Australia, New Zealand 130/90-18 69R

Miscellaneous (continued)

YZ450 front
  US, Canada, Australia, New Zealand, South Africa and France..... 80/100-21 51M
  Europe except France 80/100-21 51R
YZ450 rear
  US, Canada, Australia, New Zealand, South Africa and France..... 110/90-19 62M
  Europe (except France) .... 110/90-19 NHS
WR450 front
  US, Canada, South Africa 80/100-21 51M
  Europe, Australia, New Zealand... 90/90-21 54R
WR450 rear
  US, Canada, South Africa 110/100-18 64M
  Europe, Australia, New Zealand... 130/90-18 69R

2004 through 2007
YZ250 front
  US, Canada, Australia, New Zealand, South Africa and France..... 80/100-21 51M
  Europe (except France) 80/100-21 51R
YZ250 rear
  US, Canada, Australia, New Zealand, South Africa and France..... 100/90-19 57M
  Europe (except France) .... 100/90-19 NHS
WR250 front
  US, Canada and South Africa 80/100-21 51M
  Europe, Australia and New Zealand... 90/90-21 54R
WR250 rear
  US, Canada and South Africa 100/100-18 59M
  Europe, Australia, New Zealand... 130/90-18 69R
YZ450 front
  US, Canada, Australia, New Zealand, South Africa and France..... 80/100-21 51M
  Europe except France 80/100-21 51R
YZ450 rear
  US, Canada, Australia, New Zealand, South Africa and France..... 110/90-19 62M
  Europe (except France) .... 110/90-19 NHS
WR450 front
  US, Canada, South Africa 80/100-21 51M
  Europe, Australia, New Zealand... 90/90-21 54R
WR450 rear
  US, Canada, South Africa 110/100-18 64M
  Europe, Australia, New Zealand... 130/90-18 69R

2008
YZ250 front... 80/100-21 51M
YZ250 rear.. 100/90-19 57M
WR250 front.. 80/100-21 51M
WR250 rear... 100/100-18 59M
YZ450 front... 80/100-21 51M
YZ450 rear
  US, Canada, Australia, New Zealand and South Africa 120/80-19 63M
  Europe (except France) 110/90-19 62M
WR450 front.. 80/100-21 51M
WR450 rear... 110/100-18 64M

Drive chain slack
1998 through 2004.. 40 to 50 mm (1.6 to 2.0 inches)
2005 and later .... 48 to 58 mm (1.9 to 2.3 inches)

Torque specifications

Engine oil drain bolts
  In oil filter cover.. 10 Nm (86 inch-lbs)
  In crankcase
    1998 through 2003 .... 20 Nm (14 ft-lbs)
    2004 (YZ250, YZ450) .... 20 Nm (14 ft-lbs)
    2004 and 2005 (YZ450, WR450)
      Crankcase left side.. 10 Nm (86 inch-lbs)
      Crankcase rear corner... 20 Nm (14 ft-lbs)
    2006 and later (YZ450, WR450, left and right sides)........................... 20 Nm (14 ft-lbs)
  In frame oil tank (2005 and earlier)... 23 Nm (17 ft-lbs)
  In external oil tank (2006 and later)... 18 Nm (13 ft-lbs)
Oil filter cover bolts .... 10 Nm (86 inch-lbs)
Oil strainer (in frame oil tank)
  2000 through 2002.. 90 Nm (65 ft-lbs)

2003
    YZ250, WR250, YZ450 .... 70 Nm (50 ft-lbs)
    WR450 .... 90 Nm (65 ft-lbs)
2004 and 2005 .... 70 Nm (50 ft-lbs)
Oil strainer (in external oil tank, 2006 and later) .... 9 Nm (78 inch-lbs)
Oil gallery bolt (oil pressure check bolt)
    1998 through 2000 .... 18 Nm (13 ft-lbs)
    2001 and 2002
        All except WR250 .... 18 Nm (13 ft-lbs)
        WR250 .... 10 Nm (86 inch-lbs)
    2003 and later .... 10 Nm (86 inch-lbs)
Coolant drain bolt .... 10 Nm (86 inch-lbs)
Spark plug .... 13 Nm (113 inch-lbs)
Spark arrester bolts .... 10 Nm (86 inch-lbs)
Wheel spokes
    1998 and 1999 .... 6 Nm (51 inch-lbs)
    2000 and later .... 3 Nm (26 inch-lbs)
Rim lock locknut .... Not specified
Steering stem adjusting nut*
    Initial torque .... 38 Nm (27 ft-lbs)
    Final torque .... 7 Nm (86 inch-lbs)
*Torque setting requires a Yamaha ring nut wrench, placed on the torque wrench at a right angle.

Recommended lubricants and fluids

Fuel
    1998
        US and Australia .... Unleaded
        Canada and Europe .... Premium unleaded
        All others .... Premium
    1999
        YZ400
            US and Australia .... Unleaded
            Canada and Europe .... Premium unleaded
            All others .... Premium
        WR400
            All except Australia .... Premium unleaded, Research octane 95 or higher
            Australia .... Unleaded
    2000 and 2001
        All except Australia .... Premium unleaded, Research octane 95 or higher
        Australia .... Unleaded
    2002 through 2004
        All except South Africa .... Premium unleaded, Research octane 95 or higher
        South Africa .... Unleaded
    2005 and later .... Premium unleaded, Research octane 95 or higher
Engine/transmission oil
    Type
        API service SG or higher multigrade four-stroke oil manufactured for use in motorcycles and meeting JASO standard MA
    Viscosity (US and Canada)
        10 to 70-degrees F (-10 to 20 degrees-C) 10W-30
        50 to 110-degrees F (10 to 40-degrees C) 20W-40
        10 to 120-degrees F (-10 to 45-degrees C) 10W-50
    Viscosity (all others)
        10 to 70-degrees F (-10 to 20 degrees-C) 10W-30
        10 to 110-degrees F (-10 to 40-degrees C) 10W-40
        30 to 110-degrees F (0 to 40-degrees C) 15W-40
        50 to 110-degrees F (10 to 40-degrees C) 20W-40
        50 to 120-degrees F (10 to 45-degrees C) 20W-50
    Capacity at oil change (with filter change)*
        1998
            YZ400 .... 1.5 liters (1.59 US qt, 2.64 Imp pt)
            WR400 .... 1.6 liters (1.69 US qt, 2.82 Imp pt)
        1999 through 2001 (all models) .... 1.6 liters (1.69 US qt, 2.82 Imp pt)
        2002
            YZ250, WR250 .... 1.6 liters (1.69 US qt, 2.82 Imp pt)
            YZ426, WR426 .... 1.4 liters (1.48 US qt, 2.46 Imp pt)
        2003 through 2005
            YZ250, YZ450, WR450 .... 1.1 liters (1.16 US qt, 1.94 Imp pt)
            WR250 .... 1.3 liters (1.37 US qt, 2.28 Imp pt)

Recommended lubricants and fluids (continued)

2006
    YZ250 .... 1.25 liters (1.32 US qt, 2.20 Imp pt)
    WR250 .... 1.3 liters (1.37 US qt, 2.28 Imp pt)
    YZ450, WR450 .... 1.0 liter (1.06 US qt, 1.76 Imp pt)
2007 and later
    YZ250 .... 1.15 liters (1.22 US qt, 2.02 Imp pt)
    WR250 .... 1.2 liters (1.27 US qt, 2.12 Imp pt)
    YZ450, WR450 .... 1.0 liter (1.06 US qt, 1.76 Imp pt)

Approximate. Capacity after engine overhaul is approximately 10 to 20 percent higher. Use the dipstick or oil level window to determine the final amount.

Air cleaner element oil .... Foam filter oil
Coolant
    Type.. Ethylene glycol antifreeze compatible with aluminum engines, mixed 50/50 with water

    Capacity
    1998
        YZ400 .... 1.15 liters (1.22 US qt, 2.02 Imp pt)
        WR400 .... 1.5 liters (1.59 US qt, 2.64 Imp pt)
    1999
        YZ400 .... 1.15 liters (1.22 US qt, 2.02 Imp pt)
        WR400 .... 1.2 liters (1.27 US qt, 2.12 Imp pt)
    2000 .... 1.2 liters (1.27 US qt, 2.12 Imp pt)
    2001 and 2002
        YZ250 .... 0.9 liters (0.95 US qt, 1.58 Imp pt)
        WR250 .... 1.3 liters (1.37 US qt, 2.28 Imp pt)
        YZ426 and WR426 .... 1.2 liters (1.27 US qt, 2.12 Imp pt)
    2003 and 2004
        YZ250 .... 0.9 liters (0.95 US qt, 1.58 Imp pt)
        WR250 .... 1.3 liters (1.37 US qt, 2.28 Imp pt)
        YZ450 .... 1.2 liters (1.27 US qt, 2.12 Imp pt)
        WR450 .... 1.6 liters (1.69 US qt, 2.82 Imp pt)
    2005
        YZ250 .... 0.9 liters (0.95 US qt, 1.58 Imp pt)
        WR250 .... 1.26 liters (1.33 US qt, 2.22 Imp pt)
        YZ450 .... 1.2 liters (1.27 US qt, 2.12 Imp pt)
        WR450 .... 1.56 liters (1.65 US qt, 2.74 Imp pt)
    2006
        YZ250 .... 0.99 liters (1.05 US qt, 1.74 Imp pt)
        WR250 .... 1.26 liters (1.33 US qt, 2.22 Imp pt)
        YZ450 .... 0.99 liters (1.05 US qt, 1.74 Imp pt)
        WR450 .... SPEC TK
    2007 and later
        YZ250, WR250, YZ450 .... 0.99 liters (1.05 US qt, 1.74 Imp pt)
        WR450 .... 1.0 liters (1.06 US qt, 1.76 Imp pt)

Brake fluid .... DOT 4
Drive chain lubricant.. Chain lube or 10W-30 engine oil
Fork oil .... See Chapter 6
Miscellaneous
    Wheel bearings .... Medium weight, lithium-based multi-purpose grease
    Swingarm pivot bushings.. Molybdenum disulfide
    Cables and lever pivots... Yamaha cable lube or WD-40
    Throttle grip, cable ends .... Lightweight, lithium-based multi-purpose grease

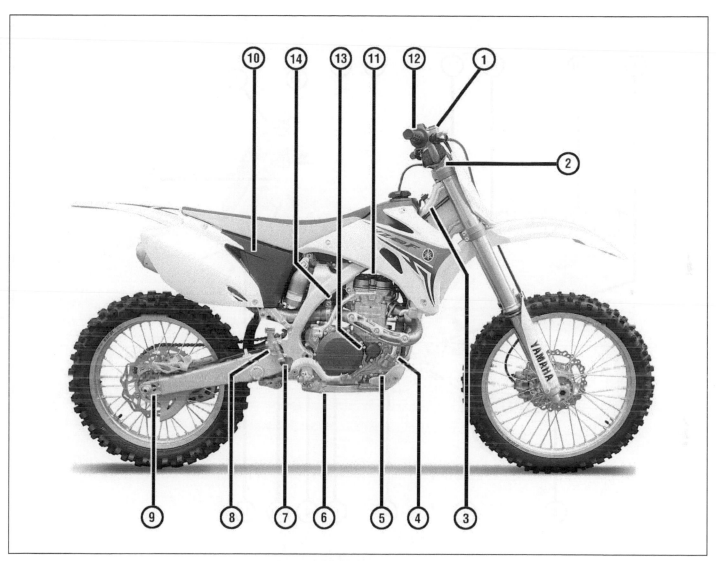

Yamaha YZ maintenance points

1 Front brake fluid reservoir
2 Steering head bearing adjuster
3 Engine oil dipstick location (models with frame oil tank)
4 Engine oil dipstick location (models with crankcase oil tank) (on left side)
5 Coolant drain bolt location
6 Crankcase drain plug (see text for specific location)
7 Rear brake pedal adjuster
8 Rear brake fluid reservoir
9 Drive chain adjuster
10 Air filter (under seat)
11 Spark plug
12 Clutch adjuster (and hot start cable adjuster, if equipped)
13 Oil filter
14 Throttle stop screw (on left side)

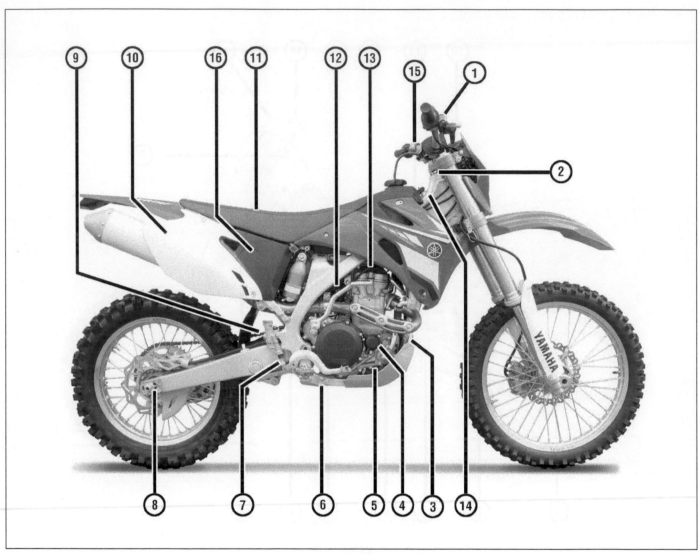

Yamaha WR maintenance points

1 Front brake fluid reservoir
2 Steering head bearing adjuster
3 Engine oil dipstick location (models with crankcase oil tank) (on left side)
4 Engine oil filter
5 Coolant drain bolt location
6 Crankcase drain plug (see text for specific location)

7 Rear brake pedal adjuster
8 Drive chain adjuster
9 Rear brake fluid reservoir
10 Coolant reservoir (inside left side panel)
11 Fuse and battery (under seat)
12 Throttle stop screw (on left side)
13 Spark plug

14 Engine oil dipstick location (models with frame oil tank)
15 Clutch adjuster (and hot start cable adjuster, if equipped)
16 Air filter (under seat on 2002 and earlier models, behind left side cover on 2003 and later models)

1 Yamaha YZF and WRF Routine maintenance intervals

Note: *This schedule was developed for your motorcycle's intended purpose, motocross or enduro competition. It's based on numbers of races. The intervals listed below are the shortest intervals recommended by the manufacturer for each particular operation during the model years covered in this manual. Your owner's manual may have different intervals for your model. If your bike is not used in competition, perform the "Before each race" and "Before every third race" inspection and maintenance procedures every 3000 km (1800 miles) or 3 months. Replace the brake fluid and coolant once a year. Periodic disassembly of the engine for inspection is not necessary.*

Before each race

- ☐ Make sure the engine kill switch works properly (see Section 3)
- ☐ Check the engine oil, brake fluid and coolant levels (see Section 4)
- ☐ Check the throttle, choke and hot start knobs/levers for smooth operation and correct freeplay (See Section 5)
- ☐ Check clutch and decompression lever operation and freeplay (See Section 6)
- ☐ Inspect and lubricate the control cables (See Section 7)
- ☐ Check the brake pads for wear (See Section 8)
- ☐ Check the operation of both brakes - check the front brake lever and rear brake pedal for correct freeplay (See Section 9)
- ☐ Check the tires for damage, the presence of foreign objects and correct air pressure (see Section 10)
- ☐ Lubricate and inspect the drive chain, sprockets and sliders (see Section 11)
- ☐ Check all fasteners, including axle nuts, for tightness (see Section 12)
- ☐ Check and adjust the steering head bearings (See Section 13)
- ☐ Inspect the front and rear suspensions (see Section 14)

- ☐ Lubricate the swingarm bearings and rear shock linkage pivot points (see Chapter 6)
- ☐ Inspect the cooling system (see Section 15)
- ☐ Clean the air filter element (see Section 16)
- ☐ Inspect and tighten the exhaust pipe and muffler (see Section 17)
- ☐ Check and adjust the carburetor; clean if necessary (see Section 18 and Chapter 4)
- ☐ Clean and inspect the frame (see Chapter 8)

Before every third race

- ☐ Clean and gap the spark plug (see Section 19)
- ☐ Check and adjust the valve clearances (see Section 20)
- ☐ Change the engine oil and filter (see Section 21)
- ☐ Inspect the fuel tank and supply valve (see Chapter 4)
- ☐ Inspect and lubricate the wheel bearings (see Chapter 7)

Before every fifth race

- ☐ Replace the glass wool in the muffler (see Section 17)
- ☐ Change the fork oil (see Chapter 6)
- ☐ Clean and lubricate the steering head bearings (see Chapter 6)
- ☐ Disassemble and inspect the engine top end (camshafts, timing chain and sprockets, cylinder head, cylinder piston, pin and rings (see Chapter 2)
- ☐ Change the coolant (see Section 15)

Every year

- ☐ Change the brake fluid (see Chapter 7)
- ☐ Replace the rear shock absorber spring seat (see Chapter 6)

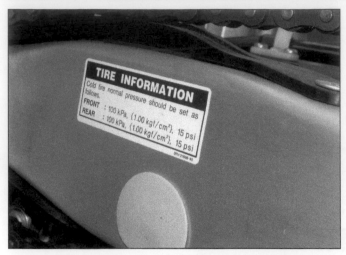

2.1 Decals on the motorcycle include maintenance and safety information

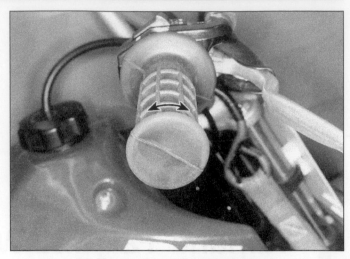

5.3 Measure throttle freeplay at the grip

2 Introduction to tune-up and routine maintenance

This Chapter covers in detail the checks and procedures necessary for the tune-up and routine maintenance of your motorcycle. Section 1 includes the routine maintenance schedule, which is designed to keep the machine in proper running condition and prevent possible problems. The remaining Sections contain detailed procedures for carrying out the items listed on the maintenance schedule, as well as additional maintenance information designed to increase reliability. Maintenance and safety information is also printed on decals, which are mounted in various locations on the motorcycle **(see illustration)**. Where information on the decals differs from that presented in this Chapter, use the decal information.

Since routine maintenance plays such an important role in the safe and efficient operation of your motorcycle, it is presented here as a comprehensive check list. For the rider who does all of the bike's maintenance, these lists outline the procedures and checks that should be done on a routine basis.

Deciding where to start or plug into the routine maintenance schedule depends on several factors. If you have owned the bike for some time but have never performed any maintenance on it, then you may want to start at the nearest interval and include some additional procedures to ensure that nothing important is overlooked. If you have just had a major engine overhaul, then you may want to start the maintenance routine from the beginning. If you have a used machine and have no knowledge of its history or maintenance record, you may desire to combine all the checks into one large service initially and then settle into the maintenance schedule prescribed.

The Sections which outline the inspection and maintenance procedures are written as step-by-step comprehensive guides to the actual performance of the work. They explain in detail each of the routine inspections and maintenance procedures on the check list. References to additional information in applicable Chapters are also included and should not be overlooked.

Before beginning any actual maintenance or repair, the machine should be cleaned thoroughly, especially around the oil filler plug, radiator cap, air filter cover, carburetor, etc. Cleaning will help ensure that dirt does not contaminate the engine and will allow you to detect wear and damage that could otherwise easily go unnoticed.

3 Engine kill switch - check

Start the engine, then use the kill switch to shut it off. If it doesn't shut off or if the engine doesn't start, refer to Chapter 5 and the wiring diagrams at the end of the book to test the switch.

4 Fluid levels - check

At the specified intervals, check the engine oil, coolant and brake fluid levels as described in *Daily checks* at the front of this manual.

5 Throttle, choke and hot start lever operation/ freeplay - check and adjustment

Throttle check

1   Make sure the throttle twistgrip moves easily from fully closed to fully open with the front wheel turned at various angles. The grip should return automatically from fully open to fully closed when released. If the throttle sticks, check the throttle cable for cracks or kinks in the housings. Also, make sure the inner cable is clean and well-lubricated.

2   Check and adjust the idle speed (see Section 18).

3   Check for a small amount of freeplay at the twistgrip **(see illustration)** and compare the freeplay to the value listed in this Chapter's Specifications.

Throttle adjustment

1998 and 1999 models

4   These models use two throttle cables. The upper cable is pulled during acceleration and the lower cable is pulled during deceleration.

5   Locate the cable adjuster in the upper and lower cables near the carburetor.

6   Loosen the locknut on the lower cable adjuster. Turn the adjuster to take up any slack in the cable, then tighten the locknut.

7   Loosen the locknut on the upper cable adjuster. Turn the adjuster to obtain the freeplay listed in this Chapter's Specifications, then tighten the locknut.

2000 through 2002 models

8   These models use two throttle cables. The lower cable is pulled during acceleration and the upper cable is pulled during deceleration.

9   Locate the cable adjuster in the upper and lower cables near the carburetor.

10   Loosen the locknut on the upper cable adjuster. Turn the adjuster to take up any slack in the cable, then tighten the locknut.

11   Loosen the locknut on the lower cable adjuster. Turn the adjuster to obtain the freeplay listed in this Chapter's Specifications, then tighten the locknut.

5.13 Make sure the choke knob (A) pulls out with a click, but operates smoothly. (B) is the throttle stop screw

5.14a Measure freeplay of the hot start lever at the lever tip - slide back the cover (arrow) for access to the adjuster . . .

2003 and later models

12 Throttle cable adjustments are made at the throttle grip end of the cable. Pull back the rubber cover from the adjuster and loosen the locknut on the cable. Turn the adjuster until the specified freeplay is obtained, then retighten the locknut or lockwheel.

Choke and hot start knobs - operation check

13 Check that the choke knob moves smoothly (see illustration). If not, refer to Chapter 4 and remove the choke plunger for inspection. On models equipped with a hot start knob (colored red and located next to the choke knob), check it also.

Hot start lever - check and adjustment

14 Check the freeplay at the lever tip (see illustration). If it's not within the value listed in this Chapter's Specifications, pull the rub-

ber cover away from the adjuster (see illustration). Loosen the locknut, turn the adjusting nut to set the freeplay and tighten the locknut.

6 Clutch and decompression levers - check and freeplay adjustment

Clutch lever

1 Operate the clutch lever and measure freeplay at the tip of the lever or in the gap between the lever and bracket (see illustration), depending on model (refer to this Chapter's Specifications for the gap and measurement point). If it's not within the range listed in this Chapter's Specifications, adjust it as follows.

2 On early models, pull back the rubber cover from the adjuster at the handlebar (see illustration). Loosen the lockwheel or lock-

5.14b . . . loosen the locknut and turn the adjuster (upper arrow) - the lower adjuster (lower arrow) is for the clutch

nut and turn the adjuster to change freeplay.

3 Some models are equipped with a mid-

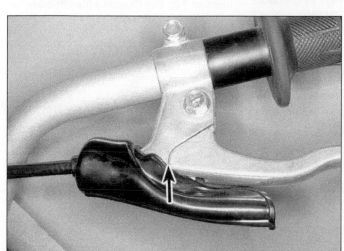

6.1 On some models, clutch freeplay is measured at the lever gap (see this Chapter's Specifications) . . .

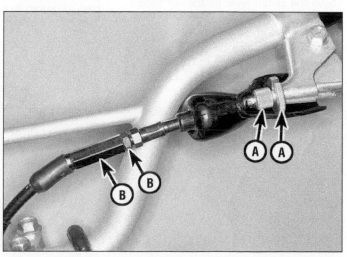

6.2 . . . to adjust it, loosen the locknut and turn the adjuster first at the handlebar adjuster (A), then at the mid-cable adjuster (if equipped) (B)

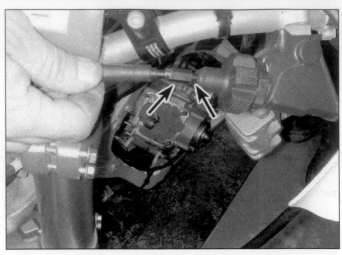

6.4a On later models, measure clutch freeplay at the lever tip; the adjusters are inside the boot (left arrow) and cap (right arrow) . . .

6.4b . . . to make initial adjustments, pull back the boot, loosen the locknut (right arrow) and turn the adjuster (left arrow) . . .

cable adjuster, which is several inches farther down the cable (see illustration 6.2). If freeplay can't be brought within specifications by using the handlebar adjuster, make adjustments at the mid-cable adjuster.

4   Later models have a fine adjuster under the cap (see illustrations). This can be used to make very small adjustments after the freeplay has been set using the locknut and adjuster.

5   If freeplay can't be adjusted to within the specified range, the cable is probably stretched and should be replaced with a new one (see Chapter 2).

Decompression lever

6   The decompression lever is used to release engine compression during startup on 1998 through 2002 models.

7   Place the cylinder at TDC on the compression stroke (see Section 21).

8   Check the freeplay at the end of the decompression lever (it's the short lever above the clutch lever on the left handlebar). If it's not within the value listed in this

Chapter's Specifications, loosen the locknut, turn the adjusting nut to set the freeplay and tighten the locknut.

9   If freeplay can't be adjusted to within the specified range, the cable is probably stretched and should be replaced with a new one (see Chapter 2).

7   Lubrication - general

1   Since the controls, cables and various other components of a motorcycle are exposed to the elements, they should be lubricated periodically to ensure safe and trouble-free operation.

2   The throttle twistgrip, brake lever, brake pedal and kickstarter pedal pivot should be lubricated frequently. In order for the lubricant to be applied where it will do the most good, the component should be disassembled. However, if chain and cable lubricant is being used, it can be applied to the pivot

joint gaps and will usually work its way into the areas where friction occurs. If motor oil or light grease is being used, apply it sparingly as it may attract dirt (which could cause the controls to bind or wear at an accelerated rate). **Note:** *One of the best lubricants for the control lever pivots is a dry-film lubricant (available from many sources by different names).*

3   The throttle and clutch cables should be removed and treated with a commercially available cable lubricant which is specially formulated for use on motorcycle control cables. Small adapters for pressure lubricating the cables with spray can lubricants are available and ensure that the cable is lubricated along its entire length (see illustration). When attaching the cable to the lever, be sure to lubricate the barrel-shaped fitting at the end with multi-purpose grease.

4   To lubricate the cables, disconnect them at the upper end, then lubricate the cable with a pressure lube adapter (see illustration 7.3). See Chapter 4 (throttle cable) or Chapter 2 (clutch cable).

6.4c . . . and if necessary, pull back the cap and turn this adjuster to make fine adjustments

6.4d Align the tab inside the cap (arrow) with the adjuster groove when you install the cap

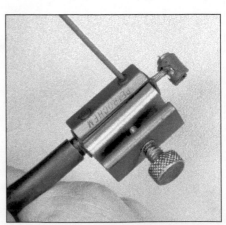

7.3 Lubricating a cable with a pressure lube adapter (make sure the tool seats around the inner cable)

8.4 The back of the caliper is open - when the pad material is worn to the wear indicator notches (arrows), it's time for new pads

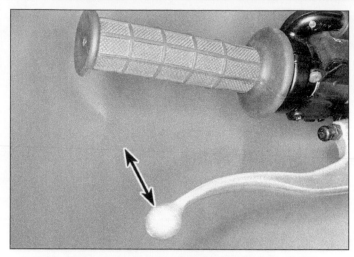

9.1 Measure the distance from the brake lever tip to the handlebar

5   Lubrication of the swingarm and rear suspension linkage pivots requires removal of the components.

6   Refer to Chapter 6 for the following lubrication procedures:

a) *Swingarm bearings, dust seals and rear suspension linkage pivots*
b) *Steering head bearings*

7   Refer to Chapter 7 for the following lubrication procedures:

a) *Front and rear wheel bearings*
b) *Brake pedal pivot*

8   Brake system - general check

1   A routine general check of the brakes will ensure that any problems are discovered and remedied before the rider's safety is jeopardized.

2   Check the brake lever and pedal for loose connections, excessive play, bends, and other damage. Replace any damaged parts with new ones (see Chapter 7).

3   Make sure all brake fasteners are tight. Check the brakes for wear as described below.

4   There's a groove on the inside of each pad next to the metal backing (see illustration). When the friction material is nearly worn down to the groove, it's time to replace the pads (even if only one pad is worn that far).

9   Brake lever and pedal - check and adjustment

Front brake lever

1   Check the distance from the lever tip to the handlebar (see illustration). If it exceeds the limit listed in this Chapter's Specifications, adjust the front brake.

2   On early models, loosen the locknut and turn the adjusting nut to adjust lever distance. Once the adjustment is correct, tighten the locknut.

3   On later models, loosen the locknut and turn the adjusting bolt (see illustration). Tighten the locknut after making the adjustment.

4   Recheck lever travel (see Step 1). If it's still not within the Specifications, refer to Chapter 7 and bleed the brakes.

Rear brake pedal

5   Measure brake pedal height in relation to the footpeg and compare it with the value listed in this Chapter's Specifications (see illustration). Adjust if necessary.

6   To adjust pedal height, loosen the locknut and turn the pushrod on the rear master

9.3 Make adjustments by loosening the locknut and turning the adjusting bolt

9.5 Measure brake pedal height (right arrow) in relation to the footpeg (left arrow) . . .

9.6 . . . to adjust it, loosen the locknut (lower arrow) and turn the adjusting nut (upper arrow) - do not over-adjust it (see text)

10.5 Make sure the spokes are tight, but don't overtighten them

cylinder **(see illustration)**. Be sure there is enough thread showing on the pushrod after the adjustment. If you have to use up all the threads to get the adjustment within Specifications, bleed the brakes or overhaul the master cylinder and caliper (see Chapter 7). Also do this if the end of the pushrod comes closer than 2 mm (0.080 inch) to the brake pedal.

7   Pedal freeplay on disc brake models is automatic and no means of manual adjustment is provided.

10 Tires/wheels - general check

1   Routine tire and wheel checks should be made with the realization that your safety depends to a great extent on their condition.
2   Check the tires carefully for cuts, tears, embedded nails or other sharp objects and excessive wear. Operation of the motorcycle with excessively worn tires is extremely hazardous, as traction and handling are directly affected. Check the tread depth at the center

of the tire and compare it to the value listed in this Chapter's Specifications. Yamaha doesn't specify a minimum tread depth for some models, but as a general rule, tires should be replaced with new ones when the tread knobs are worn to 8 mm (5/16 inch) or less for safety's sake, but for performance, you'll want to replace a tire long before it gets that worn.
3   Repair or replace punctured tires as soon as damage is noted. Do not try to patch a torn tire, as wheel balance and tire reliability may be impaired.
4   Check the tire pressures when the tires are cold and keep them properly inflated (refer to *Daily checks* at the front of this manual). Proper air pressure will increase tire life and provide maximum stability and ride comfort. Keep in mind that low tire pressures may cause the tire to slip on the rim or come off, while high tire pressures will cause abnormal tread wear and unsafe handling.
5   The wheels should be kept clean and checked periodically for cracks, bending, loose spokes and rust. Never attempt to repair damaged wheels; they must be replaced with new ones. Loose spokes can be tightened with a spoke wrench **(see illus-**

tration), but be careful not to overtighten and distort the wheel rim.
6   Check the valve stem locknuts to make sure they're tight. Also, make sure the valve stem cap is in place and tight. If it is missing, install a new one made of metal or hard plastic.
7   Check the tightness of the locknut on the rim lock **(see illustration)**. If it's loose, tighten it securely.

11 Drive chain and sprockets - check, adjustment and lubrication

1   A neglected drive chain won't last long and can quickly damage the sprockets. Routine chain adjustment isn't difficult and will ensure maximum chain and sprocket life. **Note:** *The chain should be routinely replaced at the interval listed in Section 1.*
2   To check the chain, support the bike securely with the rear wheel off the ground. Place the transmission in neutral.
3   Push down and pull up on the top run of the chain, directly above the swingarm protector mounting bolt, and measure the slack midway between the two sprockets **(see illustration)**. Compare the measurements to the value listed in this Chapter's Specifications. As wear occurs, the chain will actually stretch, which means adjustment by removing some slack from the chain. In some cases where lubrication has been neglected, corrosion and galling may cause the links to bind and kink, which effectively shortens the chain's length. If the chain is tight between the sprockets, rusty or kinked, it's time to replace it with a new one. **Note:** *Turn the back wheel to move the chain and repeat the chain slack measurement along the length of the chain - ideally, every inch or so. If you find a tight area, mark it with a felt pen or paint and repeat the measurement after the bike has been ridden. If the chain is still tight in the same areas, it may be damaged or worn.*

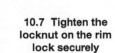

10.7 Tighten the locknut on the rim lock securely

11.3 Measure drive chain slack along the upper chain run

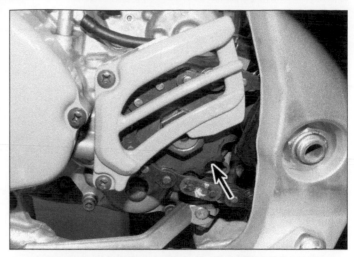

11.5 Check the engine sprocket for wear - if it's worn, replace both sprockets and the drive chain as a set

11.6a Check the chain sliders for wear; typical locations are below the swingarm . . .

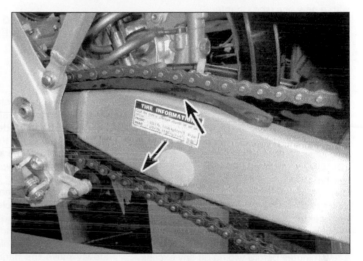

11.6b . . . wrapped around the swingarm . . .

Because a tight or kinked chain can damage the transmission countershaft bearing, it's a good idea to replace it.

4   Check the entire length of the chain for damaged rollers, loose links and loose pins.

5   Check the teeth on the engine sprocket **(see illustration)** and the rear sprocket for wear (see Chapter 6). Refer to Chapter 6 for the sprocket replacement procedure if the sprockets appear to be worn excessively.

6   Check the chain sliders and roller(s) **(see illustrations)**. If a slider is worn, measure its thickness. If it's less than the value listed in the Chapter 6 Specifications, replace it (see Chapter 6).

Adjustment

7   Rotate the rear wheel until the chain is positioned with the least amount of slack present.

8   Loosen the rear axle nut (see Chapter 7). Loosen the locknut and turn the adjuster on each side of the swingarm evenly until the proper chain tension is obtained (get

11.6c . . . and attached to the frame above the swingarm

the adjuster on the chain side close, then set the adjuster on the opposite side) **(see illustration)**. Be sure to turn the adjusters evenly to keep the wheel in alignment. If the adjust-

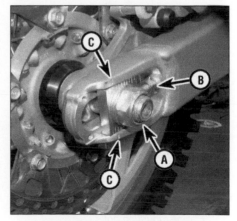

11.8 Chain adjuster details

A   Axle nut
B   Locknut and adjusting bolt
C   Indicator marks

ers reach the end of their travel, the chain is excessively worn and should be replaced

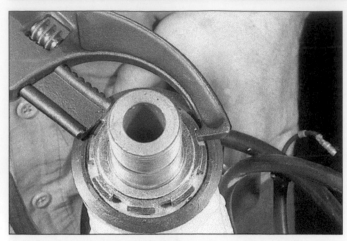

13.6 Loosen the ring nut with an adjustable spanner wrench such as this one, then tighten it slightly

14.4a Oil leakage from the fork seals (arrow) . . .

with a new one (see Chapter 6).

9   When the chain has the correct amount of slack, make sure the marks on the adjusters correspond to the same relative marks on each side of the swingarm (**see illustration 11.8**). Tighten the axle nut to the torque listed in the Chapter 7 Specifications.

Lubrication

Note: *If the chain is dirty, it should be removed and cleaned before it's lubricated (see Chapter 6).*

10   Use a good quality chain lubricant of the type listed in this Chapter's Specifications. Apply the lubricant along the top of the lower chain run, so that when the bike is ridden, centrifugal force will move the lubricant into the chain, rather than throwing it off.

11   After applying the lubricant, let it soak in a few minutes before wiping off any excess.

12 Fasteners - check

1   Since vibration of the machine tends to loosen fasteners, all nuts, bolts, screws, etc. should be periodically checked for proper tightness. Also make sure all cotter pins or other safety fasteners are correctly installed.

2   Pay particular attention to the following:
   Spark plug
   Engine and transmission oil drain plug and check bolt
   Gearshift pedal
   Brake lever and pedal
   Kickstarter pedal
   Footpegs
   Engine mounting bolts
   Steering stem locknut
   Front axle nut
   Rear axle nut
   Skid plate bolts

3   If a torque wrench is available, use it along with the torque specifications at the beginning of this, or other, Chapters.

13 Steering head bearings - check and adjustment

Inspection

1   These motorcycles are equipped with roller-and-cone type steering head bearings, which can become dented, rough or loose during normal use of the machine. In extreme cases, worn or loose steering head bearings can cause steering wobble that is potentially dangerous.

2   To check the bearings, lift up the front end of the motorcycle and place a secure support beneath the engine so the front wheel is off the ground.

3   Point the wheel straight ahead and slowly move the handlebars from side-to-side. Dents or roughness in the bearing will be felt and the bars will not move smoothly. **Note:** *Make sure any hesitation in movement is not being caused by the cables and wiring harnesses that run to the handlebars.*

4   Next, grasp the fork legs and try to move the wheel forward and backward. Any looseness in the steering head bearings will be felt. If play is felt in the bearings, adjust the steering head as follows.

Adjustment

5   Remove the handlebar and upper triple clamp (see Chapter 6).

6   Loosen the ring nut, then retighten it to the initial torque listed in this Chapter's Specifications (**see illustration**).

7   Loosen the ring nut one full turn, then tighten it to the final torque listed in this Chapter's Specifications.

8   Recheck the adjustment by turning the steering stem lock-to-lock. If it feels a little tight, loosen the lower ring nut a little bit and recheck. If it feels loose, repeat the adjustment (see Steps 6 and 7).

9   Install the upper triple clamp and handlebar (see Chapter 6).

14 Suspension - check

1   The suspension components must be maintained in top operating condition to ensure rider safety. Loose, worn or damaged suspension parts decrease the motorcycle's stability and control.

2   Lock the front brake and push on the handlebars to compress the front forks several times. See if they move up-and-down smoothly without binding. If binding is felt, the forks should be disassembled and inspected as described in Chapter 6.

3   Check the tightness of all front suspension nuts and bolts to be sure none have worked loose.

4   Check for oil leaking from the fork seals (**see illustration**). As part of this inspection, check for foreign material at the bases of the fork legs and for built-up dirt behind the fork protectors (**see illustrations**). Either of these will wear out the fork seals prematurely when the fork leg compresses, causing the seals to leak.

5   Inspect the rear shock absorber for fluid leakage and tightness of the mounting nuts and bolts. If leakage is found, the shock should be replaced.

6   Check the rear suspension linkage for loose fasteners, leaking grease and damaged components. Tighten loose fasteners to the torques listed in the Chapter 6 Specifications. If grease has been leaking, disassemble the linkage, clean and lubricate it and replace the seals (see Chapter 6). Replace any damaged components with new ones. The rear shock linkage bearings should be lubricated before each race if the bike is used in competition (see Chapter 6).

7   Support the motorcycle securely upright with its rear wheel off the ground. Grab the swingarm on each side, just ahead of the axle. Rock the swingarm from side to side - there should be no discernible movement

14.4b ... can be caused by sand or dirt at the bottom of the fork leg ...

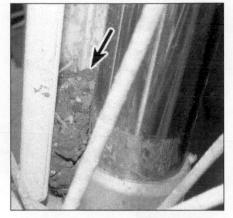

14.4c ... and built-up dirt between the fork leg and protector (arrow) can cause the same problem

15.5 If coolant has been leaking from the weep hole, the water pump should be removed and inspected

at the rear. If there's a little movement or a slight clicking can be heard, make sure the swingarm pivot shaft is tight. If the pivot shaft is tight but movement is still noticeable, the swingarm will have to be removed and the bearings replaced as described in Chapter 6.

8   Inspect the tightness of the rear suspension nuts and bolts.

15 Cooling system - inspection and coolant change

System inspection

1   Refer to Chapter 3 and remove the radiator shrouds. Clean mud, leaves or other obstructions out of the radiator fins with low-pressure water or compressed air.

2   If any fins are bent, carefully straighten them with a small screwdriver, taking care not to puncture the coolant tubes in the radiator.

3   If more than 20 percent of the radiator's surface area is blocked, replace the radiator.

4   Check the coolant hoses for swelling, cracks, burns, cuts or other defects. Replace the hoses if their condition is doubtful. Make sure the hose clamps are tight and free of corrosion. Tighten loose clamps and replace corroded or damaged ones.

5   Check for leaks at the water pump weep hole (see illustration) and gaskets. Also check for leaks at the coolant drain plug(s). Replace water pump or drain plug gaskets if they've been leaking. If coolant has been leaking from the weep hole, it's time for a new water pump seal (see Chapter 3).

Coolant change

⚠ Warning 1: Do not allow antifreeze to come in contact with your skin or painted surfaces of the vehicle. Rinse off spills immediately with plenty of water. Antifreeze is highly toxic if ingested. Never leave antifreeze lying around in an open container or in puddles on the floor; children and pets are attracted by its sweet smell and may drink it. Check with local authorities about disposing of

used antifreeze. Many communities have collection centers which will see that antifreeze is disposed of safely.

⚠ Warning 2: Don't remove the radiator cap or the drain bolt when the engine and radiator are hot. Scalding coolant and steam will be blown out under pressure, which could cause serious injury. To open the radiator cap or the drain bolts, place a thick rag, like a towel, over the radiator cap; slowly rotate the cap counterclockwise to the first stop. This procedure allows any residual pressure to escape. When the steam has stopped escaping, press down on the cap while turning it counterclockwise and remove it.

6   Place a drain pan beneath the water pump drain bolt (see illustration).

7   Remove the drain bolt. Coolant will dribble out until the radiator cap is removed, then it will flow (see illustration).

8   Once the coolant has drained completely, place a new gasket on the drain bolt and install it in the engine. Fill the cooling system with the antifreeze and water mixture

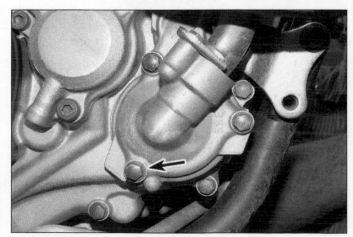

15.6 Remove the drain bolt and sealing washer (arrow) to drain the coolant

15.7 Coolant will spurt out once the radiator cap is removed, so have a container ready

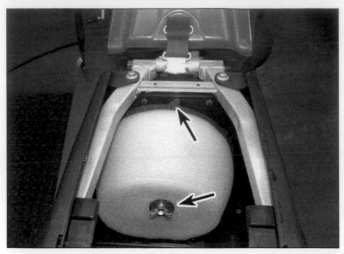

16.2 Unscrew the element bolt (lower arrow) and lift out the element; the tab (upper arrow) should be up when the element is installed

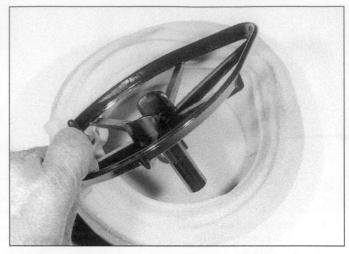

16.3 Separate the foam element from the holder for cleaning and re-oiling

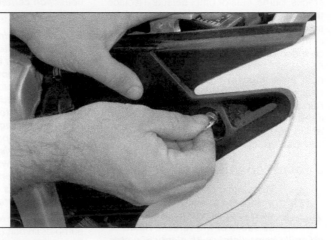

16.9 Undo the quick-release screw and take off the cover . . .

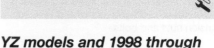

16 Air cleaner element - servicing

YZ models and 1998 through 2002 WR models

1   Remove the seat (see Chapter 8).
2   Unscrew the filter element bolt or nut **(see illustration)**. Lift out the element and pull the bolt out of it.
3   Separate the element holder from the foam element **(see illustration)**.
4   Thoroughly clean the element in high-flash point or non-flammable safety solvent. Do not use gasoline. After cleaning, squeeze out the solvent and allow the element to dry completely.
5   Coat the element with foam filter oil recommended in this Chapter's Specifications, then squeeze the element to work the oil through the foam. Once the element is

listed in this Chapter's Specifications.
9   Lean the bike about twenty-degrees to one side, then the other, several times. This will allow air trapped in the coolant passages to make its way to the top of the coolant.
10   Check coolant level. It should be up

to the bottom of the radiator filler neck. Add more antifreeze/water mixture if necessary.
11   Start the engine and check for leaks. Warm up the engine, then let it cool completely and recheck the coolant level.

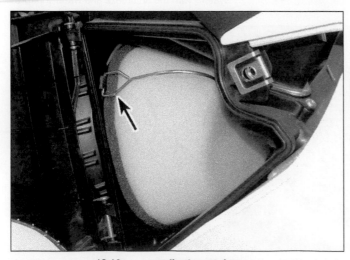

16.10a . . . unclip the retainer . . .

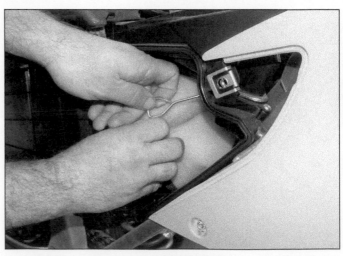

16.10b . . . pivot the retainer out of the way and remove the element

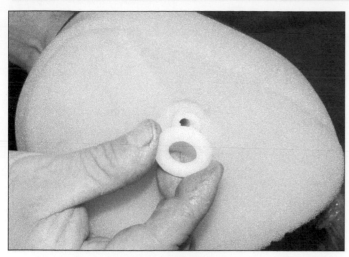

16.11 Remove the plastic washer from the element before cleaning it

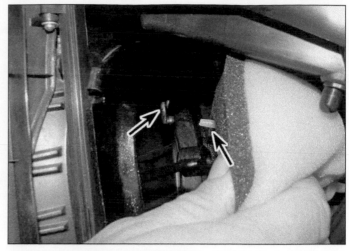

16.13 On installation, insert the plastic peg into the hole (arrows)

soaked with oil, squeeze out the excess.

6   Place the holder on the element. On some models, there's an inner tab on the end foam piece; align this with the groove in the element holder.

7   Install the element in the case with its locating tab up **(see illustration 16.2)**.

8   The remainder of installation is the reverse of the removal steps.

2003 and later WR models

9   Lift up the folding handle on the cover quick-release screw, unscrew it and pull the cover forward off of the bike **(see illustration)**.

10   Pull the element clip forward and unhook it **(see illustration)**. Fold the retainer away form the element and take the element out of the housing **(see illustration)**.

11   Take the plastic washer off the element **(see illustration)**. Take the plastic reinforcing plate off the other side of the element.

12   Clean and oil the filter as described in Steps 4 and 5.

13   Installation is the reverse of the removal steps. Place the plastic pin on the element plate in its hole in the air filter housing **(see illustration)**.

17 Exhaust system - inspection and glass wool replacement

> ⚠ **Warning: Make sure the exhaust system is cool before doing this procedure.**

1   Periodically check the exhaust system for leaks and loose fasteners (see Chapter 4).

2   If you're working on a WR model, remove and clean the spark arrester. Unbolt the trim cap (2005 and later models) and take it off the muffler. Unbolt the spark arrester and pull it out.

3   Tap the spark arrester on a hard surface to loosen any carbon. Finish cleaning it with a wire brush.

4   Install the spark arrester by reversing Step 2.

5   If you're working on a YZ model, at the specified interval, remove the muffler (see Chapter 4), clean the inner pipe and replace the glass wool in the muffler as described below.

6   Place the muffler bracket in a padded vise. Remove the bolts or drill out the pop rivets from the muffler casing, then pull out the inner pipe and glass wool insert. Pull the inner pipe out of the glass wool **(see illustrations)**.

7   Clean the small holes in the inner pipe with a wire brush.

8   Apply muffler sealant to the front and rear ends of the inner pipe where it contacts the muffler case. Install a new glass wool insert and the inner pipe in the case. Tighten the bolts or install new pop rivets.

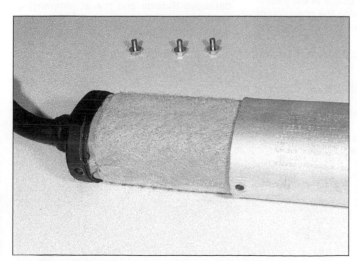

17.6a Pull the inner pipe and glass wool out of the muffler . . .

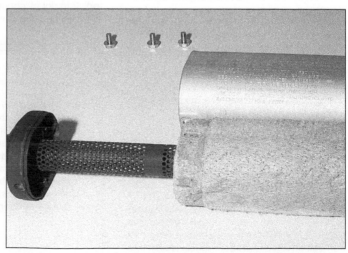

17.6b . . . and separate the inner pipe from the glass wool

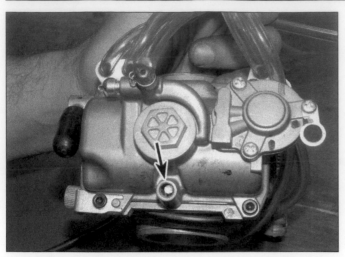

18.3a The pilot screw (arrow) is in the underside of the carburetor (carburetor removed for clarity)

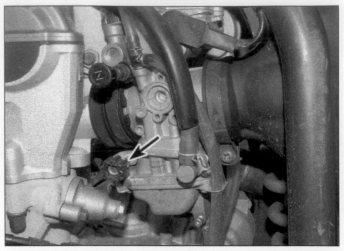

18.3b On some models, the throttle stop screw is turned with a knob (arrow)

18 Carburetor - check and adjustment

Check

1 Check around the carburetor for signs of fuel leakage. If the float chamber gasket or jet plug O-ring has been leaking, replace them (see Chapter 4).

2 Make sure the intake tubes are securely connected and in good condition. Tighten the clamps if they're loose. If the rubber has deteriorated or is damaged, remove the carburetor and replace the tubes with new ones (see Chapter 4).

Adjustment

3 Warm the engine to normal operating temperature (ten minutes of stop-and-go riding should be enough). Locate the throttle stop screw and pilot screw on the carburetor **(see 5.13 and the accompanying illustrations)**.

4 With the engine off, turn the pilot air screw in until it bottoms *lightly*. Back the screw out the number of turns listed in this Chapter's Specifications.

5 Start the engine and let it idle. Turn the throttle stop screw to set the idle to the speed listed in this Chapter's Specifications.

19 Spark plug - check and replacement

1 Remove the fuel tank (see Chapter 4). On 2002 and earlier models, detach the spark plug wire from the spark plug. The ignition coil on 2003 and later models is built into the spark plug boot. Disconnect the electrical

connector and twist the ignition coil to break it free from the plug, then pull it off **(see illustration)**.

Caution: Don't use pliers or a screwdriver to remove the coil - it's easily damaged.

2 If available, use compressed air to blow any accumulated debris from around the spark plug. Remove the plug with a spark plug socket.

3 Inspect the electrodes for wear. Both the center and side electrodes should have square edges and the side electrode should be of uniform thickness. Look for excessive deposits (especially oil fouling) and evidence of a cracked or chipped insulator around the center electrode. Compare your spark plug to the color spark plug chart on the inside of the back cover. Check the threads, the washer and the ceramic insulator body for cracks and other damage.

4 If the electrodes are not excessively worn, and if the deposits are not excessive, the plug can be regapped and reused (If no cracks or chips are visible in the insulator). If in doubt concerning the condition of the plug, replace it with a new one, as the

expense is minimal. The plug should be replaced at the interval listed in the maintenance schedule.

5 Cleaning the spark plug is not recommended.

6 Before installing a new plug, make sure it is the correct type and heat range. Check the gap between the electrodes, as it is not preset. For best results, use a wire-type gauge rather than a flat gauge to check the gap **(see illustration)**. If the gap must be adjusted, bend the side electrode only and be very careful not to chip or crack the insulator nose **(see illustration)**. Make sure the washer is in place before installing the plug.

7 Since the cylinder head is made of aluminum, which is soft and easily damaged, thread the plug into the head by hand. Slip a short length of hose over the end of the plug to use as a tool to thread it into place. The hose will grip the plug well enough to turn it, but will start to slip if the plug begins to cross-thread in the hole - this will prevent damaged threads and the accompanying repair costs.

8 Once the plug is finger tight, tighten

19.1 The ignition coil on 2003 and later models is built into the spark plug boot (arrow) - twist and pull to free it from the cylinder head, but don't use pliers or a screwdriver

19.6a Spark plug manufacturers recommend using a wire type gauge when checking the gap - if the wire doesn't slide between the electrodes with a slight drag, adjustment is required

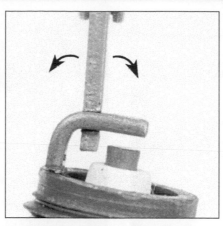

19.6b To change the gap, bend the side electrode only, as indicated by the arrows, and be very careful not to crack or chip the ceramic insulator surrounding the center electrode

19.9 Lubricate the spark plug seal with silicone spray

the plug to the torque listed in this Chapter's Specifications.
9   Lubricate the spark plug seal with silicone spray and wipe off the excess **(see illustration)**. Reconnect the spark plug cap and push it in until it clicks.

20 Engine oil and filter - change

1   Consistent routine oil changes are the single most important maintenance procedure you can perform on these models. The oil not only lubricates the internal parts of the transmission and clutch, but it also acts as a coolant, a cleaner, a sealant, and a protectant. Because of these demands,

the oil takes a terrific amount of abuse and should be replaced often with new oil of the recommended grade and type. Saving a little money on the difference in cost between a good oil and a cheap oil won't pay off if the engine is damaged.

Caution: Yamaha recommends against using oil additives. The engine oil also lubricates the clutch plates, and additives may cause the clutch to slip. Do not use oils labeled "Energy Conserving II" or "CD."

2   Before changing the oil, warm up the engine by running it for several minutes so the oil will drain easily. Be careful when draining the oil, as the exhaust pipe, the engine and the oil itself can cause severe burns.
3   Park the motorcycle over a clean drain pan. Remove the skid plate(s) under or at the front of the frame (see Chapter 8).
4   Remove the oil filler cap to vent the

crankcase and act as a reminder that there is no oil in the engine. Remove the dipstick on models so equipped.

Frame oil tank models

Note: *This step applies to 1998 through 2005 YZ400/426/450 models, 1998 through 2006 WR400/426/450 models, 2000 through 2005 YZ250 models and 2000 through 2006 WR250 models.*
5   Locate the crankcase drain plug(s). On 1998 through 2003 400/426/450 models, there's a single plug in the bottom of the crankcase, accessible through a hole in the skid plate. On 250 models, there's a single small plug, on the left side of the crankcase just below and in front of the neutral switch. On 2004 and 2005 450 models, there are two plugs, one in the left rear corner of the crankcase and one on the left side **(see illustrations)**.

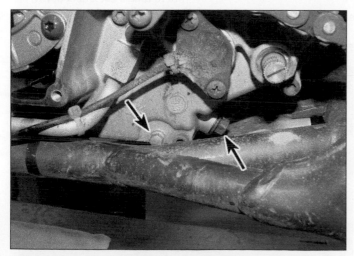

20.5a All models with a frame oil tank have a drain plug on the bottom left of the crankcase (left arrow) - 450 models have an additional plug (right arrow) . . .

20.5b . . . which is accessible through a hole in the skid plate

20.6a Remove the dipstick from the side of the frame (shown) or from the top . . .

20.6b . . . oil is added through the filler hole (arrow) or, if there isn't one, through the dipstick hole

20.7 Remove the drain plug and sealing washer from the frame oil tank

20.8 Loosen the hose clamp (lower arrow) and slide it down, then unscrew the strainer hex (upper arrow), letting the strainer turn inside the hose . . .

6   Remove the dipstick from the frame oil tank **(see illustration)**. On 1998 through 2003 models, the dipstick hole is the oil filler hole. On 2004 and 2005 models, the filler hole is in the left front corner of the crankcase **(see illustration)**.

7   Remove the drain plug from the frame oil tank **(see illustration)**.

8   Loosen the hose clamp on the right oil

hose at the frame tank **(see illustration)**. Slide the clamp down to the bottom of the hose.

9   Unscrew the oil strainer screen from the frame **(see illustration)**. Turn the strainer screen with a wrench on the hex, letting it spin inside the hose as it's unscrewed.

10   The right external oil (a combined metal pipe and hose) isn't flexible enough to allow the strainer screen to be pulled out of the frame tank while the oil line is still attached to the engine. For this reason, you'll need to detach the right oil line from the engine **(see illustration)**. Refer to Chapter 2 if necessary. Take the oil line out, with the strainer screen still in it, then separate the strainer screen from the hose portion of the oil line.

11   On 2000 YZ250 and WR250 models, as well as 1998 through 2003 400/426/450 models, remove the exhaust pipe for access to the filter cover bolts (see Chapter 4).

⚠️ *Warning: Since the engine was warmed up to drain the oil, the exhaust system may still be too hot to touch. Let it cool down enough that it won't burn you before removing it.*

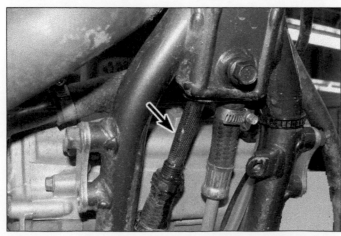

20.9 . . . lower the strainer (arrow) out of the frame . . .

20.10 . . . unbolt the oil line from the crankcase and remove the oil line together with the strainer

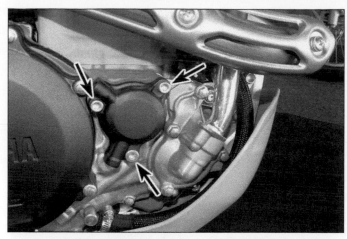

20.12 Remove the lower bolt to drain the oil from the filter housing, then remove the upper bolts . . .

20.13a . . . then remove the cover, noting the locations of the O-rings (arrows) . . .

12   Remove the lower bolt from the oil filter cover and let the oil drain (see illustration).
13   Remove the remaining cover bolts, take the cover off and remove the filter element (see illustrations).
14   Check the O-rings and drain plug sealing washers and replace them if their condition is in doubt. It's a good idea to replace them whenever the oil is changed.
15   Clean the oil strainer screen in solvent and blow it dry. If there's any clogging that can't be removed, replace the strainer. Check the strainer hose and replace it if it's cracked, brittle or deteriorated.
16   Install the strainer screen in the hose, making sure that the lower hose clamp is tight and the upper hose clamp is on the hose, but loose. Install a new sealing washer on the strainer screen. Install the oil line on the engine, using a new O-ring, positioning the strainer screen below its hole in the frame tank as you do so. Install the oil line bolt and tighten it to the torque listed in the Chapter 2 Specifications. Push the strainer screen up into the frame tank and tighten it to the torque listed in this Chapter's Specifications.

17   Install the drain bolt(s), filter element and oil filter cover, tightening them to the torque listed in this Chapter's Specifications (see illustration).
18   Add the amount of oil listed in this Chapter's Specifications to the engine through the oil filler hole. Install the filler cap or dipstick.
19   Locate the oil pressure check bolt (see illustration). On early models, this is the oil line banjo bolt; on later models, there's a separate oil pressure check bolt in the cylinder head. Loosen the bolt, but don't remove it.
20   Start the engine and let it idle. Oil should seep from the check bolt within one minute. If it doesn't, stop and find out why before continuing. Don't let the engine keep running in this condition or severe engine damage will occur.

External oil tank models
Note: This procedure applies to 2006 and later YZ250 models and 2007 and later WR250 models.
21   After running the engine to warm it up (see Step 2), shut the engine off and let the

20.13b . . . remove the filter element, noting how the post (arrow) fits inside the open end of the element

bike sit undisturbed for five minutes.
22   Remove the oil filler cap (see illustration).
23   Remove the vent bolt and washer from

20.17 The closed end of the filter element goes into the engine first

20.19 Later models have a separate oil check bolt (right arrow) - if there isn't one, use the oil line banjo bolt (left arrow)

20.22 On 250 models with an external oil tank, add oil through the filler hole (left arrow) and check the level in the sight glass (right arrow)

20.23 Remove the vent plug and sealing washer from the oil tank

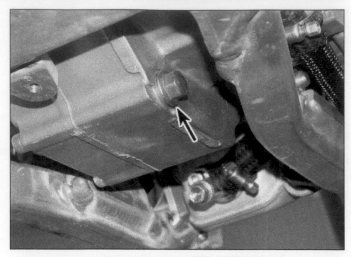

20.25 The drain plug on external oil tank models is at the right rear corner of the crankcase

the upper front of the oil tank **(see illustration)**.

24 Place a drain pan beneath the oil tank drain bolt and remove the bolt.

25 Place a drain pan beneath the crankcase drain bolt at the right rear corner of the engine and unscrew the bolt **(see illustration)**.

26 Loosen the hose clamp on the lower oil tank hose and disconnect the hose from the fitting **(see illustration)**.

27 Remove the bolt that secures the oil strainer screen to the tank and pull the screen out **(see illustration)**.

28 Remove the oil filter element as described in Steps 12 and 13 above.

29 Inspect the oil strainer as described in Step 15 above.

30 Install the strainer screen in the oil tank, using a new O-ring. Connect the hose and tighten the clamp.

31 Install the drain bolt, new filter element and oil filter cover, tightening them to the torque listed in this Chapter's Specifications **(see illustration 20.17)**.

32 Add the amount of oil listed in this Chapter's Specifications to the engine through the oil filler hole. Install the filler cap.

33 Install the oil tank vent bolt **(see illustration 20.23)**.

34 Locate the oil pressure check bolt **(see illustration 20.19)**. On early models, this

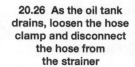

20.26 As the oil tank drains, loosen the hose clamp and disconnect the hose from the strainer

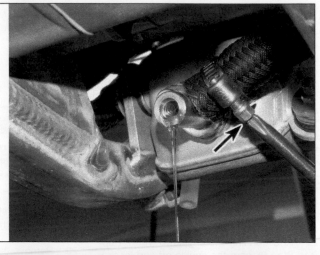

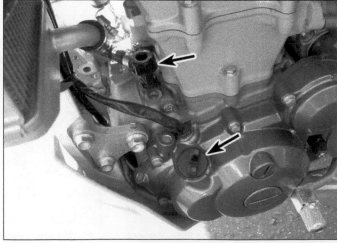

20.27 Unbolt the strainer and take it out of the tank, noting the location of the O-ring

20.38 On bikes with an integral oil tank, check the level with the dipstick (left arrow) and add oil through the filler hole (right arrow)

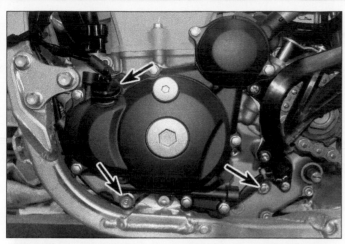

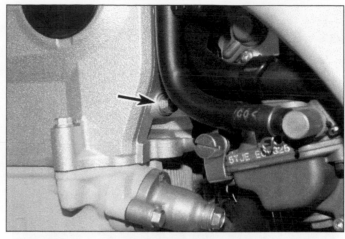

20.39 Oil filler cap (upper arrow), left drain plug (lower left arrow) and oil level check bolt (lower right arrow) (integral oil tank models)

20.44 Models without an external oil pipe have an oil pressure check bolt in the cylinder head (arrow)

is the oil line banjo bolt; on later models, there's a separate oil pressure check bolt in the cylinder head. Loosen the bolt, but don't remove it.

35   Start the engine and let it idle. Oil should seep from the check bolt within one minute. If it doesn't, stop and find out why before continuing. Don't let the engine keep running in this condition or severe engine damage will occur.

36   Check the oil level in the sight glass and adjust it as needed **(see illustration 20.22)**.

Internal oil tank models

Note: *This procedure applies to 2006 and later YZ450 models and 2007 and later WR450 models. On these bikes, the oil tank is a separate reservoir inside the front portion of the crankcase.*

37   After running the engine to warm it up (see Step 2), shut the engine off.

38   Remove the oil filler cap and dipstick **(see illustration)**.

39   Place a drain pan under the engine. Remove the drain plug at the left front corner of the crankcase **(see illustration)**. Remove the drain plug at the right rear corner of the

crankcase **(see illustration 20.25)**.

40   Remove the oil filter cover and element as described in Steps 12 and 13 above.

41   Let the oil drain from the drain plug holes until it stops.

42   Install the drain bolt, new filter element and oil filter cover, tightening them to the torque listed in this Chapter's Specifications **(see illustration 20.17)**.

43   Add the amount of oil listed in this Chapter's Specifications to the engine through the filler hole. Remove the oil level check bolt from the left side of the crankcase **(see illustration 20.39)**. If oil flows from the hole, let it flow until it stops. Reinstall the check bolt, using a new washer.

44   Locate the oil pressure check bolt in the cylinder head **(see illustration)**. Loosen the bolt, but don't remove it.

45   Start the engine and let it idle. Oil should seep from the check bolt within one minute. If it doesn't, stop and find out why before continuing. Don't let the engine keep running in this condition or severe engine damage will occur.

46   Install the filler cap and recheck the oil level with the dipstick.

All models

47   The old oil drained from the engine cannot be reused in its present state and should be disposed of. Check with your local refuse disposal company, disposal facility or environmental agency to see whether they will accept the oil for recycling. Don't pour used oil into drains or onto the ground. After the oil has cooled, it can be drained into a suitable container (capped plastic jugs, topped bottles, milk cartons, etc.) for transport to one of these disposal sites.

21 Valve clearances - check and adjustment

⚠ ***Warning: The engine must be completely cool before beginning this procedure.***

Check

1   Drain the cooling system and remove the right-hand radiator (see Section 15 and Chapter 3).

2   Remove the seat (see Chapter 8).

3   Remove the fuel tank and carburetor (see Chapter 4).

4   Remove the spark plug (see Section 19).

5   Remove the engine upper mounting bracket and valve cover (see Chapter 2).

6   Unscrew the crankshaft end plug and timing hole access plug. Use a coin gripped with pliers if you don't have a screwdriver big enough to fit the slots **(see illustration)**.

7   Make sure the transmission is in Neutral. Using a socket on the crankshaft rotation bolt, turn the engine to position the piston at Top Dead Center on the compression stroke. **Note:** *Rotate the crankshaft counterclockwise. When this occurs, the timing mark on the rotor will align with the notch in the*

21.6 Use a coin or large washer and a pair of pliers to unscrew the crankshaft bolt cover and timing hole cover directly above it

21.7a Align the timing mark with the notch (arrow). If the piston is on the compression stroke . . .

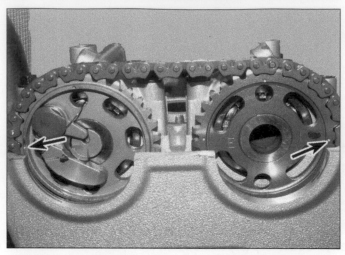

21.7b . . . the camshaft punch marks will align with the valve cover gasket surface on the cylinder head (arrows) . . .

crankcase cover **(see illustration)**. To make sure the piston is on the compression stroke, not the exhaust stroke, check the camshaft position. The punch marks on the camshafts should be 180-degrees away from each other and aligned with the gasket surface on the top of the cylinder head **(see illustration)**. In addition, the camshaft lobes should point away from each other **(see illustration)**. If the camshaft lobes and marks are out of position, rotate the crankshaft one full turn, so the crankshaft timing mark again aligns.

8   With the engine in this position, all of the valve clearances can be checked. There are five valves, three intakes and two exhausts.

9   Insert a feeler gauge of the same thickness as the valve clearance listed in this Chapter's Specifications between each of the cam lobes and the valve lifter beneath it **(see illustrations)**. Pull the feeler gauge out slowly - you should feel a light drag. If there's no drag, the clearance is too loose. If there's a heavy drag, the clearance is too tight.

21.7c . . . and the cam lobes on the right-hand side of the engine will point away from each other like this

10   Write down the locations of any valves with incorrect clearances. Recheck the clearances of these valves, trying different feeler gauges until you find the thickness that fits correctly (a light drag). Once you do, write this thickness down - you'll need this information later to select a new valve adjusting shim.

11   If any of the clearances need to be

21.9a You'll need narrow feeler gauges to check the valve clearances - they can be bought in sets like this

21.9b Slip the feeler gauge between the cam lobe and lifter to measure the clearance

21.13a The shim thickness is marked on the shim . . .

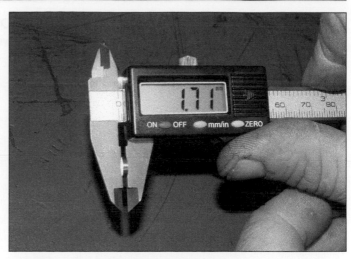

21.13b . . . but it should also be checked with a vernier caliper or micrometer

adjusted, go Step 12. If all of the clearances are within the Specifications, go to Step 18.

Adjustment

12   Remove the camshafts (see Chapter 2). Remove the lifter and shim for each of the valves that need to be adjusted.

13   Determine the thickness of the shim that was removed. It should be marked on the bottom of the shim (see illustration), but the ideal way is to measure it with a micrometer or vernier caliper (see illustration). Note: If the number on the shim does not end with zero or 5, round it off to the nearest zero or 5. For example, if the number on the shim is 228, round it off to 230. If it's 224, round it off to 225.

14   If the clearance (measured in Step 9 and written down in Step 10) was too large, you need a thicker shim. If it was too small, you need a thinner shim.

15   If the measured valve clearance in Step 9 was too great, subtract the mid-range specified clearance from the measured clearance. Write this number down, then add it to the thickness of the adjusting shim you removed. This will give you the thickness of the needed new shim. For example:

Measured clearance: 0.20 mm
Specified clearance: 0.10 to 0.15 mm
Mid-range (desired) clearance: 0.12 mm
Measured clearance minus desired clearance = 0.08 mm

So if the existing shim is numbered 180 (1.80 mm thick) the new shim should be 1.88 mm thick. The closest to this is a 200 (2.0 mm thick). This is the thickness of the new shim you will need for that valve.

16   Select new shims for any remaining valves that are not within the Specifications.

17   Install the new shims and their lifters (see Chapter 2).

18   The remainder of installation is the reverse of the removal steps. After the camshafts are installed, recheck the valve clearances to make sure they're within the Specifications.

Notes

Chapter 2
Engine, clutch and transmission

Contents

Degrees of difficulty

Easy, suitable for novice with little experience	**Fairly easy,** suitable for beginner with some experience	**Fairly difficult,** suitable for competent DIY mechanic	**Difficult,** suitable for experienced DIY mechanic	**Very difficult,** suitable for expert DIY or professional

Specifications

YZ250F, WR250F

General

Bore	77 mm (3.03 inches)
Stroke	53.6 mm (2.11 inches)
Displacement	249 cc

Camshaft

Lobe height (intake)
  Standard
    2006 and earlier .... 30.296 to 30.346 mm (1.1923 to 1.1947 inches)
    2007
      YZ250 .... 30.246 to 30.346 mm (1.1908 to 1.1947 inches)
      WR250 .... 29.65 to 29.75 mm (1.1673 to 1.1713 inches)
  Limit
    2006 and earlier .... 30.196 mm (1.188 inches)
    2007
      YZ250 .... 30.146 mm (1.1869 inches)
      WR250 .... 29.55 mm (1.1634 inches)
Lobe height (exhaust)
  Standard .... 30.399 to 30.499 mm (1.1968 to 1.2007 inches)
  Limit .... 30.196 mm (1.188 inches)
Camshaft runout limit .... 0.03 mm (0.012 inch)
Bearing oil clearance
  Standard
    2001 through 2004 .... 0.020 to 0.054 mm (0.0008 to 0.0021 inch)
    2005 and later .... 0.020 to 0.062 mm 0.0011 to 0.0024 inch)
  Limit .... 0.08 mm (0.0031 inch)

Cylinder head, valves and valve springs

Cylinder head warpage limit .... 0.05 mm 0.002 inch)
Valve stem runout .... 0.01 mm (0.0004 inch)
Valve stem diameter
  YZ250, WR450
    Intake
      Standard .... 3.975 to 3.990 mm (0.1565 to 0.1571 inch)
      Limit .... 3.945 mm (0.1553 inch)
    Exhaust
      Standard .... 4.460 to 4.475 mm (0.1756 to 0.1752 inch)
      Limit .... 4.430 mm (0.1744 inch)
Valve guide inside diameter
  Standard .... 4.000 to 4.012 mm (0.1575 to 0.1580 inch)
  Limit .... 4.050 mm (0.1594 inch)
Stem-to-guide clearance
  Intake
    Standard .... 0.010 to 0.037 mm (0.0004 to 0.0015 inch)
    Limit .... 0.08 mm (0.0031 inch)
  Exhaust
    Standard .... 0.025 to 0.052 mm (0.0010 to 0.0020 inch)
    Limit .... 0.1 mm (0.004 inch)
Valve seat width (intake and exhaust)
  Standard .... 0.9 to 1.1 mm (0.0354 to 0.0433 inch)
  Limit .... 1.6 mm (0.063 Inch)
Valve margin thickness
  2001 (intake and exhaust) .... 0.7 mm (0.276 inch)
  2002
    Intake .... 0.8 mm (0.315 inch)
    Exhaust .... 0.7 mm (0.276 inch)
Valve spring free length (intake)
  Standard
    YZ250 .... 37.81 mm (1.49 inch)
    WR250
      2001 through 2006 .... 37.81 mm (1.49 inch)
      2007 .... 36.58 mm (1.44 inches)
  Limit
    YZ250 .... 35.9 mm (1.413 inch)
    WR250
      2001 through 2003 .... 35.9 mm (1.413 inch)
      2004 and 2005 .... 36.81 mm (1.45 inch)
      2006 and later .... 35.58 mm (1.40 inch)
Valve spring free length (exhaust)
  Standard .... 37.54 mm (1.48 inch)
  Limit
    2001 through 2003 .... 35.7 mm (1.405 inch)
    2004 and later .... 36.54 mm (1.44 inch)

Valve spring bend limit
    Intake .... 1.7 mm (0.067 inch)
    Exhaust .... 1.6 mm (0.063 inch)

Cylinder
Bore diameter
    Standard .... 77.00 to 77.01 mm (3.0315 to 3.0319 inches)
    Limit .... Not specified
Out-of-round limit................................ 0.05 mm (0.002 inch)
Taper limit .... 0.05 mm (0.002 inch)
Measuring point.................................. Top, center and bottom of bore

Piston
Diameter .... 72.92 to 72.97 mm (2.8709 to 2.8728 inches)
Measuring point.................................. 8.0 mm (0.31 inch) from bottom of skirt
Piston-to-cylinder clearance
    2001 through 2005
        Standard .... 0.040 to 0.065 mm (0.0016 to 0.0026 inches)
        Limit .... 0.1 mm (0.004 inch)
    2006 and later
        Standard .... 0.030 to 0.055 mm (0.0012 to 0.0022 inches)
        Limit .... 0.1 mm (0.004 inch)
Piston pin bore
    Standard .... 16.002 to 16.013 mm (0.6300 to 0.6304 inch)
    Limit .... 16.043 mm (0.6316 inch)
Piston pin outer diameter
    Standard .... 15.991 to 16.000 mm (0.6296 to 0.6299 inch)
    Limit .... 15.971 mm (0.6288 inch)
Piston pin-to-piston clearance .... Not specified
Ring side clearance
    Top ring
        Standard .... 0.030 to 0.065 mm (0.0012 to 0.0026 inch)
        Limit .... 0.12 mm (0.0047 inch)
    Second ring
        Standard .... 0.020 to 0.055 mm (0.0008 to 0.0022 inch)
        Limit .... 0.12 mm (0.0047 inch)
    Oil ring.. Not specified
Ring end gap
    Top ring
        Standard .... 0.15 to 0.25 mm (0.006 to 0.010 inch)
        Limit .... 0.5 mm (0.20 inch)
    Second ring
        Standard .... 0.30 to 0.45 mm (0.012 to 0.018 inch)
        Limit .... 0.8 mm (0.31 inch)
    Oil ring
        Standard .... 0.10 to 0.40 mm (0.004 to 0.016 inch)
        Limit .... Not specified

Clutch
Spring free length
    YZ250, 2001 WR250
        Standard .... 40.4 mm (1.59 inches)
        Limit .... 39.4 mm (1.55 inches)
    2002 and later WR250
        Standard .... 37.0 mm (1.46 inches)
        Limit .... 36.0 mm (1.42 inches)
Metal plate thickness .... 1.1 to 1.3 mm (0.043 to 0.067 inch)
Friction plate thickness
    Standard .... 2.9 to 3.1 mm (0.114 to 0.122 inch)
    Limit .... 2.7 mm (0.11 inch)
Friction and metal plate warpage limit .... 0.1 mm (0.004 inch)

Oil pump
Outer rotor-to-body clearance .... 0.03 to 0.10 mm (0.0012 to 0.0039 inch)
Inner-to-outer rotor clearance
    Standard .... 0.12 mm (0.005 inch) or less
    Limit .... 0.20 mm (0.008 inch)

Oil pump (continued)

Rotor to straightedge clearance
Standard .... 0.09 to 0.17 mm (0.0035 to 0.0067 inch)
Limit .... 0.24 mm (0.009 inch)

Transmission

Shift fork shaft bend limit .... 0.05 mm (0.002 inch)
Mainshaft and countershaft runout limit......................... Not specified

Crankshaft and connecting rod

Runout limit .... 0.03 mm (0.0012 inch)
Assembly width .... 55.95 to 56.00 mm (2.203 to 2.205 inches)
Connecting rod big-end side clearance
Standard .... 0.15 to 0.45 mm (0.006 to 0.018 inch)
Limit .... 0.50 mm (0.020 inch)
Connecting rod small-end freeplay
Standard .... 0.4 to 1.0 mm (0.016 to 0.039 inch)
Limit .... 2.0 mm (0.08 inch)

YZ400F, WR400F

General

Bore .... 92 mm (3.62 inches)
Stroke .... 60.1 mm (2.37 inches)
Displacement.. 399 cc

Camshaft

Lobe height (intake)
Standard .... 31.7 to 31.8 mm (1.248 to 1.252 inches)
Limit .... 31.6 mm (1.244 inches)
Lobe height (exhaust)
Standard .... 31.2 to 31.3 mm (1.2283 to 1.2323 inches)
Limit .... 31.1 mm (1.224 inches)
Camshaft runout limit .... 0.03 mm (0.012 inch)
Bearing oil clearance .... 0.020 to 0.051 mm (0.0008 to 0.0020 inch)

Cylinder head, valves and valve springs

Cylinder head warpage limit.. 0.05 mm 0.002 inch)
Valve stem runout .... 0.01 mm (0.0004 inch)
Valve stem diameter
Intake .... 4.475 to 4.490 mm (0.1762 to 0.1768 inch)
Exhaust .... 4.960 to 4.975 mm (0.1953 to 0.1959 inch)
Valve guide inside diameter
Intake .... 4.500 to 4.512 mm (0.1772 to 0.1776 inch)
Exhaust .... 5.000 to 5.012 mm (0.1969 to 0.1973 inch)
Stem-to-guide clearance
Intake
Standard .... 0.010 to 0.037 mm (0.0004 to 0.0015 inch)
Limit .... 0.08 mm (0.0031 inch)
Exhaust
Standard .... 0.025 to 0.052 mm (0.0010 to 0.0020 inch)
Limit .... 0.1 mm (0.004 inch)
Valve seat width (intake and exhaust)
Standard .... 0.9 to 1.1 mm (0.0354 to 0.0433 inch)
Limit .... 1.5 mm (0.059 inch)
Valve margin thickness (intake and exhaust)
Standard .... 1.0 mm (0.039 inch)
Limit .... 0.85 mm (0.033 inch)
Valve spring free length (intake)
All except 2000 WR400
Standard .... 39.42 mm (1.55 inch)
Limit .... 37.5 mm (1.48 inch)
2000 WR400
Standard .... 40.37 mm (1.59 inch)
Limit .... 39.4 mm (1.55 inch)
Valve spring free length (exhaust)
All except 2000 WR400
Standard .... 40.77 mm (1.61 inch)
Limit .... 38.7 mm (1.52 inch)

2000 WR400
    Standard .... 42.66 mm (1.68 inch)
    Limit .... 41.7 mm (1.64 inch)
Valve spring bend limit
  All except 2000 WR400
    Intake .... 1.7 mm (0.067 inch)
    Exhaust .... 1.8 mm (0.071 inch)
  2000 WR400
    Intake .... 1.8 mm (0.071 inch)
    Exhaust .... 1.9 mm (0.075 inch)

Cylinder

Bore diameter
    Standard .... 92.00 to 92.01 mm (3.6220 to 3.6224 inches)
    Limit .... Not specified
Out-of-round limit .... 0.05 mm (0.002 inch)
Taper limit .... 0.05 mm (0.002 inch)
Measuring point .... Top, center and bottom of bore

Piston

Diameter
    1998 .... 92.00 to 92.01 mm (3.6220 to 3.6224 inches)
    1999 and 2000 .... 91.945 to 91.960 mm (3.6199 to 3.6205 inches)
Measuring point .... 9.0 mm (0.35 inch) from bottom of skirt
Piston-to-cylinder clearance
  1998
    Standard .... 0.072 to 0.085 mm (0.0028 to 0.0033 inches)
    Limit .... 0.15 mm (0.006 inch)
  1999 and 2000
    Standard .... 0.040 to 065 mm (0.001 to 0.0026 inch)
    Limit .... 0.1 mm (0.004 inch)
Piston pin bore .... 18.004 to 18.015 mm (0.7088 to 0.7093 inch)
Piston pin outer diameter .... 17.991 to 18.000 mm (0.7083 to 0.7097 inch)
Piston pin-to-piston clearance .... Not specified
Ring side clearance
  Top ring
    Standard .... 0.030 to 0.065 mm (0.0012 to 0.0026 inch)
    Limit .... 0.13 mm (0.005 inch)
  Second ring
    Standard .... 0.020 to 0.055 mm (0.0008 to 0.0022 inch)
    Limit .... 0.13 mm (0.005 inch)
  Oil ring .... Not specified
Ring end gap
  Top ring
    Standard .... 0.20 to 0.35 mm (0.008 to 0.014 inch)
    Limit .... 0.7 mm (0.28 inch)
  Second ring
    Standard .... 0.40 to 0.55 mm (0.016 to 0.022 inch)
    Limit .... 0.8 mm (0.31 inch)
  Oil ring
    Standard .... 0.20 to 0.50 mm (0.01 to 0.02 inch)
    Limit .... Not specified

Clutch

Spring free length
    Standard .... 44 mm (1.73 inches)
    Limit .... 43 mm (1.69 inches)
Metal plate thickness .... 1.1 to 1.3 mm (0.043 to 0.067 inch)
Friction plate thickness
  Thinner plates
    Standard .... 2.72 to 2.88 mm (0.107 to 0.113 inch)
    Limit .... 2.62 mm (0.103 inch)
  Thick plate
    Standard .... 2.92 to 3.08 mm (0.115 to 0.121 inch)
    Limit .... 2.7 mm (0.106 inch)
Friction plate warpage limit .... Not specified
Metal plate warpage limit .... 0.2 mm (0.008 inch)

Oil pump
Outer rotor-to-body clearance
  All except 2000 WR400
    Standard .... 0.03 to 0.08 mm (0.0012 to 0.0031 inch)
    Limit .... 0.15 mm (0.006 inch)
  2000 WR400
    Standard .... 0.09 to 0.15 mm (0.0035 to 0.0059 inch)
    Limit .... 0.22 mm (0.009 inch)
Inner-to-outer rotor clearance
  Standard .... 0.07 to 0.12 mm (0.0028 to 0.005 inch)
  Limit .... 0.15 mm (0.006 inch)
Rotor-to-straightedge clearance
  Standard .... 0.03 to 0.08 mm (0.0012 to 0.0031 inch)
  Limit .... 0.15 mm (0.006 inch)

Transmission
Shift fork shaft bend limit .... 0.05 mm (0.002 inch)
Mainshaft and countershaft runout limit.......................... Not specified

Crankshaft and connecting rod
Runout limit .... 0.03 mm (0.0012 inch)
Assembly width .... 61.95 to 62.00 mm (2.439 to 2.441 inches)
Connecting rod big-end side clearance
  Standard .... 0.15 to 0.45 mm (0.006 to 0.018 inch)
  Limit .... 0.50 mm (0.020 inch)
Connecting rod small-end freeplay
  Standard
    1998... 0.8 to 1.0 mm (0.032 to 0.039 inch)
    1999 and 2000.. 0.4 to 1.0 mm (0.02 to 0.04 inch)
  Limit .... 2.0 mm (0.08 inch)

YZ426F, WR426F

General
Bore .... 95 mm (3.74 inches)
Stroke .... 60.1 mm (2.37 inches)
Displacement.. 426 cc

Camshaft
Lobe height (intake)
  Standard .... 31.7 to 31.8 mm (1.248 to 1.252 inches)
  Limit .... 31.6 mm (1.244 inches)
Lobe height (exhaust)
  Standard .... 31.2 to 31.3 mm (1.2283 to 1.2323 inches)
  Limit .... 31.1 mm (1.224 inches)
Camshaft runout limit... 0.03 mm (0.012 inch)
Bearing oil clearance .... 0.020 to 0.051 mm (0.0008 to 0.0020 inch)

Cylinder head, valves and valve springs
Cylinder head warpage limit... 0.05 mm 0.002 inch)
Valve stem runout.. 0.01 mm (0.0004 inch)
Valve stem diameter
  Intake .... 4.475 to 4.490 mm (0.1762 to 0.1768 inch)
  Exhaust
    2000 YZ426 .... 4.960 to 4.975 mm (0.1953 to 0.1959 inch)
    All except 2000 YZ426 .... 4.965 to 4.980 mm (0.1955 to 0.1961 inch)
Valve guide inside diameter
  Intake .... 4.500 to 4.512 mm (0.1772 to 0.1776 inch)
  Exhaust .... 5.000 to 5.012 mm (0.1969 to 0.1973 inch)
Stem-to-guide clearance
  Intake
    Standard.. 0.010 to 0.037 mm (0.0004 to 0.0015 inch)
    Limit... 0.08 mm (0.0031 inch)
  Exhaust
    2000 YZ426
      Standard... 0.025 to 0.052 mm (0.0010 to 0.0020 inch)
      Limit.. 0.1 mm (0.004 inch)
    All except 2000 YZ426
      Standard... 0.020 to 0.047 mm (0.0008 to 0.0019 inch)
      Limit.. 0.1 mm (0.004 inch)

Valve seat width (intake and exhaust)
    Standard .... 0.9 to 1.1 mm (0.0354 to 0.0433 inch)
    Limit .... 1.5 mm (0.059 inch)
Valve margin thickness (intake and exhaust)
    Standard .... 1.0 mm (0.039 inch)
    Limit .... 0.85 mm (0.033 inch)
Valve spring free length (intake)
    2000 YZ426
        Standard .... 39.42 mm (1.55 inch)
        Limit .... 37.5 mm (1.48 inch)
    All except 2000 YZ426
        Standard .... 41.36 mm (1.63 inch)
        Limit .... 39.3 mm (1.55 inch)
Valve spring free length (exhaust)
    2000 YZ426
        Standard .... 41.80 mm (1.65 inch)
        Limit .... 39.8 mm (1.57 inch)
    All except 2000 YZ426
        Standard .... 43.60 mm (1.72 inch)
        Limit .... 41.4 mm (1.63 inch)
Valve spring bend limit
    2000 YZ426
        Intake .... 1.7 mm (0.067 inch)
        Exhaust .... 1.8 mm (0.071 inch)
    All except 2000 YZ426
        Intake .... 1.8 mm (0.071 inch)
        Exhaust .... 1.9 mm (0.075 inch)

Cylinder
Bore diameter
    Standard .... 95.00 to 95.01 mm (3.7402 to 3.7406 inches)
    Limit .... Not specified
Out-of-round limit .... 0.05 mm (0.002 inch)
Taper limit .... 0.05 mm (0.002 inch)
Measuring point .... Top, center and bottom of bore

Piston
Diameter .... 94.945 to 94.960 mm (3.738 to 3.739 inches)
Measuring point .... 9.0 mm (0.35 inch) from bottom of skirt
Piston-to-cylinder clearance
    Standard .... 0.040 to 065 mm (0.001 to 0.0026 inch)
    Limit .... 0.1 mm (0.004 inch)
Piston pin bore
    Standard .... 19.004 to 19.015 mm (0.7482 to 0.7486 inch)
    Limit .... 19.045 mm (0.7498 inch)
Piston pin outer diameter
    Standard .... 18.991 to 19.000 mm (0.7477 to 0.7480 inch)
    Limit .... 18.971 mm (0.7469 inch)
Piston pin-to-piston clearance .... Not specified
Ring side clearance
    Top ring
        Standard .... 0.030 to 0.065 mm (0.0012 to 0.0026 inch)
        Limit .... 0.13 mm (0.005 inch)
    Second ring
        Standard .... 0.020 to 0.030 mm (0.0008 to 0.0012 inch)
        Limit .... 0.55 mm (0.22 inch)
    Oil ring .... Not specified
Ring end gap
    Top ring
        Standard .... 0.20 to 0.35 mm (0.008 to 0.014 inch)
        Limit .... 0.7 mm (0.28 inch)
    Second ring
        Standard .... 0.35 to 0.50 mm (0.014 to 0.020 inch)
        Limit .... 0.85 mm (0.33 inch)
    Oil ring
        Standard .... 0.20 to 0.50 mm (0.01 to 0.02 inch)
        Limit .... Not specified

Clutch

Spring free length
  2000 YZ426F
    Standard .... 44 mm (1.73 inches)
    Limit .... 43 mm (1.69 inches)
  2001 and 2002 YZ426F
    Standard .... 50 mm (1.97 inches)
    Limit .... 49 mm (1.93 inches)
  2001 and 2002 WR426F
    Standard .... 48.4 mm (1.91 inches)
    Limit .... 47.4 mm (1.87 inches)
Metal plate thickness .... 1.1 to 1.3 mm (0.043 to 0.067 inch)
Friction plate thickness
  Standard .... 2.9 to 3.1 mm (0.114 to 0.122 inch)
  Limit .... 2.7 mm (0.106 inch)
Friction plate warpage limit .... Not specified
Metal plate warpage limit .... 0.2 mm (0.008 inch)

Oil pump

2000 YZ426
  Outer rotor-to-body clearance
    Standard .... 0.07 to 0.012 mm (0.0028 to 0.0047 inch)
    Limit .... 0.15 mm (0.006 inch)
  Inner-to-outer rotor clearance
    Standard .... 0.07 to 0.12 mm (0.0028 to 0.005 inch)
    Limit .... 0.15 mm (0.006 inch)
  Rotor-to-straightedge clearance
    Standard .... 0.03 to 0.08 mm (0.0012 to 0.0031 inch)
    Limit .... 0.15 mm (0.006 inch)
All except 2000 YZ426
  Outer rotor-to-body clearance
    Standard .... 0.09 to 0.017 mm (0.00235 to 0.0067 inch)
    Limit .... 0.24 mm (0.009 inch)
  Inner-to-outer rotor clearance
    Standard .... 0.12 mm or less (0.0047 inch or less)
    Limit .... 0.20 mm (0.008 inch)
  Rotor-to-straightedge clearance
    Standard .... 0.03 to 0.10 mm (0.0012 to 0.0039 inch)
    Limit .... 0.17 mm (0.007 inch)

Transmission

Shift fork shaft bend limit .... 0.05 mm (0.002 inch)
Mainshaft and countershaft runout limit .... Not specified

Crankshaft and connecting rod

Runout limit .... 0.03 mm (0.0012 inch)
Assembly width .... 61.95 to 62.00 mm (2.439 to 2.441 inches)
Connecting rod big-end side clearance
  Standard .... 0.15 to 0.45 mm (0.006 to 0.018 inch)
  Limit .... 0.50 mm (0.020 inch)
Connecting rod small-end freeplay
  Standard .... 0.4 to 1.0 mm (0.02 to 0.04 inch)
  Limit .... 2.0 mm (0.08 inch)

YZ450F, WR450F

General

Bore .... 95 mm (3.74 inches)
Stroke .... 63.4 mm (2.50 inches)
Displacement .... 450 cc

Camshaft

Lobe height (intake)
  YZ450F, 2003 through 2006 WR450F
    Standard .... 31.2 to 31.3 mm (1.2283 to 1.2323 inches)
    Limit .... 31.1 mm (1.2244 inches)
  WR450F (2007 and later)
    Standard .... 30.1 to 30.2 mm (1.1850 to 1.1890 inches)
    Limit .... 30.0 mm (1.1811 inches)

Lobe height (exhaust)
   YZ450F (2007 and earlier)
      Standard .... 30.95 to 31.05 mm (1.2185 to 1.2224 inches)
      Limit .... 30.85 mm (1.2146 inches)
   YZ450F (2008)
      Standard .... 30.90 to 31.00 mm (1.2165 to 1.2205 inches)
      Limit .... 30.85 mm (1.2146 inches)
   WR450F (2006 and earlier)
      Standard .... 30.95 to 31.05 mm (1.2185 to 1.2224 inches)
      Limit .... 30.85 mm (1.2146 inches)
   WR450F (2007 and later)
      Standard .... 30.3 to 30.3 mm (1.1890 to 1.1929 inches)
      Limit .... 30.1 mm (1.1850 inches)
Camshaft runout limit .... 0.03 mm (0.012 inch)
Bearing oil clearance
   Standard .... 0.020 to 0.054 mm (0.0008 to 0.0021 inch)
   Limit .... 0.08 mm (0.003 inch)

Cylinder head, valves and valve springs

Cylinder head warpage limit... 0.05 mm 0.002 inch)
Valve stem runout... 0.01 mm (0.0004 inch)
Valve stem diameter
   Intake .... 4.475 to 4.490 mm (0.1762 to 0.1768 inch)
   Exhaust .... 4.965 to 4.980 mm (0.1955 to 0.1961 inch)
Valve guide inside diameter
   Intake .... 4.500 to 4.512 mm (0.1772 to 0.1776 inch)
   Exhaust .... 5.000 to 5.012 mm (0.1969 to 0.1973 inch)
Stem-to-guide clearance
   Intake
      Standard .... 0.010 to 0.037 mm (0.0004 to 0.0015 inch)
      Limit .... 0.08 mm (0.0031 Inch)
   Exhaust
      Standard .... 0.020 to 0.047 mm (0.0008 to 0.0019 Inch)
      Limit .... 0.1 mm (0.004 inch)
Valve seat width (intake and exhaust)
   Standard .... 0.9 to 1.1 mm (0.0354 to 0.0433 inch)
   Limit .... 1.6 mm (0.063 inch)
Valve margin thickness (intake and exhaust)
   Standard .... 1.0 mm (0.039 inch)
   Limit .... 0.85 mm (0.033 inch)
Valve spring free length (intake)
   Standard .... 37.03 mm (1.46 inch)
   Limit .... 36.03 mm (1.42 inch)
Valve spring free length (exhaust)
   Standard .... 37.68 mm (1.48 Inch)
   Limit .... 36.68 mm (1.44 inch)
Valve spring bend limit .... Not specified

Cylinder

Bore diameter
   Standard .... 95.00 to 95.01 mm (3.7402 to 3.7406 inches)
   Limit .... Not specified
Out-of-round limit... 0.05 mm (0.002 inch)
Taper limit .... 0.05 mm (0.002 inch)
Measuring point.. Top, center and bottom of bore

Piston

Diameter
   2003 through 2007.. 94.945 to 94.960 mm (3.738 to 3.739 inches)
   2008 .... 94.965 to 94.980 mm (3.7388 to 3.7394 inches)
Measuring point.. 8.0 mm (0.315 inch) from bottom of skirt
Piston-to-cylinder clearance
   2003 through 2007
      Standard .... 0.040 to 065 mm (0.001 to 0.0026 inch)
      Limit .... 0.1 mm (0.004 inch)
   2008
      Standard .... 0.020 to 0.045 mm (0.0008 to 0.0018 inch)
      Limit .... 0.1 mm (0.004 inch)

Piston (continued)

Piston pin bore
   Standard .... 18.004 to 18.015 mm (0.7088 to 0.70936 inch)
   Limit .... 18.045 mm (0.7104 inch)
Piston pin outer diameter
   Standard .... 17.991 to 18.000 mm (0.7083 to 0.7087 inch)
   Limit .... 17.971 mm (0.7075 inch)
Piston pin-to-piston clearance .... Not specified
Ring side clearance
   Top ring
      Standard .... 0.030 to 0.065 mm (0.0012 to 0.0026 inch)
      Limit .... 0.12 mm (0.005 inch)
   Second ring
      Standard .... 0.020 to 0.055 mm (0.0008 to 0.0022 inch)
      Limit .... 0.12 mm (0.005 inch)
   Oil ring .... Not specified
Ring end gap
   Top ring
      Standard .... 0.20 to 0.30 mm (0.008 to 0.012 inch)
      Limit .... 0.55 mm (0.22 inch)
   Second ring
      Standard .... 0.35 to 0.50 mm (0.014 to 0.020 inch)
      Limit .... 0.85 mm (0.33 inch)
   Oil ring
      Standard .... 0.20 to 0.50 mm (0.01 to 0.02 inch)
      Limit .... Not specified

Clutch

Spring free length
   YZ450F
      2003 and 2004
         Standard .... 50 mm (1.97 inches)
         Limit .... 49 mm (1.93 inches)
      2005 and 2006
         Standard .... 47.8 mm (1. inches)
         Limit .... 47.4 mm (1.87 inches)
      2007 and later
         Standard .... 50 mm (1.97 inches)
         Limit .... 49 mm (1.93 inches)
   WR450F
      2003 and 2004
         Standard .... 48.4 mm (1.91 inches)
         Limit .... 47.4 mm (1.87 inches)
      2005
         Standard .... 37.0 mm (1.46 inches)
         Limit .... 36.0 m (1.42 inches)
      2006 and later
         Standard .... 50 mm (1.97 inches)
         Limit .... 49 mm (1.93 inches)
Metal plate thickness
   2003 through 2006 YZ450F, 2003 through 2005 WR450F .... 1.1 to 1.3 mm (0.043 to 0.067 inch)
   2007 and later YZ450F .... 1.5 to 1.7 mm (0.059 to 0.067 inch)
   2006 and later WR450F
      Thick plates (4)... 1.9 to 2.1 mm (0.075 to 0.083 inch)
      Thin plates (3) .... 1.5 to 1.7 mm (0.059 to 0.067 inch)
Friction plate thickness
   Standard .... 2.92 to 3.08 mm (0.115 to 0.121 inch)
   Limit .... 2.8 mm (0.110 inch)
Friction plate warpage limit .... Not specified
Metal plate warpage limit .... 0.2 mm (0.008 inch)

Oil pump

Outer rotor-to-body clearance
   Standard .... 0.09 to 0.017 mm (0.00235 to 0.0067 inch)
   Limit .... 0.24 mm (0.009 inch)
Inner-to-outer rotor clearance
   Standard .... 0.12 mm or less (0.0047 inch or less)
   Limit .... 0.20 mm (0.008 inch)

Rotor-to-straightedge clearance
Standard .... 0.03 to 0.10 mm (0.0012 to 0.0039 inch)
Limit .... 0.17 mm (0.007 inch)

Transmission

Shift fork shaft bend limit .... 0.05 mm (0.002 inch)
Mainshaft and countershaft runout limit............................ Not specified

Crankshaft and connecting rod

Runout limit .... 0.03 mm (0.0012 inch)
Assembly width .... 61.95 to 62.00 mm (2.439 to 2.441 inches)
Connecting rod big-end side clearance
Standard .... 0.15 to 0.45 mm (0.006 to 0.018 inch)
Limit .... 0.50 mm (0.020 inch)
Connecting rod small-end freeplay
Standard .... 0.4 to 1.0 mm (0.02 to 0.04 inch)
Limit .... 2.0 mm (0.08 inch)

Torque specifications

YZ250, WR250

Valve cover bolts .... 10 Nm (86 inch-lbs)
Decompression shaft bolt (2001 through 2003 only) .... 7 Nm (61 inch-lbs)
Decompression bracket bolt (2001 through 2003 only) .... 10 Nm (86 inch-lbs)
Cylinder head small bolts or nuts...................................... 10 Nm (86 inch-lbs)
Cylinder head main bolts.. 38 Nm (27 ft-lbs) (1)
Oil check bolt .... See Chapter 1
Oil strainer to frame (all except 2006 and later YZ250) .... See Chapter 1
Oil strainer to tank (2006 and later YZ250)........................ See Chapter 1
Oil line union bolts (three-fitting metal line)
M10 thread .... 20 Nm (14 ft-lbs)
M8 thread .... 18 Nm (13 ft-lbs)
Oil hose mounting bolts .... 8 Nm (70 inch-lbs)
Oil tube (outside of right crankcase) .... 10 Nm (86 inch-lbs)
Camshaft bearing cap bolts .... 10 Nm (86 inch-lbs)
Cam chain guide bolts .... 10 Nm (86 inch-lbs) (2)
Cam chain tensioner body bolts .... 10 Nm (86 inch-lbs)
Cam chain tensioner cap bolt .... 8 Nm (70 inch-lbs)
Cylinder base bolt .... 10 Nm (86 inch-lbs)
Crankcase bolts .... 12 Nm (104 inch-lbs)
Crankcase cover bolts.. 10 Nm (86 inch-lbs)
Oil pump bolts .... 10 Nm (86 inch-lbs)
Clutch boss nut .... 60 Nm (43 ft-lbs) (3)
Clutch spring bolts
All except 2006 and later YZ250 .... 8 Nm (70 inch-lbs)
2006 and later YZ250 .... 10 Nm (86 inch-lbs)
Shift drum segment retaining pin .... 30 Nm (22 ft-lbs)
External shift linkage stopper lever bolt .... 10 Nm (86 inch-lbs)
Primary drive gear nut .... 75 Nm (54 ft-lbs)
Balancer nut .... 50 Nm (36 ft-lbs)
Engine mounting bolts/nuts
Upper rear bracket to frame .... 34 Nm (24 ft-lbs)
Upper rear bracket to engine.. 55 Nm (40 ft-lbs)
Front bracket to frame .... 34 Nm (24 ft-lbs)
Front bracket to engine... 69 Nm (50 ft-lbs)

1 *Apply molybdenum disulfide oil to the threads, bolt seating surfaces and upper and lower sides of the washers.*
2 *Apply non-permanent thread locking agent to the threads.*
3 *Use a new lockwasher.*

YZ400F, WR400F, YZ426F, WR426F, YZ450F, WR450F

Valve cover bolts .... 10 Nm (86 inch-lbs)
Decompression shaft bolt .... 7 Nm (61 inch-lbs)
Decompression bracket bolt .... 10 Nm (86 inch-lbs)
Cylinder head underside nuts
At timing chain end of head .... 10 Nm (86 inch-lbs)
At front and rear ends of head .... 20 Nm (14 ft-lbs)

YZ400F, WR400F, YZ426F, WR426F, YZ450F, WR450F (continued)

Cylinder head main bolts	
1998 through 2002..	44 Nm (32 ft-lbs) (1)
2003 and later	
First step..	30 Nm (22 ft-lbs)
Second step	Loosen completely
Third step...	20 Nm (14 ft-lbs)
Final step...	180-degrees
Oil check bolt	See Chapter 1
Oil strainer to frame...	See Chapter 1
Oil line union bolts (three-fitting metal line)	
M10 thread..	20 Nm (14 ft-lbs)
M8 thread...	18 Nm (13 ft-lbs)
Oil hose mounting bolts	
1998 through 2003...	10 Nm (86 inch-lbs)
2004	8 Nm (70 inch-lbs)
Oil tube (outside of right crankcase)	10 Nm (86 inch-lbs)
Camshaft bearing cap bolts	10 Nm (86 inch-lbs)
Cam chain guide bolts	10 Nm (86 inch-lbs) (2)
Cam chain tensioner body bolts	10 Nm (86 inch-lbs)
Cam chain tensioner cap bolt	
1998 and 1999	10 Nm (86 inch-lbs)
2000	7 Nm (61 inch-lbs)
Cylinder base bolt	10 Nm (86 inch-lbs)
Crankcase bolts	12 Nm (104 inch-lbs)
Crankcase cover bolts	
1998 through 2005...	10 Nm (86 inch-lbs)
2006 and later	
Hex head	10 Nm (86 inch-lbs)
Allen head	12 Nm (104 inch-lbs)
Oil pump bolts	10 Nm (86 inch-lbs)
Clutch boss nut	75 Nm (54 ft-lbs) (3)
Clutch spring bolts	8 Nm (70 inch-lbs)
Shift drum segment retaining pin	30 Nm (22 ft-lbs)
External shift linkage stopper lever bolt	10 Nm (86 inch-lbs)
Primary drive gear nut	
1998 through 2003..	75 Nm (54 ft-lbs)
2004	85 Nm (61 ft-lbs)
2005 and later	110 Nm (80 ft-lbs)
Balancer driven gear nut	50 Nm (36 ft-lbs)
Balancer weight plate	10 Nm (86 inch-lbs)
Kickstarter pedal bolt	33 Nm (24 ft-lbs)
Balancer weight (2006 and later).............................	45 Nm (32 ft-lbs)
Shift pedal bolt	
1998 and 1999	10 Nm (86 inch-lbs)
2000 through 2002...	Not specified
2003 and later	12 Nm (104 inch-lbs)
Engine mounting bolts/nuts	
Upper rear bracket to frame	34 Nm (24 ft-lbs)
Upper rear bracket to engine	69 Nm (50 ft-lbs)
Front bracket to frame and engine	69 Nm (50 ft-lbs)

1 *Apply molybdenum disulfide oil to the threads, bolt seating surfaces and upper and lower sides of the washers.*
2 *Apply non-permanent thread locking agent to the threads.*
3 *Use a new lockwasher.*

1 General information

The engine/transmission unit is of the liquid-cooled, single-cylinder four-stroke design.

All models have five valves (three intake and two exhaust). The valves are operated by dual overhead camshafts, which are chain driven off the crankshaft.

The engine/transmission assembly is constructed from aluminum alloy. The crankcase is divided vertically.

The crankcase incorporates a dry sump, pressure-fed lubrication system which uses a gear-driven rotor-type oil pump and an oil filter. The bike's frame is used as an oil tank on all models except 2006 and later YZ250s, which use an external oil tank. Early models have a separate strainer screen.

On all models, a wet, multi-plate clutch connects the crankshaft to the transmission. Engagement and disengagement of the clutch is controlled by a lever on the left handlebar through a cable.

The transmission has five forward gears.

A kickstarter is used on all models. 2002 and earlier models are equipped with a decompression lever that makes it easier to crank the engine with the kickstarter. 2003 and later models use electric starters as well as a kickstarter.

2 Operations possible with the engine in the frame

The components and assemblies listed below can be removed without having to remove the engine from the frame. If, however, a number of areas require attention at the same time, removal of the engine is recommended.

Kickstarter (if equipped)
Starter motor (if equipped)
Starter reduction gears (if equipped)
Starter clutch (if equipped)
Magneto rotor and stator
Clutch
External shift mechanism
Cam chain tensioner
Camshafts and lifters
Cylinder head
Cylinder and piston
Oil pump
Balancer gears (if equipped)

3 Operations requiring engine removal

It is necessary to remove the engine/transmission assembly from the frame and separate the crankcase halves to gain access to the following components:

Crankshaft and connecting rod
Transmission shafts
Shift drum and forks

4 Major engine repair - general note

1   It is not always easy to determine when or if an engine should be completely overhauled, as a number of factors must be considered.
2   High mileage is not necessarily an indication that an overhaul is needed, while low mileage, on the other hand, does not preclude the need for an overhaul. Frequency of servicing is probably the single most important consideration. An engine that has regular and frequent oil and filter changes, as well as other required maintenance, will most likely give many miles of reliable service. Conversely, a neglected engine, or one which has not been broken in properly, may require an overhaul very early in its life.
3   Exhaust smoke and excessive oil consumption are both indications that piston rings and/or valve guides are in need of attention. Make sure oil leaks are not responsible before deciding that the rings and guides are bad. Perform a cylinder compression check

5.9 Slide back the clip and disconnect the breather hose (right arrow) from the top of the engine - some models also have an oil tank vent hose (left arrow)

to determine for certain the nature and extent of the work required.
4   If the engine is making obvious knocking or rumbling noises, the connecting rod and/or main bearings are probably at fault.
5   Loss of power, rough running, excessive valve train noise and high fuel consumption rates may also point to the need for an overhaul, especially if they are all present at the same time. If a complete tune-up does not remedy the situation, major mechanical work is the only solution.
6   An engine overhaul generally involves restoring the internal parts to the specifications of a new engine. During an overhaul the piston rings are replaced and the cylinder walls are bored and/or honed. If a rebore is done, then a new piston is also required. Generally the valves are serviced as well, since they are usually in less than perfect condition at this point. While the engine is being overhauled, other components such as the carburetor and the starter motor (if equipped) can be rebuilt also. The end result should be a like-new engine that will give as many trouble-free miles as the original.
7   Before beginning the engine overhaul, read through all of the related procedures to familiarize yourself with the scope and requirements of the job. Overhauling an engine is not all that difficult, but it is time consuming. Check on the availability of parts and make sure that any necessary special tools, equipment and supplies are obtained in advance.
8   Most work can be done with typical shop hand tools, although a number of precision measuring tools are required for inspecting parts to determine if they must be replaced. Often a dealer service department or repair shop will handle the inspection of parts and offer advice concerning reconditioning and replacement. As a general rule, time is the primary cost of an overhaul, so it doesn't pay to install worn or substandard parts.
9   As a final note, to ensure maximum life and minimum trouble from a rebuilt engine, everything must be assembled with care in a spotlessly clean environment.

5 Engine - removal and installation

 Warning: Engine removal and installation should be done with the aid of an assistant to avoid damage or injury that could occur if the engine is dropped.

Removal

1   Drain the engine oil (see Chapter 1).
2   If the motorcycle has a battery, disconnect the ground cable from the engine and disconnect both battery cables (negative cable first).
3   Remove the seat. Remove the air scoops and side covers (if equipped). If the bike has guard bars or a skid plate, remove it (see Chapter 8).
4   Remove the fuel tank, carburetor and exhaust system (see Chapter 4). On 2006 and later WR models, remove the air induction system.
5   Disconnect the clutch cable from the engine (see Section 18).
6   Disconnect the odometer/speedometer/tachometer cable if equipped).
7   Remove the starter motor (if equipped) (see Chapter 5).
8   Detach the rear shock absorber gas reservoir from the frame and set it out of the way.
9   Disconnect the breather hose(s) from the valve cover **(see illustration)**. Disconnect the air intake system hose (if equipped) (see Chapter 4).
10   Disconnect the spark plug wire (see Chapter 1).
11   Label and disconnect the following wires (see Chapter 5 for component locations if necessary):
     CDI magneto and alternator
     Neutral switch
12   Remove the shift pedal (see Section 24).
13   On 2005 and earlier models, disconnect the external oil hoses from the engine (see Section 16).
14   Remove the brake pedal (see Chapter 7).

5.17a On early models, remove the top-mount-to-frame bolts on both sides (upper arrows) and remove the through-bolt (lower arrow)

5.17b On later models, remove four top mount bolts from the right side of the bike (arrows) . . .

15   Remove the drive sprocket from the engine (see Chapter 6).
16   Support the engine securely from below.

17   Remove the engine mounting bolts, nuts and brackets at the lower front, lower rear and upper rear **(see illustrations)**.
18   Remove the swingarm pivot bolt part-

way (see Chapter 6). **Note:** *The swingarm pivot bolt acts as an engine mounting bolt. Pull out the bolt just far enough to free the engine, but leave it in far enough to support one side of the swingarm* **(see illustration)**.
19   Have an assistant help you support the engine. Remove it from the right side.
20   Slowly lower the engine to a suitable work surface.

Installation
21   Check the engine supports for wear or damage and replace them if necessary before installing the engine.
22   Make sure the motorcycle is securely supported so it can't be knocked over during the remainder of this procedure.
23   With the help of an assistant, lift the engine up into the frame. Install the mounting nuts and bolts at the rear, front and top. Be sure the mount brackets are installed on the

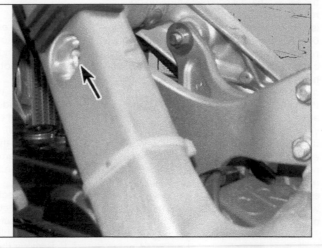

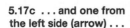
5.17c . . . and one from the left side (arrow) . . .

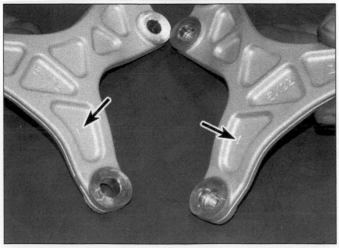

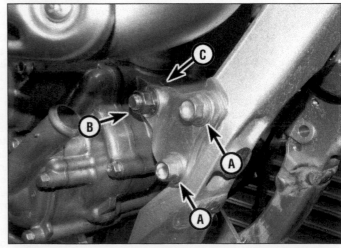

5.17d . . . and remove the top mount brackets - they're marked L and R for left and right sides of the bike

5.17e Remove two bolts with washers (A) and the through-bolt nut (B) from the right front mount - note the location of the tabbed washer (C)

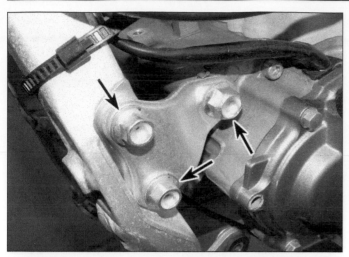

5.17f Pull out the through-bolt, noting the location of the tabbed washer (right arrows) - unscrew the bolts and washer (left arrows) and remove the front mount brackets . . .

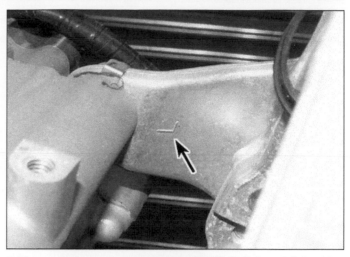

5.17g . . . the brackets are marked L and R for left and right sides of the bike (arrow)

5.17h Remove the through-bolt nut from the lower mount and pull out the through-bolt

5.18 Remove the swingarm pivot just far enough to clear the engine

correct sides of the bike (refer to their L and R marks). Finger-tighten the mounting bolts, but don't tighten them to the specified torque yet.

24 Tighten the engine mounting bolts and nuts evenly to the torques listed in this Chapter's Specifications.

25 The remainder of installation is the reverse of the removal steps, with the following additions:

a) Use new gaskets at all exhaust pipe connections.

b) Adjust the throttle cable and clutch cable following the procedures in Chapter 1.

c) Fill the engine with oil and coolant, also following the procedures in Chapter 1. Run the engine and check for oil, coolant and exhaust leaks.

6 Engine disassembly and reassembly - general information

1 Before disassembling the engine, clean the exterior with a degreaser and rinse it with water. A clean engine will make the job easier and prevent the possibility of getting dirt into the internal areas of the engine.

2 In addition to the precision measuring tools mentioned earlier, you will need a torque wrench, a valve spring compressor, oil gallery brushes (see illustration), a piston ring removal and installation tool, and a piston ring compressor. Some new, clean engine oil of the correct grade and type, some engine assembly lube (or moly-based grease) and a tube of RTV (silicone) sealant will also be required.

6.2 A selection of brushes is required for cleaning holes and passages in the engine components

6.3 An engine stand can be made from short lengths of lumber and lag bolts or nails

7.6 Remove the valve cover Allen bolts

3   An engine support stand made from short lengths of 2 x 4's bolted together will facilitate the disassembly and reassembly procedures **(see illustration)**. If you have an automotive-type engine stand, an adapter plate can be made from a piece of plate, some angle iron and some nuts and bolts.

4   When disassembling the engine, keep mated parts together (including gears, rocker arms and shafts, etc.) that have been in contact with each other during engine operation. These mated parts must be reused or replaced as an assembly.

5   Engine/transmission disassembly should be done in the following general order with reference to the appropriate Sections.

Remove the cam chain tensioner
Remove the camshafts and lifters
Remove the cylinder head
Remove the cylinder
Remove the piston
Remove the clutch
Remove the balancer gears
Remove the oil pump
Remove the external shift mechanism
Remove the alternator rotor (flywheel)
Remove the starter reduction gears (if equipped)
Remove the balancer and gears
Separate the crankcase halves
Remove the shift drum and forks
Remove the transmission gears and shafts
Remove the crankshaft and connecting rod

6   Reassembly is accomplished by reversing the general disassembly sequence.

7 Valve cover - removal and installation

1   Remove the seat (see Chapter 8).
2   Remove the fuel tank (and air induction system if equipped) (see Chapter 4).
3   If you're working on a 2006 or later model, remove the engine upper mounting bracket **(see illustrations 5.17b through 5.17d)**.
4   Remove the spark plug (see Chapter 1).
5   Disconnect the breather hose from the cylinder head **(see illustration 5.9)**.
6   Unscrew the valve cover mounting bolts **(see illustration)**.
7   Lift the valve cover off the cylinder head **(see illustration)**. If it's stuck, don't attempt to pry it off - tap around the sides of it with a plastic hammer to dislodge it. Remove the sealing washers **(see illustration)**.
8   Work the gasket free of the cylinder head and remove it **(see illustration)**.
9   Check the valve cover gasket for damage or deterioration and replace it as needed. It's a good idea to replace the bolt sealing washers whenever they're removed.
10   Installation is the reverse of the removal steps, with the following additions:

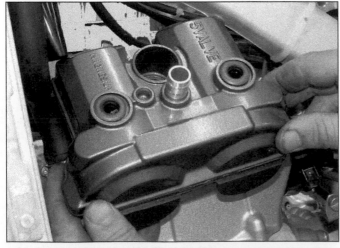

7.7a Tilt the valve cover sharply to the left and remove it from the left side

7.7b Remove the sealing washers, noting the direction they face

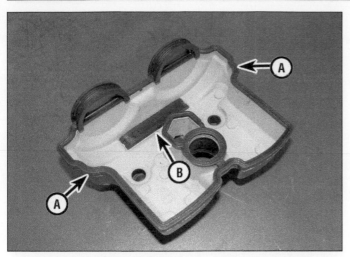

7.8 Apply non-hardening sealant to the lower side of the gasket, in the area between the arrows (A) (including the semi-circular cutouts) - check the timing chain guide (B) for wear

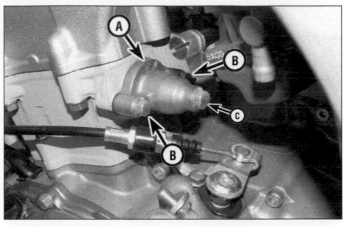

8.1 Remove the tensioner cap bolt and copper washer

A   UP mark
B   Tensioner bolts (inner bolt has a copper washer on 2007 450 models)
C   Tensioner cap bolt

a) Coat the area of the valve cover gasket that fits into the cylinder head cutouts with non-hardening sealer (see illustration 7.8).
b) Install new sealing washers and tighten the valve cover bolts evenly to the torque listed in this Chapter's Specifications.

8   Cam chain tensioner - removal, inspection and installation

Removal

Caution: Once you start to remove the tensioner bolts you must remove the tensioner all the way and reset it before tightening the bolts. The tensioner extends and locks in place, so if you loosen the bolts partway and then tighten them, the

tensioner or cam chain will be damaged.

1   Unscrew the tensioner cap bolt and remove the sealing washer (see illustration).
2   Turn the tensioner piston clockwise with a small screwdriver to retract it (see illustration). When the piston is fully retracted, there will be slack in the upper run of the cam chain (see illustration).

TOOL TIP *If you don't have a small enough screwdriver, you can make one by bending a piece of coat hanger into an L shape and grinding the tip of the long end into a screwdriver shape (see illustration).*

3   Remove the tensioner mounting bolts and detach the tensioner from the cylinder (see illustration 8.1).

Installation

4   Clean all old gasket material from the tensioner body and engine.

8.2a Insert a thin screwdriver into the tensioner and turn it clockwise to retract the piston - it will lock when fully retracted

8.2b As the tensioner piston retracts, the top run of the cam chain will develop slack (arrow)

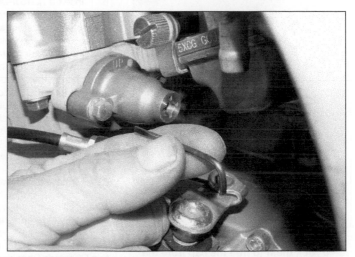

8.2c A tensioner retracting tool can be made from a length of coat hanger

8.5 Place a new gasket on the tensioner body

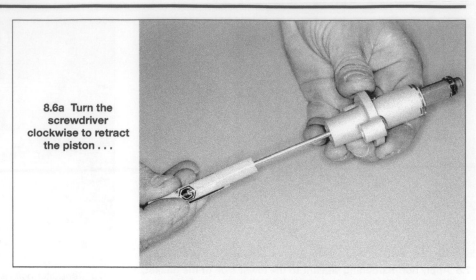

8.6a Turn the screwdriver clockwise to retract the piston . . .

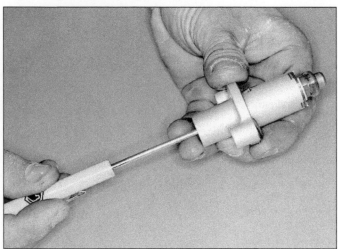

8.6b . . . until it's in this position; it should lock and stay by itself, but if not, then hold the screwdriver

9.5a Camshaft bolt TIGHTENING sequence

9.5b Lift up two of the bearing cap bolts and rock them to free the cap from the cylinder head

5   Place a new gasket on the tensioner body **(see illustration)**.

6   Insert a narrow-bladed screwdriver into the tensioner and rotate it clockwise to retract the tensioner piston **(see illustrations)**. The tensioner piston should lock in the retracted position if you turn it far enough. If not, hold the screwdriver in position while installing the tensioner on the cylinder.

7   Position the tensioner body on the cylinder, making sure the UP mark is upward **(see illustration 8.1)**. Install the bolts, tightening them to the torque listed in this Chapter's Specifications. **Note:** *On later 450 models, the bolt nearest the center of the engine has a copper washer.*

8   Remove the screwdriver so the tensioner piston can extend. When this happens, the slack will disappear from the top run of the cam chain **(see illustration 8.2b)**.

9   Install the cap bolt with a new sealing washer and tighten it to the torque listed in this Chapter's Specifications.

9 Camshafts and lifters - removal, inspection and installation

Removal

Camshafts

1   Remove the valve cover (see Section 7).

2   Refer to *Valve clearances - check and adjustment* in Chapter 1 and place the engine at Top Dead Center on the compression stroke.

3   Remove the cam chain tensioner (see Section 8).

4   Tie the cam chain up with wire so it can't fall into the chain cavity.

5   Loosen the camshaft cap bolts in several stages, in the reverse of the tightening sequence **(see illustration)**. On the intake cap, work from the inner bolts to the outer bolts. On the exhaust cap, loosen the bolts in a criss-cross pattern. Take off the caps and dowels **(see illustrations)**.

9.5c Lift off the bearing caps and remove the retaining ring (arrow); there's one for each cap

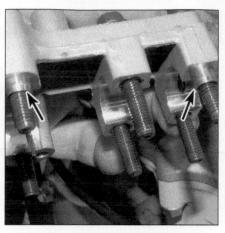

9.5d Here are the intake cap dowels (arrows) . . .

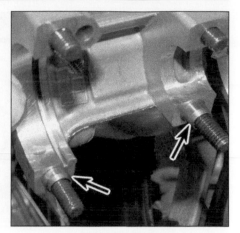

9.5e . . . and the exhaust cap dowels (arrows)

6   Lift the intake camshaft out of its saddles and disengage the sprocket from the chain (see illustrations). Remove the exhaust camshaft in the same way.

HAYNES HiNT *If you're removing the camshafts to adjust the valves, don't remove them all the way - you can just roll them out of their saddles (one at a time), leaving them attached to the chain as shown. That way, you won't need to realign the timing marks when you install the camshafts.*

Lifters

7   Stuff a clean shop rag into the timing chain cavity so the valve adjusting shims can't fall into it when they're removed.
8   Remove the camshafts following the procedure given above. Be sure to keep tension on the cam chain.
9   Make a holder for each lifter and its adjusting shim (an egg carton or box will work). Label the sections according to

9.6a If you're removing the camshafts for valve adjustment, you can roll them out of position while leaving them engaged with the chain . . .

whether the lifter belongs with the intake or exhaust camshaft, and left or right valve (or center valve on the intake camshaft). The lifters form a wear pattern with their bores and

must be returned to their original locations if reused.
10   Label each lifter and pull each lifter out of the bore, using a magnet or suction cup

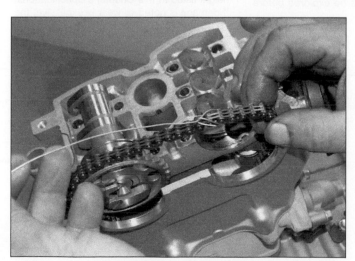

9.6b . . . to remove the camshafts completely, disengage the intake sprocket and support the chain . . .

9.6c . . . disengage the exhaust cam sprocket from the chain and remove the camshaft

9.10a Before removing the lifters, label them (intake or exhaust, left, center and right) - these are the intake lifters

9.10b Pull the lifters out of their bores with a magnet . . .

(see illustrations). Make sure the shims stay with their lifters. The shims are inside the lifters, so be careful not to let them fall as you lift the lifters out **(see illustration)**.

Timing chain and guides

11   The front (exhaust) cam chain guide is held in position by the cylinder head. If inspection procedures below indicate a problem, you'll need to remove the head to lift the guide out of the cylinder (see Sections 10 and 19). The rear (intake) guide is bolted at the bottom, so the left crankcase cover and alternator rotor will have to be removed for access if the guide or the cam chain need to be removed (see Section 19 and Chapter 5).

12   Stuff clean rags into the cam chain opening so dirt, small parts or tools can't fall into it.

Inspection

Camshaft, chain and guides

Note: *Before replacing camshafts or the cylinder head because of damage, check with local machine shops specializing in motorcycle engine work. In the case of the camshaft, it may be possible for cam lobes to be welded, reground and hardened, at a*

9.10c . . . the valve adjusting shim should come up with the lifter, but it may stick in the valve spring retainer

cost far lower than that of a new camshaft. If the bearing surfaces in the cylinder head are damaged, it may be possible for them to be bored out to accept bearing inserts. Due to the cost of a new cylinder head it is recommended that all options be explored before condemning it as trash!

9.10d Make sure each valve adjusting shim stays with its lifter

13   Check the camshaft lobes for heat discoloration (blue appearance), score marks, chipped areas, flat spots and spalling **(see illustration)**. Measure the height of each lobe with a micrometer **(see illustration)** and compare the results to the minimum lobe height listed in this Chapter's Specifications. If damage is noted or wear is excessive, the

9.13a Check the cam lobes for wear - here's a good example of damage which will require replacement (or repair) of the camshaft

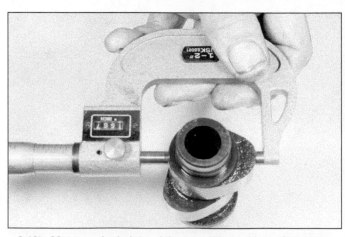

9.13b Measure the height of the cam lobes with a micrometer

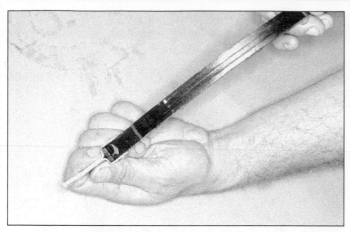

9.16 Check the chain guides for wear, especially on the friction surface

9.24 Place the valve adjusting shims in the spring retainers with the number mark upward (arrow)

camshaft must be replaced. Check the bearing surfaces for scoring or wear. Also, be sure to check the condition of the lifters, as described below.

14 Except in cases of oil starvation, the camshaft chain wears very little. If the chain has stretched excessively, which makes it difficult to maintain proper tension, replace it with a new one. To remove the chain from the crankshaft sprocket, remove the alternator rotor (see Chapter 5).

15 Check the sprockets for wear, cracks and other damage, replacing it if necessary. If a sprocket is worn, the chain is also worn, and possibly the sprocket on the crankshaft. If wear this severe is apparent, the entire engine should be disassembled for inspection. The sprockets are permanently attached to the camshaft, so if the sprockets must be replaced, the camshafts must be replaced as well.

16 Check the chain guides for wear or damage, especially along the friction surfaces **(see illustration)**. Use a flashlight to look down the cam chain tunnel. If they are worn or damaged, replace them (see Section 19).

17 Check the camshaft bearing oil clearances with Plastigage, referring to *Tools and*

Workshop Tips at the end of this manual.

18 Compare the results to this Chapter's Specifications.

19 If oil clearance is greater than specified, measure the diameter of the cam bearing journal with a micrometer. If the journal diameter is less than the specified limit, replace the camshaft with a new one and recheck the clearance.

20 If the clearance is still too great, replace the cylinder head and bearing caps with new parts (see the **Note** that precedes Step 13).

21 On 2003 and later models, check the automatic compression release in the left camshaft. Operate the lever on the sprocket end of the camshaft by hand. It should move smoothly, causing the plunger in the camshaft to extend. When released, it should return by itself. If the compression release doesn't work properly, replace the camshaft.

Lifters

22 Check the lifters and their bores for wear, scuff marks, scratches and other damage. Check the camshaft contact surface, as well as the outer surface that rides in the bore. Replace the lifters if they're visibly worn or damaged.

Installation

23 Make sure the piston is still at Top Dead Center on the compression stroke (refer to the valve adjustment procedure in Chapter 1 if necessary).

24 Coat the lifters and their bores with clean engine oil. Apply a small amount of moly-based grease to the shims and stick them to their respective valve stems with the thickness number upward **(see illustration)**.

25 Slide the lifters into their bores, taking care not to knock the valve shims out of position. When the lifters are correctly installed, it should be possible to rotate them with a finger.

26 Coat the camshaft contact surfaces of the lifters and the bearing surfaces of the camshafts with moly-based grease.

27 Install the exhaust camshaft, then the intake camshaft in the cylinder head, engaging the sprocket with the chain as you do so, and make sure the sprockets are in the correct positions (see Chapter 1, *Valve clearances - check and adjustment*). Install the cap dowels in their holes (if they were removed).

28 Install the camshaft bearing caps **(see illustrations)**. Tighten the cap bolts in stages to the torque listed in this Chapter's Speci-

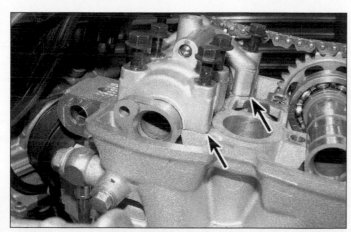

9.28a The camshaft caps should press on easily and make even contact with the cylinder head (arrows) - if not, the bearing retaining clip may have slipped out of position

9.28b Be sure to tighten the camshaft caps in the correct sequence with an accurate torque wrench

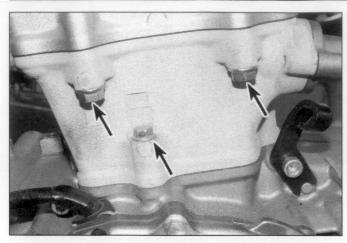

10.5 Remove the two small bolts or nuts the secure the underside of the cylinder head (upper arrows) - the single bolt (lower arrow) secures the cylinder to the crankcase . . .

10.6 . . . then loosen the four main bolts evenly in a criss-cross pattern

fications, following the tightening sequence **(see illustration 9.5a). Caution:** *The caps must be tightened in the proper sequence with an accurate torque wrench, or the camshafts may seize.*

29   Install the timing chain tensioner (see Section 8). Release the tensioner so its piston presses against the chain.

30   Recheck the crankshaft timing mark in the timing hole cover and the match marks on both camshafts. If they are not still aligned, stop and find out why before continuing. **Caution:** *Don't run the engine with the marks misaligned or the valves may strike the pistons, bending the valves.*

31   Rotate the crankshaft two full turns and make sure the timing marks still line up correctly.

32   Check the valve clearances (see Chapter 1). This is necessary to make sure none of the shims has slipped out of position.

33   Change the engine oil (see Chapter 1).

34   The remainder of installation is the reverse of removal.

10 Cylinder head - removal, inspection and installation

Removal

1   Drain the cooling system and remove the radiators (see Chapters 1 and 3).

2   Remove the seat (see Chapter 8).

3   Remove the fuel tank, carburetor, air induction system (if equipped) and exhaust system (see Chapter 4).

4   Remove the valve cover and camshafts (see Sections 7 and 9).

5   Remove the two nuts or small bolts that secure the left-hand side of the cylinder head **(see illustration)**.

6   Loosen the four main head bolts in several stages, in a criss-cross pattern, until they're completely loose **(see illustration)**. Lift out the bolts and their washers.

7   Lift the cylinder head off the cylinder **(see illustration)**. If it's stuck, don't attempt to pry it off - tap around the sides of it with a plastic hammer to dislodge it.

8   Locate the dowels **(see illustration 10.7)**. There are two of them, one in each of the head bolt holes nearest the timing chain. The dowels may be in the cylinder or they may have come off with the head.

9   Remove the old head gasket from the cylinder or head.

Inspection

10   Check the cylinder head gasket and the mating surfaces on the cylinder head and cylinder for leakage, which could indicate warpage.

11   Refer to Section 12 and check the flatness of the cylinder head.

12   Clean all traces of old gasket material from the cylinder head and cylinder. Be careful not to let any of the gasket material fall into the crankcase, the cylinder bore or the bolt holes.

Installation

13   Install the two dowel pins, then place the new head gasket on the cylinder **(see illustration)**. Never reuse the old gasket and

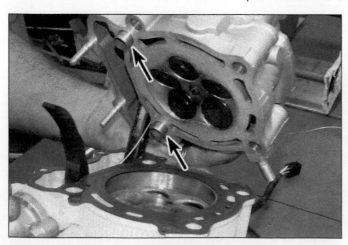

10.7 Lift the head off the cylinder and locate the dowels (arrows)

10.13 Make sure these coolant passages in the gasket and head line up correctly - if they don't, the gasket is on upside down

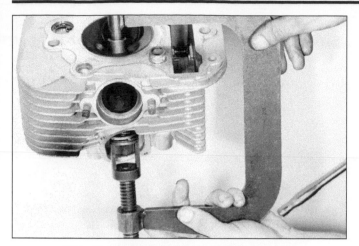

12.6a Compress the valve springs with a valve spring compressor

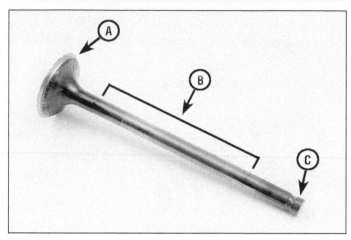

12.6b Check the valve face (A), stem (B) and keeper groove (C) for wear and damage

don't use any type of gasket sealant. **Caution:** *The head gasket will almost line up if it's installed upside down. Make sure the coolant passages and bolt holes line up exactly.*

14 Install the exhaust side cam chain guide, fitting the lower end and the middle guide into their notches **(see illustration 19.2b)**. **Note:** *Don't forget to install the chain guide at this point. It won't be possible to install it once the cylinder head is installed.*

15 Carefully lower the cylinder head over the dowels, guiding the cam chain through the slot in the cylinder head. It's helpful to have an assistant support the cam chain with a piece of wire so it doesn't fall and become kinked or detached from the crankshaft. When the head is resting on the cylinder, wire the cam chain to another component to keep tension on it.

16 Lubricate the threads and seating surfaces of the four main cylinder head bolts with molybdenum disulfide grease. Lubricate the upper and lower sides of the head bolt washers with the same grease.

17 Install the washers on the four main head bolts and install the bolts finger-tight. Tighten the four bolts in a criss-cross pattern, in several stages, to the initial torque listed in this Chapter's Specifications.

18 Loosen the bolts all the way, again in a criss-cross pattern. Retighten them to the second-step torque setting listed in this Chapter's Specifications, then tighten them exactly 1/2 turn further.

19 After the main nuts or bolts are tightened, tighten the two Allen bolts to the torque listed in this Chapter's Specifications.

20 The remainder of installation is the reverse of the removal steps.

11 Valves/valve seats/valve guides - servicing

1 Because of the complex nature of this job and the special tools and equipment required, servicing of the valves, the valve seats and the valve guides (commonly known as a valve job) is best left to a professional.

2 The home mechanic can, however, remove and disassemble the head, do the initial cleaning and inspection, then reassemble and deliver the head to a dealer service department or properly equipped repair shop for the actual valve servicing. Refer to Section 12 for those procedures.

3 The dealer service department will remove the valves and springs, recondition or replace the valves and valve seats, replace the valve guides, check and replace the valve springs, spring retainers and keepers (as necessary), replace the valve seals with new ones and reassemble the valve components.

4 After the valve job has been performed, the head will be in like-new condition. When the head is returned, be sure to clean it again very thoroughly before installation on the engine to remove any metal particles or abrasive grit that may still be present from the valve service operations. Use compressed air, if available, to blow out all the holes and passages.

12 Cylinder head and valves - disassembly, inspection and reassembly

1 As mentioned in the previous Section, valve servicing and valve guide replacement should be left to a dealer service department or other repair shop. However, disassembly, cleaning and inspection of the valves and related components can be done (if the necessary special tools are available) by the home mechanic. This way no expense is incurred if the inspection reveals that service work is not required at this time.

2 To properly disassemble the valve components without the risk of damaging them, a valve spring compressor is absolutely necessary. If the special tool is not available, have a dealer service department or other repair

shop handle the entire process of disassembly, inspection, service or repair (if required) and reassembly of the valves.

Disassembly

3 Before the valves are removed, scrape away any traces of gasket material from the head gasket sealing surface. Work slowly and do not nick or gouge the soft aluminum of the head. Gasket removing solvents, which work very well, are available at most motorcycle shops and auto parts stores.

4 Carefully scrape all carbon deposits out of the combustion chamber area. A hand held wire brush or a piece of fine emery cloth can be used once most of the deposits have been scraped away. Do not use a wire brush mounted in a drill motor, or one with extremely stiff bristles, as the head material is soft and may be eroded away or scratched by the wire brush.

5 Before proceeding, arrange to label and store the valves along with their related components so they can be kept separate and reinstalled in the same valve guides from which they are removed (plastic bags work well for this).

6 Compress the valve spring(s) on the first valve with a spring compressor, then remove the keepers and the retainer from the valve assembly **(see illustration)**. Do not compress the spring(s) any more than is absolutely necessary. Carefully release the valve spring compressor and remove the spring(s), spring seat and valve from the head. If the valve binds in the guide (won't pull through), push it back into the head and deburr the area around the keeper groove with a very fine file or whetstone **(see illustration)**.

7 Repeat the procedure for the remaining valve. Remember to keep the parts for each valve together so they can be reinstalled in the same location.

8 Once the valves have been removed and labeled, pull off the valve stem seals with pliers and discard them (the old seals should never be reused).

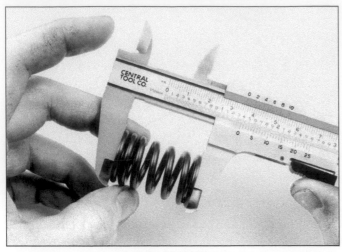

12.18a Measuring the free length of the valve springs

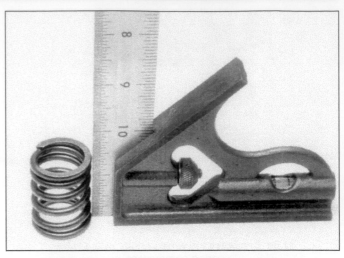

12.18b Checking the valve springs for squareness

9   Next, clean the cylinder head with sol-vent and dry it thoroughly. Compressed air will speed the drying process and ensure that all holes and recessed areas are clean.

10   Clean all of the valve springs, keepers, retainers and spring seats with solvent and dry them thoroughly. Clean the parts from one valve at a time so that no mixing of parts between valves occurs.

11   Scrape off any deposits that may have formed on the valve, then use a motorized wire brush to remove deposits from the valve heads and stems. Again, make sure the valves do not get mixed up.

Inspection

12   Inspect the head very carefully for cracks and other damage. If cracks are found, a new head will be required. Check the cam bearing surfaces for wear and evi-dence of seizure. Check the camshaft for wear as well (see Section 9).

13   Using a precision straightedge and a feeler gauge, check the head gasket mating surface for warpage as described in *Tools and Workshop Tips* at the end of this manual. If the head is warped it must either be machined or, if warpage is excessive, replaced with a new one.

14   Examine the valve seats in each of the combustion chambers. If they are pitted, cracked or burned, the head will require valve service that is beyond the scope of the home mechanic. Measure the valve seat width and compare it to this Chapter's Specifications. If it is not within the specified range, or if it var-ies around its circumference, valve service work is required.

15   Clean the valve guides to remove any carbon buildup, then measure the inside diameters of the guides (at both ends and the center of the guide) as described in *Tools and Workshop Tips* at the end of this manual. Record the measurements for future refer-ence. The guides are measured at the ends and at the center to determine if they are worn in a bell-mouth pattern (more wear at

the ends). If they are, guide replacement is an absolute must.

16   Carefully inspect each valve face for cracks, pits and burned spots. Check the valve stem and the keeper groove area for cracks **(see illustration 12.6b)**. Rotate the valve and check for any obvious indication that it is bent. Check the end of the stem for pitting and excessive wear and make sure the bevel is the specified width. The pres-ence of any of the above conditions indicates the need for valve servicing.

17   Measure the valve stem diameter with a micrometer. If the diameter is less than listed in this Chapter's Specifications, the valves will have to be replaced with new ones. Also check the valve stem for bending. Set the valve in a V-block with a dial indicator touch-ing the middle of the stem. Rotate the valve and look for a reading on the gauge (which indicates a bent stem). If the stem is bent, replace the valve.

18   Check the end of each valve spring for wear and pitting. Measure the free length **(see illustration)** and compare it to this Chapter's Specifications. Any springs that are shorter than specified have sagged and should not be reused. Stand the spring on a flat surface and check it for squareness **(see illustration)**.

19   Check the spring retainers and keepers for obvious wear and cracks. Any question-able parts should not be reused, as exten-sive damage will occur in the event of failure during engine operation.

20   If the inspection indicates that no ser-vice work is required, the valve components can be reinstalled in the head.

Reassembly

21   If the valve seats have been ground, the valves and seats should be lapped before installing the valves in the head to ensure a positive seal between the valves and seats. This procedure requires coarse and fine valve lapping compound (available at auto parts stores) and a valve lapping tool. If a

lapping tool is not available, a piece of rub-ber or plastic hose can be slipped over the valve stem (after the valve has been installed in the guide) and used to turn the valve.

22   Apply a small amount of coarse lapping compound to the valve face **(see illustra-tion)**, then slip the valve into the guide. **Note:** *Make sure the valve is installed in the correct guide and be careful not to get any lapping compound on the valve stem.*

23   Attach the lapping tool (or hose) to the valve and rotate the tool between the palms of your hands. Use a back-and-forth motion rather than a circular motion. Lift the valve off the seat and turn it at regular intervals to dis-tribute the lapping compound properly. Con-tinue the lapping procedure until the valve face and seat contact area is of uniform width and unbroken around the entire cir-cumference of the valve face and seat **(see illustration)**. Once this is accomplished, lap the valve again with fine lapping compound.

24   Carefully remove the valve from the guide and wipe off all traces of lapping com-pound. Use solvent to clean the valve and wipe the seat area thoroughly with a solvent soaked cloth. Repeat the procedure for the remaining valves.

12.22 Apply the lapping compound very sparingly, in small dabs, to the valve face only

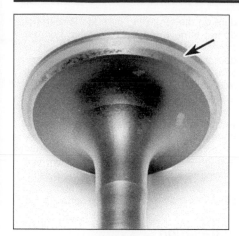

12.23 After lapping, the valve face should exhibit a uniform, unbroken contact pattern (arrow)

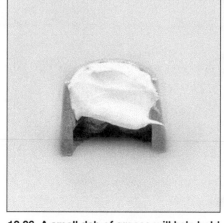

12.26 A small dab of grease will help hold the keepers in place on the valve while the spring compressor is released

13.4 Lift the cylinder straight off and locate the dowels - they may stay in the crankcase like this (arrow) . . .

25   Lay the spring seat in place in the cylinder head, then install new valve stem seals on both of the guides. Use an appropriate size deep socket to push the seals into place until they are properly seated. Don't twist or cock them, or they will not seal properly against the valve stems. Also, don't remove them again or they will be damaged.

26   Coat the valve stems with assembly lube or moly-based grease, then install one of them into its guide. Next, install the spring seat, springs and retainers, compress the springs and install the keepers. **Note:** *Install the springs with the tightly wound coils at the bottom (next to the spring seat).* When compressing the springs with the valve spring compressor, depress them only as far as is absolutely necessary to slip the keepers into place. Apply a small amount of grease to the keepers **(see illustration)** to help hold them in place as the pressure is released from the springs. Make certain that the keepers are securely locked in their retaining grooves.

27   Support the cylinder head on blocks so

the valves can't contact the workbench top, then very gently tap each of the valve stems with a soft-faced hammer. This will help seat the keepers in their grooves.

28   Once all of the valves have been installed in the head, check for proper valve sealing by pouring a small amount of solvent into each of the valve ports. If the solvent leaks past the valve(s) into the combustion chamber area, disassemble the valve(s) and repeat the lapping procedure, then reinstall the valve(s) and repeat the check. Repeat the procedure until a satisfactory seal is obtained.

13 Cylinder - removal, inspection and installation

Removal

1   Following the procedure given in Section 10, remove the cylinder head. Make sure

the crankshaft is positioned at Top Dead Center (TDC).

2   Lift out the cam chain front guide.

3   Remove the bolt securing the base of the cylinder to the crankcase **(see illustration 10.5)**.

4   Lift the cylinder straight up to remove it **(see illustration)**. If it's stuck, tap around its perimeter with a soft-faced hammer. Don't attempt to pry between the cylinder and the crankcase, as you'll ruin the sealing surfaces.

5   Locate the dowel pins (they may have come off with the cylinder or still be in the crankcase) **(see illustrations)**. 2007 YZ450 and WR450 models have two dowels, located opposite the timing chain cavity. All others have three dowels, two across the back of the cylinder and one near the front; the front dowel, which is larger, has an O-ring. Be careful not to let these drop into the engine. Stuff rags around the piston and remove the gasket and all traces of old gasket material from the surfaces of the cylinder and the crankcase.

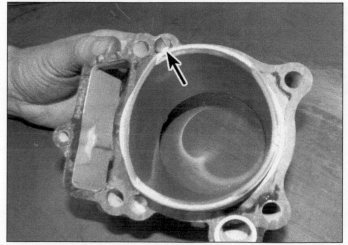

13.5a . . . or come off with the cylinder (arrow) - remove all traces of old base gasket (arrow) and install a new one on assembly

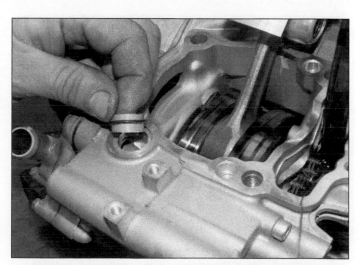

13.5b The large dowel (not used on 2007 450 models) has an O-ring

13.16 If you're experienced and very careful, the cylinder can be installed over the rings without a ring compressor, but a compressor is recommended

Inspection

6   Don't attempt to separate the liner from the cylinder.

7   Check the cylinder walls carefully for scratches and score marks.

8   Using the appropriate precision measuring tools, check the cylinder's diameter. Measure parallel to the crankshaft axis and across the crankshaft axis, at the depth from the top of the cylinder listed in this Chapter's Specifications. Average the two measurements and compare the results to this Chapter's Specifications. If the cylinder walls are tapered, out-of-round, worn beyond the specified limits, or badly scuffed or scored, have the cylinder rebored and honed by a dealer service department or a motorcycle repair shop. If a rebore is done, an oversize piston and rings will be required as well. Check with your dealer service department about available oversizes.

9   As an alternative, if the precision measuring tools are not available, a dealer service department or repair shop will make the measurements and offer advice concerning servicing of the cylinder.

10   If it's in reasonably good condition and not worn to the outside of the limits, and if the piston-to-cylinder clearance can be maintained properly, then the cylinder does not have to be rebored; honing is all that is necessary.

11   To perform the honing operation you will need the proper size flexible hone with fine stones as shown in *Maintenance techniques, tools and working facilities* at the front of this book, or a bottle brush type hone, plenty of light oil or honing oil, some shop towels and an electric drill motor. Hold the cylinder block in a vise (cushioned with soft jaws or wood blocks) when performing the honing operation. Mount the hone in the drill motor, compress the stones and slip the hone into the cylinder. Lubricate the cylinder thoroughly, turn on the drill and move the hone up and down in the cylinder at a pace which will produce a fine crosshatch pattern on the cylinder wal, with the crosshatch lines intersecting at approximately a 60-degree angle. Be sure to use plenty of lubricant and do not take off any more material than is absolutely necessary to produce the desired effect. Do not withdraw the hone from the cylinder while it is running. Instead, shut off the drill and continue moving the hone up and down in the cylinder until it comes to a complete stop, then compress the stones and withdraw the hone. Wipe the oil out of the cylinder and repeat the procedure on the remaining cylinder. Remember, do not remove too much material from the cylinder wall. If you do not have the tools, or do not desire to perform the honing operation, a dealer service department or vehicle repair shop will generally do it for a reasonable fee.

12   Next, the cylinder must be thoroughly washed with warm soapy water to remove all traces of the abrasive grit produced during the honing operation. Be sure to run a brush through the bolt holes and flush them with running water. After rinsing, dry the cylinder thoroughly and apply a coat of light, rust-preventative oil to all machined surfaces.

Installation

13   Lubricate the cylinder bore with plenty of clean engine oil. Apply a thin film of moly-based grease to the piston skirt.

14   Install the dowel pins (and O-ring on all except 2007 540 models), then slip a new cylinder base gasket over them **(see illustrations 13.5a and 13.5b)**.

15   Attach a piston ring compressor to the piston and compress the piston rings. A large hose clamp can be used instead - just make sure it doesn't scratch the piston, and don't tighten it too much.

16   Install the cylinder and carefully lower it down until the piston crown fits into the cylinder liner **(see illustration)**. While doing this, pull the camshaft chain up, using a hooked tool or a piece of stiff wire. Push down on the cylinder, making sure the piston doesn't get cocked sideways, until the bottom of the cylinder liner slides down past the piston rings. A wood or plastic hammer handle can be used to gently tap the cylinder down, but don't use too much force or the piston will be damaged.

17   Remove the piston ring compressor or hose clamp, being careful not to scratch the piston.

18   The remainder of installation is the reverse of the removal steps.

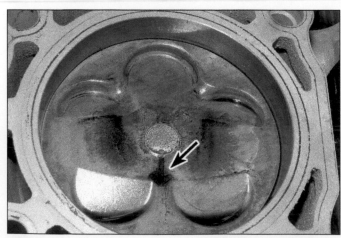

14.3a The arrow mark on top of the piston (arrow) faces the exhaust (front) side of the engine - if you can't see it, install the piston with two valve notches toward the front and three valve notches toward the rear

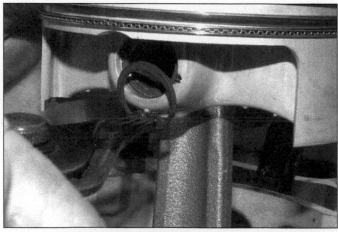

14.3b Wear eye protection and remove the snap-ring from its groove with snap-ring pliers

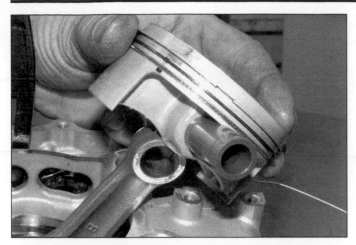

14.4a The piston pin should come out with hand pressure . . .

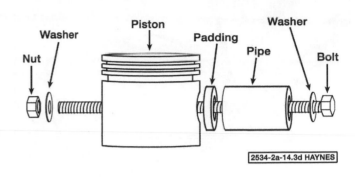

2534-2a-14.3d HAYNES

14.4b . . . if it doesn't, this removal tool can be fabricated from readily available parts

14 Piston - removal, inspection and installation

1   The piston is attached to the connecting rod with a piston pin that's a slip fit in the piston and rod.
2   Before removing the piston from the rod, stuff a clean shop towel into the crankcase hole, around the connecting rod. This will prevent the snap-rings from falling into the crankcase if they are inadvertently dropped.

Removal

3   The piston should have an arrow mark on its crown that points toward the exhaust (front) side of the engine (see illustration). If this mark is not visible due to carbon buildup, you can refer to the valve notches. There are two notches on the exhaust side of the piston and three on the intake side. Support the piston and remove the snap-ring with snap-ring pliers (see illustration).
4   Push the piston pin out from the opposite end to free the piston from the rod (see illustration). You may have to deburr the area around the groove to enable the pin to

slide out (use a triangular file for this procedure). If the pin won't come out, you can fabricate a piston pin removal tool from a long bolt, a nut, a piece of tubing and washers (see illustration).

Inspection

5   Before the inspection process can be carried out, the piston must be cleaned and the old piston rings removed.
6   Using a piston ring removal and installation tool, carefully remove the rings from the piston (see illustration). Do not nick or gouge the piston in the process.
7   Scrape all traces of carbon from the top of the piston. A hand-held wire brush or a piece of fine emery cloth can be used once the majority of the deposits have been scraped away. Do not, under any circumstances, use a wire brush mounted in a drill motor to remove deposits from the piston; the piston material is soft and will be eroded away by the wire brush.
8   Use a piston ring groove cleaning tool to remove any carbon deposits from the ring grooves. If a tool is not available, a piece broken off the old ring will do the job. Be very careful to remove only the carbon deposits.

Do not remove any metal and do not nick or gouge the sides of the ring grooves.
9   Once the deposits have been removed, clean the piston with solvent and dry them thoroughly. Make sure the oil return holes below the oil ring grooves are clear.
10   If the piston is not damaged or worn excessively and if the cylinder is not rebored, a new piston will not be necessary. Normal piston wear appears as even, vertical wear on the thrust surfaces of the piston and slight looseness of the top ring in its groove. New piston rings, on the other hand, should always be used when an engine is rebuilt.
11   Carefully inspect each piston for cracks around the skirt, at the pin bosses and at the ring lands (see illustration).
12   Look for scoring and scuffing on the thrust faces of the skirt, holes in the piston crown and burned areas at the edge of the crown. If the skirt is scored or scuffed, the engine may have been suffering from overheating and/or abnormal combustion, which caused excessively high operating temperatures. The oil pump should be checked thoroughly. A hole in the piston crown, an extreme to be sure, is an indication that abnormal combustion (pre-ignition) was

14.6 Remove the piston rings with a ring removal and installation tool

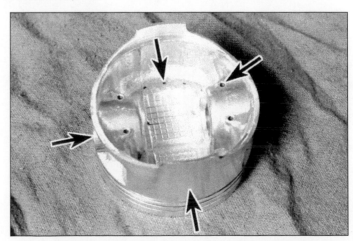

14.11 Check the piston pin bore and the piston skirt for wear, and make sure the internal holes are clear (arrows)

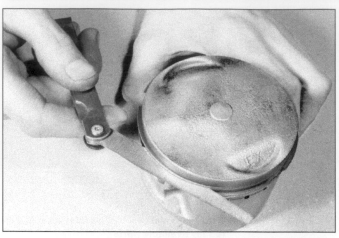

14.13 Measure the piston ring-to-groove clearance with a feeler gauge

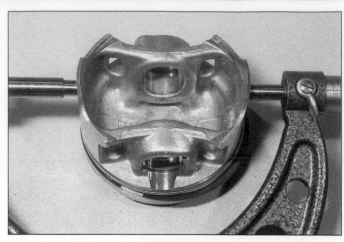

14.14 Measure the piston diameter with a micrometer

occurring. Burned areas at the edge of the piston crown are usually evidence of spark knock (detonation). If any of the above problems exist, the causes must be corrected or the damage will occur again.

13 Measure the piston ring-to-groove clearance (side clearance) by laying a new piston ring in the ring groove and slipping a feeler gauge in beside it **(see illustration)**. Check the clearance at three or four locations around the groove. Be sure to use the correct ring for each groove; they are different. If the clearance is greater then specified, a new piston will have to be used when the engine is reassembled.

14 Check the piston-to-bore clearance by measuring the bore (see Section 13) and the piston diameter **(see illustration)**. Measure the piston across the skirt on the thrust faces at a 90-degree angle to the piston pin, at the specified distance up from the bottom of the skirt. Subtract the piston diameter from the bore diameter to obtain the clearance. If it is greater than specified, the cylinder will have to be rebored and a new oversized piston and rings installed. If the appropriate precision measuring tools are not available, the piston-to-cylinder clearance can be obtained, though not quite as accurately, using feeler gauge stock. Feeler gauge stock comes in 12-inch lengths and various thicknesses and is generally available at auto parts stores. To check the clearance, slip a piece of feeler gauge stock of the same thickness as the specified piston clearance into the cylinder along with the appropriate piston. The cylinder should be upside down and the piston must be positioned exactly as it normally would be. Place the feeler gauge between the piston and cylinder on one of the thrust faces (90-degrees to the piston pin bore). The piston should slip through the cylinder (with the feeler gauge in place) with moderate pressure. If it falls through, or slides through easily, the clearance is excessive and a new piston will be required. If the piston binds at the lower end of the cylinder and is loose toward the top, the cylinder is tapered, and if tight spots are encountered as the piston/feeler gauge is rotated in the cylinder, the cylinder is out-of-round. Be sure to have the cylinder and piston checked by a dealer service department or a repair shop to confirm your findings before purchasing new parts.

15 Apply clean engine oil to the pin, insert it into the piston and check for freeplay by rocking the pin back-and-forth. If the pin is loose, a new piston and possibly new pin must be installed.

16 Repeat Step 15, this time inserting the piston pin into the connecting rod **(see illustration)**. If the pin is loose, measure the pin diameter and the pin bore in the rod (or have this done by a dealer or repair shop). A worn pin can be replaced separately; if the rod bore is worn, the rod and crankshaft must be replaced as an assembly.

17 Refer to Section 15 and install the rings on the piston.

Installation

18 Install the piston with its arrow mark toward the exhaust side (front) of the engine. Lubricate the pin and the rod bore with moly-based grease. Install a new snap-ring in the groove in one side of the piston (don't reuse the old snap-rings). Push the pin into position from the opposite side and install another new snap-ring. Compress the snap-rings only enough for them to fit in the piston. Make sure the circlips are properly seated in the grooves **(see illustration)**.

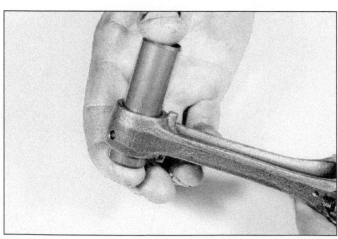

14.16 Slip the piston pin into the rod and try to rock it back-and-forth to check for looseness

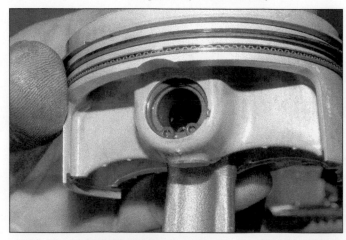

14.18 Make sure both piston pin snap-rings are securely seated in the piston grooves

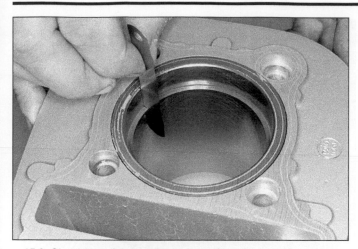

15.2 Check the piston ring end gap with a feeler gauge at the bottom of the ring travel area

15.4 If the end gap is too small, clamp a file in a vise and file the ring ends (from the outside in only) to enlarge the gap slightly

15 Piston rings - installation

1   Before installing the new piston rings, the ring end gaps must be checked.

2   Insert the top (No. 1) ring into the bottom of the first cylinder and square it up with the cylinder walls by pushing it in with the top of the piston. The ring should be about one-half inch above the bottom edge of the cylinder. To measure the end gap, slip a feeler gauge between the ends of the ring (see illustration) and compare the measurement to the Specifications.

3   If the gap is larger or smaller than specified, double check to make sure that you have the correct rings before proceeding.

4   If the gap is too small, it must be enlarged or the ring ends may come in contact with each other during engine operation, which can cause serious damage. The end gap can be increased by filing the ring ends very carefully with a fine file (see illustration). When performing this operation, file

only from the outside in.

5   Repeat the procedure for the second compression ring (ring gap is not specified for the oil ring rails or spacer).

6   Once the ring end gaps have been checked/corrected, the rings can be installed on the piston.

7   The oil control ring (lowest on the piston) is installed first. It is composed of three separate components. Slip the spacer into the groove, then install the upper side rail (see Illustrations). Do not use a piston ring installation tool on the oil ring side rails as they may be damaged. Instead, place one end of the side rail into the groove between the spacer expander and the ring land. Hold it firmly in place and slide a finger around the piston while pushing the rail into the groove (taking care not to cut your fingers on the sharp edges). Next, install the lower side rail in the same manner.

8   After the three oil ring components have been installed, check to make sure that both the upper and lower side rails can be turned smoothly in the ring groove.

9   Install the no. 2 (middle) ring next with

15.7a Installing the oil ring expander - make sure the ends don't overlap

its identification mark facing up (see illustration). Do not mix the top and middle rings; their profiles are slightly different, but the difference can be hard to see. The most important indicator is the ring thickness. The top

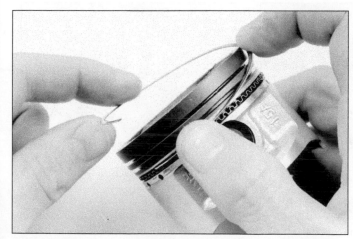

15.7b Installing an oil ring side rail - don't use a ring installation tool to do this

15.9 Install the middle ring with its identification mark up

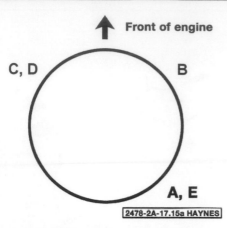

Front of engine

C, D   B

A, E

2478-2A-17.15a HAYNES

15.12 Arrange the ring gaps like this

A   Oil ring spacer   D   Second compression ring
B   Oil ring upper rail   E   Top compression ring
C   Oil ring lower rail

16.5a Here are the right side oil line/hose (upper arrow) and oil tank vent hose on a later 250 with external oil tank (lower arrow)

ring is thicker than the second ring. On a new piston, the top ring will not fit into the second ring's groove. If you're not sure which ring is which, measure their thicknesses with a micrometer.

10   To avoid breaking the ring, use a piston ring installation tool and make sure that the identification mark is facing up. Fit the ring into the middle groove on the piston. Do not expand the ring any more than is necessary to slide it into place.

11   Finally, install the no. 1 (top) ring in the same manner. Make sure the identifying mark is facing up. Be very careful not to confuse the top and second rings.

12   Once the rings have been properly installed, stagger the end gaps, including those of the oil ring side rails **(see illustration)**.

16 External oil tank and lines - removal and installation

Oil lines

Removal

1   1998 through 2005 YZ400/426/540 models, as well as 1998 through 2006 WR400/426/450 models, have the following tank and line components:
   a) Combined oil pipe/hose on each side of the engine
   b) Oil tank built into the front upper portion of the frame
   c) External three-fitting oil tube at the right rear corner of the engine

2   2006 and later YZ450 models, as well as 2007 WR450 models, have the following oil tank and line components:
   a) Oil tank breather hose on the left side of the engine
   b) Oil tank built into the front of the crank-case

3   2001 through 2005 YZ250 and 2001 through 2006 WR250 models have the following tank and line components:

16.5b Here's a left side oil line/hose; its bolted to the engine and external oil tank (arrows)

a) Combined oil pipe/hose on each side of the engine
b) Oil tank built into the front upper portion of the frame

16.5c The bolted connections at the engine use an O-ring and dowel (arrow)

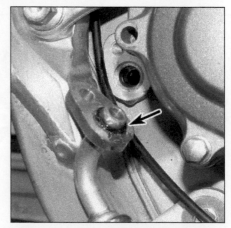

16.5d The bolted connections at the external oil tank use an O-ring on the fitting (arrow)

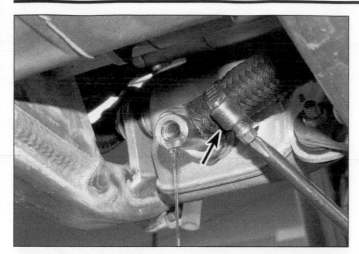

16.5e The lower oil tank hose on models with an external oil tank is secured by a clamp

16.5f Hoses on models with a frame oil tank are clamped to the bottom of the tank

c) *External three-fitting oil tube at the right rear corner of the engine*

4   2006 and later YZ250 models, as well as 2007 WR250 models, have the following tank and line components:

a) *Combined oil pipe/hose on each side of the engine*

b) *External oil tank between the lower front frame members*

c) *External three-fitting oil tube at the right rear corner of the engine*

5   To disconnect an oil pipe/hose from the engine, remove its mounting bolt **(see illustrations)**. The connection at the engine has a dowel and O-ring **(see illustration)**. The connection at the oil tank has a bolted fitting with an O-ring **(see illustrations)**, or is secured by a clamp **(see illustrations)**.

6   To disconnect the frame oil tank vent hose (if equipped), loosen its clamps and carefully pry it off the fittings **(see illustration)**.

7   To remove the three-branched oil pipe, remove the banjo bolts and sealing washers from each of the three fittings **(see illustration)**. Take the pipe off.

16.6 The frame oil tank vent hose runs from the valve cover to the tank (arrow)

8   All models have an oil tube in the left crankcase half. To remove it, remove the left crankcase cover, alternator rotor and stator (see Section 17 and Chapter 5). Remove

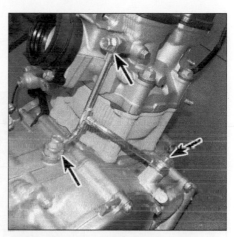

16.7 The external oil pipe (used on all except 2007 450 models) is attached to the crankcase and cylinder head with union bolts

the tube retaining bolt and pull it out of the crankcase, together with its O-rings **(see illustrations)**.

16.8a Unbolt the oil pipe (arrow) . . . .

16.8b . . . and pull it out, together with its three O-rings (arrows)

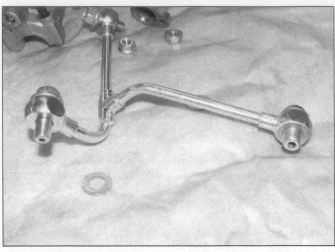

16.9 The oil pipe union bolts are different sizes

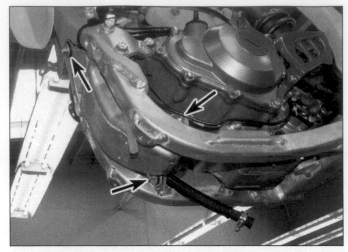

16.13a Remove the oil tank mounting bolts (arrows) . . .

Installation

9   Installation is the reverse of the removal steps, with the following additions:

a)   Replace the sealing washers whenever the union bolts or hose fittings are loos-ened. Note that the three-branched pipe's right union bolt is larger than the others (see illustration).

b)   Replace O-rings whenever they are removed.

c)   Tighten the union bolts and tube retaining bolt to the torques listed in this Chapter's Specifications.

External oil tank

10   An external oil tank is used on 2006 and later YZ250 models. On all other models, the oil tank is in the frame or engine.

Removal

11   Refer to the oil change procedure in Chapter 1 and drain the oil tank.

12   Disconnect the hoses from the tank as described above.

13   Remove the tank mounting bolts and lift the tank off the frame (see illustrations).

Installation

14   Check the mounting bushings for damage or deterioration and replace them as needed (see illustration).

15   Position the tank on the frame and tighten the mounting bolts. Connect the hoses as described above.

16   Refer to Chapter 1 and change the engine oil.

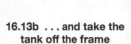
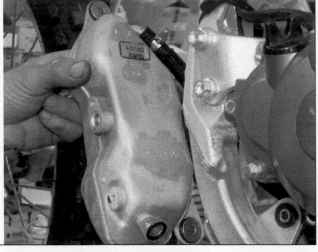

16.13b . . . and take the tank off the frame

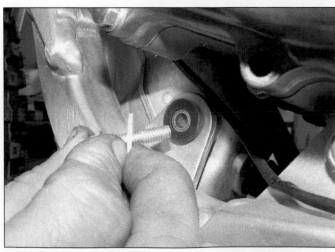

16.14 Check the mounting grommets for wear and deterioration

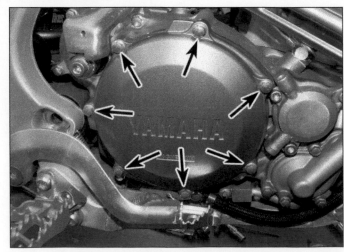

17.1 Remove the outer cover mounting bolts . . .

17.2 . . . take the cover off and locate the dowel pins (arrows)

17.8 Remove the inner cover mounting bolts (arrows) - on all 450 models, there's an additional bolt at location A, and on later 450 models, Allen bolts are used at locations B. . .

17.10 . . . take the cover off and locate the dowel pins (arrows) (lower dowel pin/O-ring not used on 2006 and later YZ450 models or 2007 and later WR450 models)

the cover off or the gasket surfaces will be damaged.
10   Locate the cover dowels and O-rings **(see illustration)**. They may have stayed in the crankcase or come off with the cover.
11   Remove all traces of the old gasket from the cover and crankcase.

Installation

12   Installation is the reverse of the removal steps. When installing the inner cover, use a new gasket, coated on both sides with gasket sealer. Be sure the dowels and O-rings are installed.
13   Tighten the cover bolts evenly to the torque listed in this Chapter's Specifications.

Left crankcase cover

Removal

14   If you're working on a WR model with an electric starter, remove the starter torque limiter (see Chapter 5). One of the crankcase cover bolts is hidden behind the torque limiter.
15   Remove the shift pedal (see Section 24).
16   Remove the cover bolts **(see illustrations)**. Loosen the bolts evenly in a criss-cross pattern, then remove them.

17 Crankcase covers - removal and installation

Right crankcase cover

Outer cover removal

1   Remove the cover mounting bolts **(see illustration)**.
2   Tap the cover gently with a soft-faced mallet to free it and take it off the engine. Locate the dowels and remove the old gasket **(see illustration)**.
3   Installation is the reverse of the removal steps. Tighten the cover bolts evenly, in a criss-cross pattern, to the torque listed in this Chapter's Specifications.

Inner cover removal

4   Remove the oil filter and drain the cooling system (see Chapter 1). Remove the outer cover as described above.
5   Remove the exhaust system and brake pedal (Chapters 4 and 7).

6   Remove the coolant tube from the cover (see Chapter 3).
7   Disconnect the oil hose(s) from the cover (see Section 16).
8   Remove the cover bolts **(see illustration)**. Loosen the bolts evenly in a criss-cross pattern, then remove them.
9   Pull the cover off. Tap it with a rubber mallet if it won't come off evenly. Don't pry

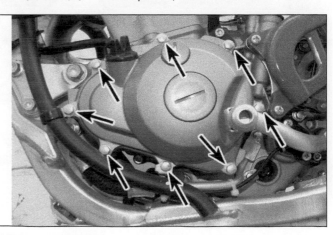

17.16a Remove the cover bolts (arrows) - this is a YZ . . .

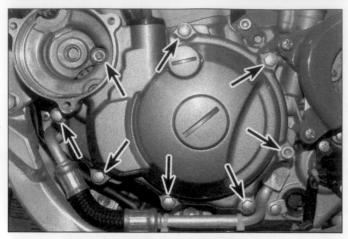

17.16b ... this is a WR250 with electric start - one of the Allen bolts is hidden behind the slipper clutch (A) ...

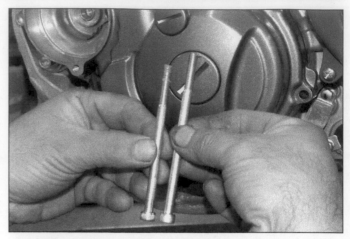

17.16c ... the rear Allen bolt is longer than the front one

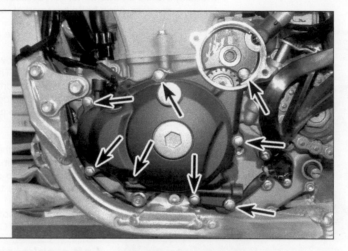

17.16d This is a WR450 - the upper rear cover bolt is hidden behind the slipper clutch

sides with gasket sealer. Be sure the dowels are installed.

21   Tighten the cover bolts evenly to the torque listed in this Chapter's Specifications.

18 Clutch and release mechanism - removal, inspection and installation

Cable and lever

Removal

1   Loosen the midline cable adjuster or handlebar cable adjuster all the way to create slack (see Chapter 1). Rotate the cable so the inner cable aligns with the slot in the lever, then slip the cable and fitting out of the lever **(see illustration)**.

2   Slip the end of the cable through the lever slot on the engine to detach it from the lever **(see illustration)**.

3   Loosen the locknut at the cable bracket on the engine and pull the cable out of the

17   Pull the cover off. Tap it with a rubber mallet if it won't come evenly. Don't pry the cover off or the gasket surfaces will be damaged.

18   Locate the cover dowels **(see illustration)**. They may have stayed in the crankcase or come off with the cover. Note that the front dowel on 2006 and later YZ450 models, as

well as 2007 WR450 models, has an O-ring **(see illustration)**.

19   Remove all traces of the old gasket from the cover and crankcase.

Installation

20   Installation is the reverse of the removal steps. Use a new gasket, coated on both

17.18a Locate the cover dowels (arrows) ...

17.18b ... on 2006 Z450 models and all 2007 and later 450 models, the front dowel has an O-ring (arrow)

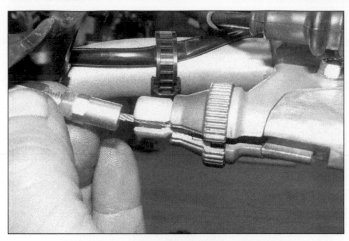

18.1 Slip the cable end out of the gap (arrow)

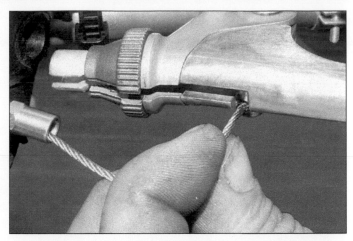

18.2 Rotate the cable and lower it away from the lever

18.3a Loosen the locknut (arrow) all the way (on some models, it's on the other side of the bracket) . . .

18.3b . . . and pull the cable out of the bracket . . .

bracket **(see illustrations)**. Turn the cable to align with the slot in the lever and lift the cable end out of the lever. Free the cable from any retainers and take it off the bike **(see illustration)**.

4   If necessary, detach the cable bracket from the engine.
5   To remove the lever from the handlebar, undo the clamp screws **(see illustration)**.

Inspection

6   Slide the inner cable back and forth in the housing and make sure it moves freely. If it doesn't, try lubricating it as described in Chapter 1. If that doesn't help, replace the cable.

18.3c . . . then turn the cable to align with the slot in the lever (arrow) and lift the cable end out of the lever

18.5 Remove the clamp screws (arrows) and detach the lever from the handlebar

18.10a Remove the spring bolts, washers and springs . . .

18.10b . . . take off the pressure plate and push piece . . .

18.10c . . . slide the plates out of the clutch housing . . .

18.10d . . . the last friction plate is narrower than the others; a damper spring (arrow) fits within it

Installation

7   Installation is the reverse of the removal steps. Refer to Chapter 1 and adjust the clutch freeplay.

Clutch

Removal

8   If you're just going to remove the clutch plates, remove the clutch cover from the right crankcase cover (see Section 17). If you're going to remove the clutch center or housing, remove the right crankcase cover (see Section 17). If you're not sure how much clutch work will be needed, start by remov-

18.10e The metal plates with blue tabs go on first and last

18.10f The damper spring (left arrow) has a spring seat behind it (right arrow) - the concave side of the damper spring faces away from the engine

18.10g The clutch center and housing are secured by a nut and lockwasher (arrow)

18.10h Bend back the lockwasher tabs

18.10i Slip a copper washer between the primary drive gear and the primary driven gear on the back of the clutch housing (arrow) to lock the gears so they won't turn, then unscrew the nut . . .

18.10j . . . remove the lockwasher and pull off the clutch center (arrow) . . .

ing the outer cover and plates. This will make it possible to inspect the center and housing to see if they need to be removed.

9   Hold the clutch from turning with a holding tool. If you don't have one, you can make your own holding tool from steel strap. **Note:** *If you've removed the right crankcase cover, you can wedge a copper washer or penny between the primary drive gear and clutch housing driven gear to keep the clutch hous-* *ing from turning while you loosen the clutch spring bolts and clutch housing nut.*

10   Refer to the accompanying illustrations to remove the clutch **(see illustration 17.2 and the accompanying illustrations).**

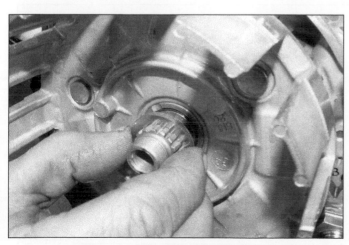

18.10k . . . remove the thrust washer . . .

18.10l . . . then remove the clutch housing

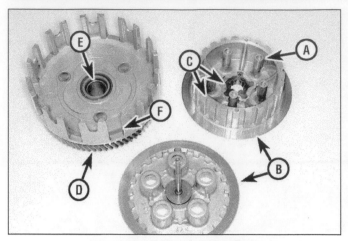

18.12a Clutch inspection points

A	Spring posts	D	Primary driven gear
B	Friction surface	E	Clutch housing bushing
C	Splines	F	Clutch housing slots

18.12b Grip the clutch housing and try to rotate the primary gear; if there's any play, replace the housing

Inspection

11   Check the bolt posts and the friction surface on the pressure plate for damaged threads, scoring or wear. Replace the pressure plate if any defects are found.

12   Check the edges of the slots in the clutch housing for indentations made by the friction plate tabs **(see illustration)**. If the indentations are deep they can prevent clutch release, so the housing should be replaced with a new one. If the indentations can be removed easily with a file, the life of the housing can be prolonged to an extent. Check the bushing surface in the center of the clutch housing for score marks, scratches and excessive wear. Also, check the driven gear teeth for cracks, chips and excessive wear. If the bushing or gear is worn or damaged, the clutch housing must be replaced with a new one. Check the primary driven gear for play **(see illustration)**. If there is any, replace the clutch housing.

13   Check the splines of the clutch boss for indentations made by the tabs on the metal plates. Check the clutch boss friction surface for wear or scoring. Replace the clutch boss if problems are found.

14   Measure the free length of the clutch springs **(see illustration)** and compare the results to this Chapter's Specifications. If the springs have sagged, or if cracks are noted, replace them with new ones as a set.

15   If the lining material of the friction plates smells burnt or if it is glazed, new parts are required. If the metal clutch plates are scored or discolored, they must be replaced with new ones. Measure the thickness of the friction plates **(see illustration)** and replace with new parts any friction plates that are worn.

16   Lay the metal plates, one at a time, on a perfectly flat surface (such as a piece of plate glass) and check for warpage by trying to slip a feeler gauge between the flat surface and the plate **(see illustration)**. The feeler gauge should be the same thickness as the maximum warp listed in this Chapter's Specifications. Do this at several places around the plate's circumference. If the feeler gauge can

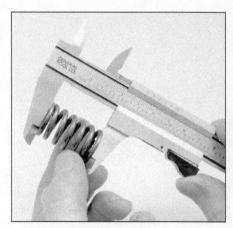

18.14 Measure the clutch spring free length

be slipped under the plate, it is warped and should be replaced with a new one.

17   Check the tabs on the friction plates for excessive wear and mushroomed edges.

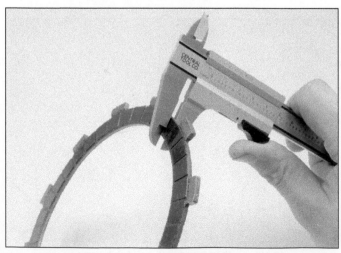

18.15 Measure the thickness of the friction plates

18.16 Check the metal plates for warpage

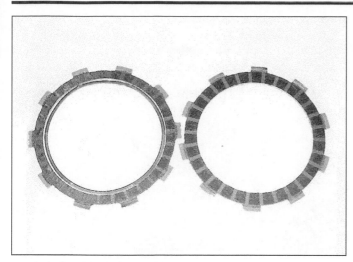

18.19 On YZ/WR250 models, the narrower friction plate and damper (left) go into the clutch housing first

18.21a Pull the ball out of the transmission shaft with a magnet . . .

They can be cleaned up with a file if the deformation is not severe. Check the friction plates for warpage as described in Step 16.
18 Check the thrust washer for score marks, heat discoloration and evidence of excessive wear.

Installation

19 Installation is the reverse of the removal steps, with the following additions:

a) Install a new lockwasher and position its tabs between the ribs of the clutch center. Tighten the clutch nut to the torque listed in this Chapter's Specifications, then bend the lockwasher tabs against two of the flats on the nut.

b) Coat the friction plates with clean engine oil before you install them.

c) Install a friction plate, then the remaining metal and friction plates until they're all installed. Friction plates go on first and last, so the friction material contacts the metal surfaces of the clutch center and the pressure plate. On YZ/WR250 mod-

els, note the location of the narrower friction plate and the damper that fits inside it (see illustration).

d) Apply grease to the ends of the clutch pushrod, the steel ball and the end of the adjuster rod.

Lifter lever and pushrod

Removal

20 Remove the right crankcase cover (see Section 17) . Remove the clutch pressure plate and push piece as described above.
21 Remove the ball and pushrod (see illustrations)
22 Remove the retaining bolt and pull the lifter lever out of the crankcase (see illustration 18.3a or 18.3c).
23 Check for visible wear or damage at the contact points of the lifter lever and pushrod (see illustration). Replace any parts that show problems.
24 Remove the snap-ring and pry the lifter shaft seal out of the crankcase. If the needle

bearing is worn or damaged, drive it out with a shouldered drift the same diameter as the bearing, then use the same tool to drive in a new one. Pack the needle bearing with grease and press in a new seal.
25 Installation is the reverse of the removal steps. Engage the notch in the lever shaft with the pushrod and hook the spring to the crankcase.
26 Refer to Chapter 1 and adjust clutch freeplay.

19 Camshaft chain and guides - removal, inspection and installation

1 If inspection procedures in Section 9 indicate problems with the chain or guides, remove them for further inspection.
2 To remove the exhaust side chain guide, remove the cylinder head (see Section 10). Lift the guide out of its pockets in the cylin-

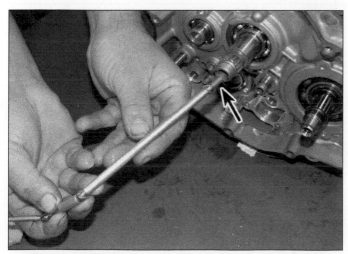

18.21b . . . then pull out the shaft - the long end of the shaft (arrow) goes into the transmission shaft first on installation

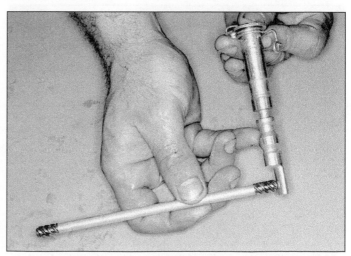

18.23 The pushrod engages the lifter lever like this when they're installed

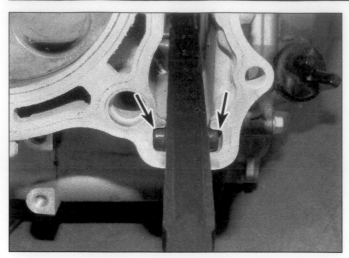

19.2a The upper end of the chain guide fits in these pockets (arrows) - be sure to install the chain guide before installing the cylinder head

19.2b The lower end of the front chain guide fits in this pocket (left arrow) - the lower end of the rear chain guide is bolted to the crankcase (arrows)

der and crankcase (see illustrations).

3   To remove the intake side guide or the chain, remove the left crankcase cover and alternator rotor (see Section 17 and Chapter 5). Unbolt the guide and slip the chain off the crankshaft sprocket (see illustration 19.2b).

4   Installation is the reverse of removal.

20 Kickstarter - removal, inspection and installation

Kickstarter

Removal

1   Remove the right crankcase cover (see Section 17).

2   Note how the kickstarter spring ends and the stopper arm on the kick axle assembly engage the crankcase, then take the kickstarter assembly out of the crankcase (see illustrations).

3   Remove the snap-ring, thrust washer and kick idle gear (see illustration).

Inspection

4   Remove the components from the kickstarter shaft. From the outer end, remove the spring guide and torsion spring. From the inner end, remove the thrust washer (if you haven't already done so). Remove the snap-ring, washer, spring and ratchet wheel. Remove the remaining snap-ring, then remove the washer and kickstarter gear. Lay the parts in order so they can be reinstalled correctly.

5   Check all parts for visible wear and damage and replace any that show problems.

6   Lubricate the parts with clean engine oil and put them back on the shaft.

Installation

7   Installation is the reverse of the removal

20.2a Note how the spring ends engage with the crankcase and collar . . .

steps. Be sure the spring and stopper arm engage the crankcase correctly.

20.2b . . . then remove the kick axle and thrust washer (center arrow) from the crankcase - on installation, hook the ratchet wheel arm (lower arrow) behind the stop (upper arrow)

20.3 Remove the snap-ring and take the kick idle gear off its shaft

22.2 The alignment marks on the balancer gears (arrows) must be aligned on installation

22.4 The balancer weight on 2007 450 models is removable - bend back the lockwasher tabs and unscrew the nut

21 Compression release (2002 and earlier models) - removal, inspection and installation

Note: *2003 and later models have an automatic compression release built into the exhaust camshaft. It's removed and inspected as part of the camshaft removal procedure (see Section 9).*

Removal

1 Loosen the cable locknut to provide slack in the cable. Follow the cable from the decompressor lever at the left handlebar grip to the cam lever at the cylinder head, removing any ties that attach it to the frame.
2 At the cylinder head, remove the bolt that attaches the cable bracket to the engine.
3 Rotate one end of the cable to align it with its lever slot, then slip the cable end plug out of the lever. Do the same thing at the other end of the cable, then take the cable off the motorcycle.
4 At the cylinder head, remove the lockbolt directly above the decompression cam. Slip the decompression cam and spring out of the engine.

Inspection

5 Slide the inner cable back and forth in its housing. If it doesn't move freely, lubricate it (see Chapter 1). If that doesn't help, replace the cable.
6 Check the lever and cam components for wear and damage and replace any that show problems.
7 Pry the seal out of the cylinder head. Press in a new one, using a socket the same diameter as the seal.

Installation

8 Installation is the reverse of the removal steps. Refer to Chapter 1 and adjust the decompressor cable freeplay.

22 Primary drive gear and balancer - removal, inspection and installation

Removal

1 If you're just planning to remove the gears, remove the right crankcase cover and clutch (see Sections 17 and 18). If you're planning to remove the balancer weight or shaft, remove the left crankcase cover (see Section 17).

2 Turn the crankshaft so the match marks on the balancer drive and driven gears align **(see illustration)**.
3 On all except 2007 450 models, bend back the tabs on the primary drive gear and balancer driven gear lockwashers **(see illustration 22.2)**. On 2007 450 models, bend back the tabs on the balancer driven gear lockwasher. The drive gear has a conical lockwasher with no tabs.
4 Wedge a copper washer or penny between the teeth of the balancer drive and driven gears to prevent them from turning. Loosen the driven gear nut. If you plan to remove the primary drive gear or balancer drive gear, wedge the gears from the other side and loosen the primary drive gear nut. If you're working on a 2007 450 model, bend back the tab on the balancer lockwasher and loosen the balancer weight nut (it's on the left side of the engine inside the crankcase cover) **(see illustration)**.
5 Unscrew the nuts and remove the lockwashers **(see illustrations)**.
6 Slide off the primary drive gear. Note the location of the short spline, then remove the

22.5a Unscrew the nuts, noting the direction they face . . .

22.5b . . . and remove the lockwashers - on installation, place the lockwashers tabs in the gear slots (arrow)

22.6a Remove the primary drive gear - the short spline arrow allows the gear to be installed only one way . . .

22.6b . . . remove the balancer drive gear - its punch mark aligns with the short spline (arrow)

22.6c The shouldered side of the gear faces the engine

22.7 Remove the balancer driven gear/weight - its punch mark aligns with the short spline (arrow)

22.8a On all except 2007 450 models, turn the balancer shaft so the weight is toward the front of the engine as shown . . .

balancer drive gear (see illustrations).

7   Note the location of the short spline, then remove the balancer driven gear (see illustration).

8   If you're working on a 2006 or earlier 450 model or any 250 model, pull the balancer shaft out of the engine from the left side (see illustrations).

9   If you're working on a 2007 450 model, remove the balancer weight from the shaft

(see illustration 22.4). The crankcase halves must be separated to remove the balancer shaft (see Section 25). Pull the shaft out of the engine.

Inspection

10   Check the gears for worn or damaged teeth and replace them as a set if problems are found.

11   Check the ball bearings for wear, loose-

ness or rough movement. If any problems are found, replace the bearings as described in Section 26.

12   Check the remaining components for wear and damage and replace any worn or damaged parts. Replace the lockwasher with a new one whenever it's removed.

13   Inspect the balancer and crankshaft ball bearings to the extent possible without disassembling the crankcase. If wear, looseness or roughness can be detected, the crankcase will have to be disassembled to replace the bearings.

Installation

14   If you removed the balancer shaft, install it.

15   Install the balancer drive and driven gears so their short splines align with the short splines on the shafts and the match marks align with each other (see illustrations 22.7, 22.6b and 22.6a).

16   Install the primary drive gear next to the balancer drive gear.

17   Install the lockwashers and nuts on the balancer shaft and crankshaft. Engage the

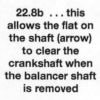

22.8b . . . this allows the flat on the shaft (arrow) to clear the crankshaft when the balancer shaft is removed

23.2 Remove the snap-ring and take the oil pump drive gear off its shaft

23.3a If you plan to disassemble the pump, loosen the assembly screw now, while the pump is still bolted to the engine

A Assembly screw
B Light colored Allen bolts
C Dark colored Allen bolt (longer)

tab in the driven gear locknut with the hole in the drive gear (see illustration 22.5b).

18 Wedge the gears as described in Step 4 and tighten the nuts to the torque listed in this Chapter's Specifications (on 2007 450 models, this includes the balancer weight on the opposite side of the engine from the gears). On tabbed lockwashers, bend the lockwasher tabs to secure the nut(s).

19 The remainder of installation is the reverse of the removal steps

23 Oil pump - removal, inspection and installation

Removal

1 Remove the right crankcase cover and clutch (see Sections 17 and 18).
2 Remove the snap-ring and washer and take the oil pump drive gear off its shaft (see illustration).
3 Remove the oil pump mounting bolts

23.3b Lift the pump off the engine and locate the dowel (upper arrow; 450 models have two dowels) - the no. 2 outer rotor may stay in the engine (lower arrow)

and take it off the engine (see illustration). The outer rotor of the no. 2 rotor set may remain in the engine (see illustration). If so, remove it.

4 Locate the pump dowel(s) (see illustration 23.3b). They may have come off with the pump or stayed in the engine.

Inspection

5 Remove the snap-ring, no. 2 inner rotor and drive pin from the pump shaft (see illustrations).
6 Remove the assembly screw from the oil pump cover. Take the cover off and remove

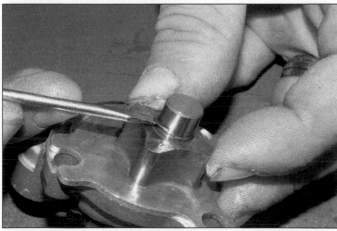

23.5a Remove the circlip from the pump shaft, slide the no. 2 inner rotor off . . .

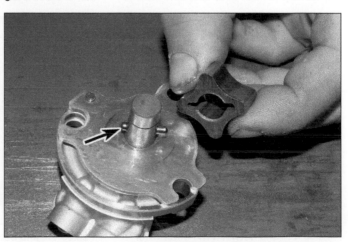

23.5b . . . and remove the drive pin (arrow)

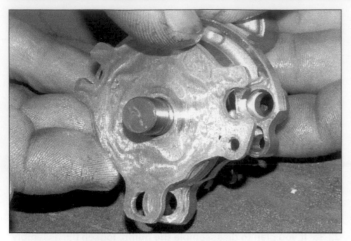

23.6a Remove the pump cover . . .

23.6b . . . outer rotor (arrow) . . .

23.6c . . . then lift the inner rotor off the drive pin and remove it . . .

23.6d . . . and take the thrust washer out of the case

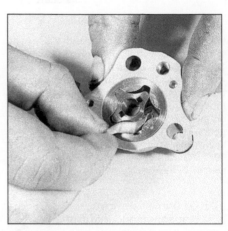

23.8a Measure the gap between the inner and outer rotors . . .

the drive pin, shaft and rotors **(see illustrations)**.

7   Wash all the components in solvent, then dry them off. Check the pump body, the rotors and the cover for scoring and wear. If any damage or uneven or excessive wear is

evident, replace the pump. If you are rebuilding the engine, it's a good idea to install a new oil pump.

8   Place the rotors in the pump cover. Measure the clearance between the outer rotor and body, and between the inner and

outer rotors, with a feeler gauge **(see illustrations)**. Place a straightedge across the pump body and rotors and measure the gap with a feeler gauge **(see illustration)**. If any of the clearances are beyond the limits listed in this Chapter's Specifications, replace the pump.

9   Reassemble the pump by reversing the disassembly steps, with the following additions:

  a) *Before installing the cover, pack the cavities between the rotors with petroleum jelly - this will ensure the pump develops suction quickly and begins oil circulation as soon as the engine is started.*
  b) *Make sure the drive pin is in position.*
  c) *Tighten the cover screw to the torque listed in this Chapter's Specifications.*

Installation

10   Installation is the reverse of removal, with the following additions:

  a) *Make sure the pump dowels are in position.*
  b) *Tighten the oil pump mounting screws to the torque listed in this Chapter's Specifications.*

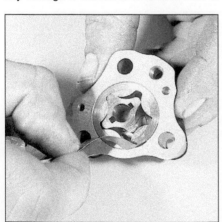

23.8b . . . and between the outer rotor and body . . .

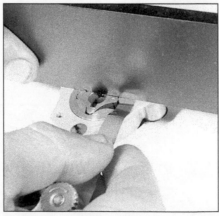

23.8c . . . and between the rotors and a straightedge

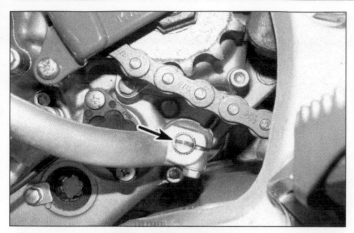

24.1 Make an alignment mark on the end of the shift shaft, then remove the pedal bolt all the way

24.7a Pull the shift shaft out of the crankcase, noting how the return spring ends fit over their pin (arrow)

24 External shift mechanism - removal, inspection and installation

Shift pedal

Removal

1   Look for alignment marks on the end of the shift pedal and shift shaft (see illustration). If they aren't visible, make your own marks with a felt pen or sharp punch. Remove the shift pedal pinch bolt completely (it fits in a groove) and slide the pedal off the shaft.

Inspection

2   Check the shift pedal for wear or damage such as bending. Check the splines on the shift pedal and shaft for stripping or step wear. Replace the pedal or shaft if these problems are found.
3   Check the shift shaft seal in the alternator cover for signs of leakage. If the seal has been leaking, remove the left crankcase cover (see Section 17). Pry the seal out of the cover, then tap in a new one with a seal driver or socket the same diameter as the seal.

24.7b Take the roller off the pawl holder, noting which direction it faces

Installation

4   Install the shift pedal or shift arm. Line up its punch marks and tighten the pinch bolt to the torque listed in this Chapter's Specifications.

External shift linkage

Removal

5   Remove the shift pedal as described above.

6   Remove the right crankcase cover and the clutch (see Sections 17 and 18).
7   Refer to the accompanying illustrations to remove the linkage (see illustrations).

Inspection

8   Check the shift shaft for bends and damage to the splines. If the shaft is bent, you can attempt to straighten it, but if the splines are damaged it will have to be replaced. Check

24.7c Remove the shift guide bolts (arrows)

24.7d Take the pawl assembly and shift guide off . . .

24.7e The pawl assembly goes together like this - note the directions the rounded ends of the springs and pawls face

24.7f . . . separate the pawl holder from the shift guide . . .

24.7g . . . and remove the pins and springs

24.7h If the punch mark on the shift drum segment (arrow) is in the position shown . . .

24.7i . . . turn it all the way counterclockwise, turning the transmission shaft (arrow) by hand at the same time to prevent damage to the shift forks . . .

24.7j . . . until the segment stops with its mark in this position

24.7k Unscrew the segment bolt . . .

24.7l . . . pry the stopper lever out of the way (arrow) and take the segment off . . .

24.7m . . . and locate the dowel pin (arrow) - align the pin and notch (arrow) on installation

24.7n Note how the spring engages the stopper arm and crankcase, then unbolt the stopper arm and take it off

the condition of the return spring, shift arm and the pawl spring. Replace the shift shaft if they're worn, cracked or distorted.

Installation

9 Installation is the reverse of the removal steps, with the following additions:
a) Tighten the stopper arm bolt and shift drum segment bolt to the torques listed in this Chapter's Specifications.
b) Check the engine oil level and add some, if necessary (see Chapter 1).

25 Crankcase - disassembly and reassembly

1 To examine and repair or replace the crankshaft, connecting rod, bearings and transmission components, the crankcase must be split into two parts.

Disassembly

2 Remove the engine from the vehicle (see Section 5).
3 Remove the CDI magneto and starter clutch (if equipped) (see Chapter 4).
4 Remove the right crankcase cover and clutch (see Sections 17 and 18).
5 Remove the external shift mechanism (see Section 24).
6 Remove the valve cover, cam chain tensioner, camshafts, cylinder head, cylinder, piston, external oil lines and crankcase oil tube (see Sections 7, 8, 9, 10, 13, 14 and 16).
7 Remove the timing chain and intake side guide (see Section 19).
8 Remove the oil pump (see Section 23).
9 Remove the balancer gears (see Section 22). On all except 2007 450 models, remove the balancer shaft. On 2007 450 models, remove the balancer weight (the shaft is removed after the crankcase has been disassembled).
10 Check carefully to make sure there aren't any remaining components that attach the upper and lower halves of the crankcase together.
11 Loosen the crankcase bolts in two or three stages, in a criss-cross pattern (see illustration).
12 Place the engine on blocks so the transmission shafts and crankshaft can extend downward. Set up a three-legged puller against the end of the crankshaft so it can pull the upper case half off the crankshaft

25.11 Crankcase bolts (YZ250; others similar)

25.12a Set up a puller like this to push against the crankshaft while pulling the upper case half off

25.12b Watch the crankcase seam to make sure the halves are separating evenly

(see illustration). Tap gently on the ends of the transmission shafts, balancer shaft and crankshaft as the case halves are being separated. Make sure the case halves separate evenly (see illustration). Carefully pry the crankcase apart at the pry points. Don't pry against the mating surfaces or they'll develop leaks.

13   Lift the right crankcase half off the left half (see illustration).

14   Locate the crankcase dowels (see illustration 25.13). If they aren't secure in their holes, remove them and set them aside for safekeeping.

15   Refer to Sections 26 through 28 for information on the internal components of the crankcase.

Reassembly

16   Remove all traces of old gasket and sealant from the crankcase mating surfaces with a sharpening stone or similar tool. Be careful not to let any fall into the case as this is done and be careful not to damage the mating surfaces.

17   Check to make sure the dowel pins are in place in their holes in the mating surface of the crankcase. Be sure to install a new O-ring on the center dowel (see illustration 25.13).

18   Coat both crankcase mating surfaces with Yamaha Quick Gasket (ACC-11001-05-01) or equivalent sealant.

19   Pour some engine oil over the transmission gears, balancer shaft (2007 450 models) and crankshaft bearing surfaces and the shift drum. Don't get any oil on the crankcase mating surfaces.

20   Carefully place the removed crankcase half onto the other crankcase half. While doing this, make sure the transmission shafts, shift drum, crankshaft and balancer (2007 450 models) fit into their bearings in the upper crankcase half.

21   Install the crankcase half bolts or screws in the correct holes and tighten them so they are just snug. Then tighten them in two or three stages, in a criss-cross pattern, to the torque listed in this Chapter's Specifications.

22   Turn the transmission shafts to make sure they turn freely. Also make sure the crankshaft and balancer shaft turn freely.

23   The remainder of installation is the reverse of removal.

26 Crankcase components - inspection and servicing

1   Separate the crankcase and remove the following:
   a) Transmission shafts and gears
   b) Balancer shaft (2007 450 models)
   c) Crankshaft and main bearings
   d) Shift drums and forks

2   Clean the crankcase halves thoroughly with new solvent and dry them with compressed air. All oil passages should be blown out with compressed air and all traces of old gasket sealant should be removed from the mating surfaces. Caution: Be very careful not to nick or gouge the crankcase mating surfaces or leaks will result. Check both crankcase sections very carefully for cracks and other damage.

3   Check the bearings in the case halves (see illustration). If they don't turn smoothly, replace them, referring to Tools and Workshop Tips at the end of this manual.

4   Check the oil strainer screen for clog-

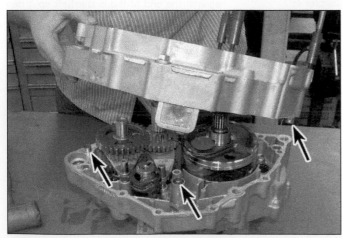

25.13 Lift the left case half off the right half and locate the dowels (arrows) - the center dowel has an O-ring

26.3 Check the case bearings for roughness, looseness or noise (YZ250 shown)

26.4a Unbolt the oil strainer . . .

26.4b . . . and lift it off - 450 models use a dowel

27.2a Transmission shafts and forks - assembled view

27.2b Disengage the shift drum from the fork pins and take it out of the crankcase, noting how it's installed - it's possible to install the shift drum backwards

ging or damage (see illustration). If problems are found, remove it for cleaning or replacement. It's a good idea to remove the screen and check its oil passage in the crankcase whenever the crankcase is disassembled (see illustration).

5   If any damage is found that can't be repaired, replace the crankcase halves as a set.

6   Assemble the case halves (see Section 25) and check to make sure the crankshaft and the transmission shafts turn freely.

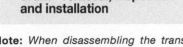

27 Transmission shafts, balancer shaft and shift drum - removal, inspection and installation

Note: *When disassembling the transmission shafts, place the parts on a long rod or thread a wire through them to keep them in order and facing the proper direction.*

Removal

1   Remove the engine, then separate the case halves (see Sections 5 and 25). The balancer shaft (2007 450 models) and transmission components (all models) remain in one case half when the case is separated.

2   Remove the shift drum, shift forks and transmission shafts from the crankcase (see illustrations).

3   On 2007 450 models, lift the balancer shaft out of the crankcase.

27.2c The fork pins (arrows) point in the directions shown when they're installed correctly

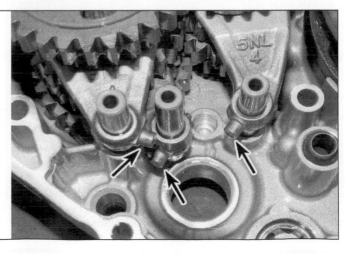

27.2d Remove the first wheel gear

27.2e Remove the mainshaft, together with its shift fork, noting how the fork fits in the third pinion gear

27.2f The letter/number markings on the forks face the right side of the crankcase

27.2g Pull the fork shafts out of their holes in the crankcase and remove the forks, together with the shafts

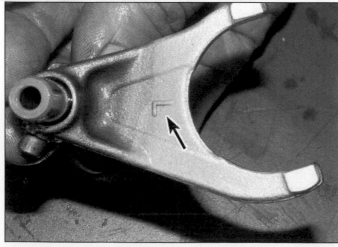

27.2h The position marks on the forks - L (left), C (center) and R (right) - face the left side of the crankcase

27.2i The center shift fork engages the third pinion gear on the mainshaft like this

27.2j Remove the right shift fork and fifth wheel gear from the countershaft

27.2k Remove the countershaft from the crankcase

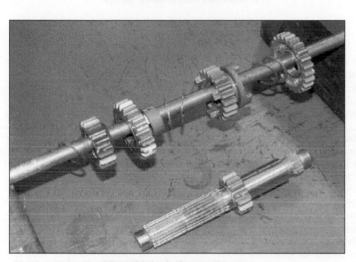

27.4a Mainshaft components

27.4b Countershaft components

4   Using snap-ring pliers, remove the snap-rings and take the gears off the shafts. Place the gears in order on a coat hanger or dowel so they won't be mixed up **(see illustrations)**.

Inspection

5   Wash all of the components in clean solvent and dry them off.

6   Inspect the shift fork grooves in the gears. If a groove is worn or scored, replace the affected part and inspect its corresponding shift fork.

7   Check the shift forks for distortion and wear, especially at the fork ears. If they are discolored or severely worn they are probably bent. Inspect the guide pins for excessive wear and distortion and replace any defective parts with new ones.

8   Check the shift fork guide bars for evidence of wear, galling and other damage. Make sure the shift forks move smoothly on the guide bars. If the shafts are worn or bent, replace them with new ones.

9   Check the edges of the grooves in the shift drums for signs of excessive wear.

10   Hold the inner race of the shift drum

bearing with your fingers and spin the outer race. Replace the bearing if it's rough, loose or noisy. Replace the shift drum segment if it's worn or damaged (see Section 24).

11   Check the gear teeth for cracking and other obvious damage. Check the bushing surface in the inner diameter of the free-wheeling gears for scoring or heat discoloration. Replace damaged parts.

12   Inspect the engagement dogs and dog holes (on gears so equipped) for excessive wear or rounding off. Replace the paired gears as a set if necessary.

13   Check the transmission shaft bearings in the crankcase for wear or heat discoloration and replace them if necessary (see Section 26).

Installation

14   Installation is basically the reverse of the removal procedure, but take note of the following points:

a) Use new snap-rings.

b) Lubricate the components with engine oil before assembling them.

c) After assembly, check the gears to

make sure they're installed correctly **(see illustration)**. Move the shift drums through the gear positions and rotate the gears to make sure they mesh and shift correctly.

27.14 The gears mesh like this when they're installed

28.2 Use a puller like this to push the crankshaft out of the case half

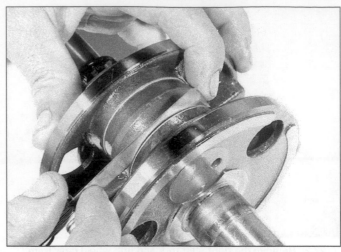

28.3 Measure the gap between the connecting rod and the crankshaft with a feeler gauge

28 Crankshaft and connecting rod - removal, inspection and installation

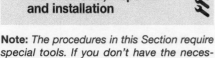

Note: *The procedures in this Section require special tools. If you don't have the necessary equipment or suitable substitutes, have the crankshaft removed and installed by a Yamaha dealer.*

Removal

1   Remove the engine and separate the crankcase halves (see Sections 5, 25 and 26). The transmission shafts need not be removed.

2   The crankshaft may be loose enough in its bearing that you can lift it out of the left crankcase half. If not, push it out with tool YU-A9642 or equivalent **(see illustration)**.

Inspection

3   Measure the side clearance between connecting rod and crankshaft with a feeler gauge **(see illustration)**. If it's more than the limit listed in this Chapter's Specifications, replace the crankshaft and connecting rod as an assembly.

4   Set up the crankshaft in V-blocks with a dial indicator contacting the big end of the connecting rod. Move the connecting rod side-to-side against the indicator pointer and compare the reading to the value listed in this Chapter's Specifications. If it's beyond the limit, the crankshaft can be disassembled and the needle roller bearing replaced. However, this is a specialized job that should be done by a Yamaha dealer or qualified machine shop.

5   Check the crankshaft splines, the cam chain sprocket, the ball bearing at the sprocket end of the crankshaft and the

bearing journals for visible wear or damage **(see illustration)**. Yamaha lists the ball bearing end of the crankshaft as a separately available part, but check with your dealer first; it may be more practical to replace the entire crankshaft if the ball bearing or cam sprocket is worn or damaged. Replace the crankshaft if any of the other conditions are found.

6   Set the crankshaft in a lathe or a pair of V-blocks, with a dial indicator contacting each end. Rotate the crankshaft and note the runout. If the runout at either end is beyond the limit listed in this Chapter's Specifications, replace the crankshaft and connecting rod as an assembly **(see illustration)**.

7   Measure the assembly width of the crankshaft (from the outside of one crank throw to the outside of the other crank throw). If it exceeds the limit listed in this Chapter's Specifications, replace the crankshaft.

28.5 Check the cam chain sprocket and the ball bearing on the end of the crankshaft

28.6 Measure runout on each side of the crankshaft (A); if the assembly width (B) is greater than specified, replace the crankshaft

Installation

8   Start the crankshaft into the case half. If it doesn't go in easily, pull it in the rest of the way with Yamaha tools YU-90050, YM-01383 and YM-91044 (see illustration).

9   The remainder of installation is the reverse of the removal steps.

28.8 These tools are used to pull the crankshaft into the left case half

29 Initial start-up after overhaul

1   Make sure the engine oil level is correct, then remove the spark plug from the engine. Place the engine kill switch in the Off position and unplug the primary (low tension) wires from the coil.

2   Crank the engine over several times to build up oil pressure. Reinstall the spark plug, connect the wires and turn the switch to On.

3   Make sure there is fuel in the tank, then operate the choke.

4   Refer to the oil change procedure in Chapter 1 to check oil pressure at the check bolt. Caution: If oil doesn't seep from the check bolt within one minute, stop the engine immediately and locate the problem before running it further. Once you've made sure that there is oil pressure, allow the engine to run at a moderately fast idle until it reaches operating temperature.

5   Check carefully for oil leaks and make sure the transmission and controls, especially the brakes, function properly before road testing the machine. Refer to Section 30 for the recommended break-in procedure.

6   Upon completion of the road test, and after the engine has cooled down completely, recheck the valve clearances (see Chapter 1).

30 Recommended break-in procedure

1   Any rebuilt engine needs time to break-in, even if parts have been installed in their original locations. For this reason, treat the machine gently for the first few miles to make sure oil has circulated throughout the engine and any new parts installed have started to seat.

2   Even greater care is necessary if the cylinder has been rebored or a new crankshaft has been installed. In the case of a rebore, the engine will have to be broken in as if the machine were new. This means greater use of the transmission and a restraining hand on the throttle for the first few operating days. There's no point in keeping to any set speed limit - the main idea is to vary the engine speed, keep from lugging (laboring) the engine and to avoid full-throttle operation. These recommendations can be lessened to an extent when only a new crankshaft is installed. Experience is the best guide, since it's easy to tell when an engine is running freely.

3   If a lubrication failure is suspected (oil doesn't seep from the check bolt, or the engine makes noise), stop the engine immediately and try to find the cause. If an engine is run without oil, even for a short period of time, irreparable damage will occur.

Notes

Chapter 3
Cooling system

Contents

Degrees of difficulty

Easy, suitable for novice with little experience	**Fairly easy,** suitable for beginner with some experience	**Fairly difficult,** suitable for competent DIY mechanic	**Difficult,** suitable for experienced DIY mechanic	**Very difficult,** suitable for expert DIY or professional

Specifications

General
Radiator cap relief pressure
   1998 .... 95 to 125 kPa (14 to 18 psi)
   1999 and later .... 110 kPa (16 psi)

Torque specifications
Water pump impeller .... 14 Nm (120 inch-lbs)
Water pump bolts .... 10 Nm (86 inch-lbs)
Radiator mounting bolts .... 10 Nm (86 inch-lbs)

3.3a The water pump-to-radiator hose is on the right side (late YZ250 shown)

3.3b There's either a single hose connecting between the radiators at the bottom, as shown here . . .

3.3c . . . or, on later models, a Y-fitting connects the water pump hose directly to both radiators

have two radiators, one on each side of the frame. The coolant is pumped through the radiators where it is cooled, then out of the radiators and through the cylinder head and cylinder.

Because these bikes are intended for competition, the cooling system is a very basic one, without a temperature gauge or light, fan or thermostat. WR models have a reservoir tank or catch tank for expanding coolant, YZ models do not.

2 Radiator cap - check

If problems such as overheating or loss of coolant occur, check the entire system as described in Chapter 1. The radiator cap opening pressure should be checked by a dealer service department or service station equipped with the special tester required to do the job. If the cap is defective, replace it with a new one.

1 General information

The motorcycles covered by this manual are equipped with a liquid cooling system which utilizes a water/antifreeze mixture to carry away excess heat produced during combustion. The combustion chamber and cylinder are surrounded by a water jacket, through which the coolant is circulated by the water pump. The pump is mounted to the right side of the crankcase near the front and is driven by a gear. All models

3.3d The radiator-to-engine hose is on the left side

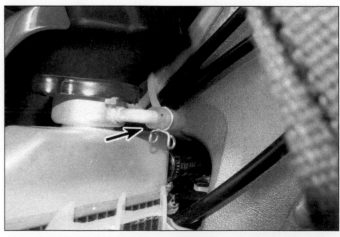

3.3e On WR models, there's a catch tank or reservoir tank hose at the bottom of the filler neck

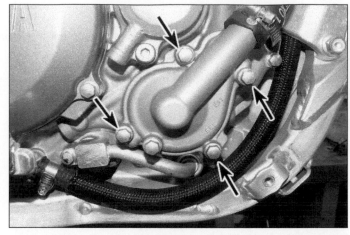

4.2 Flex the grille to free its retaining pins (left arrows) and remove the grille bolts (right arrows) - on early models, also remove the rear mounting bolt

5.3 The water pump is bolted to the lower right part of the engine (YZ250F shown)

3 Coolant hoses - removal and installation

⚠ *Warning: The engine must be completely cool before beginning this procedure.*

1   The coolant hoses are all secured by screw-type clamps to fittings on the engine and radiator.
2   Drain the cooling system (see Chapter 1).
3   To remove a hose, loosen its clamp and carefully pry it off the fitting **(see illustrations)**.
4   If the hose is stuck, pry the edge up slightly with a pointed tool and spray brake or electrical contact cleaner into the gap. Work the tool around the fitting, lifting the edge of the hose and spraying into the gap until the hose comes free of the fitting.
5   In extreme cases, you may have to slit the hose and cut it off the fitting with a knife. Make sure you can get a replacement hose before doing this.
6   Refill the cooling system (see Chapter 1).

4 Radiators - removal and installation

⚠ *Warning: The engine must be completely cool before beginning this procedure.*

1   Support the bike securely upright. Remove the radiator shroud(s) (see Chapter 8) and drain the cooling system (see Chapter 1).
2   Carefully bend the grille, pulling its pins out of the radiator **(see illustration)**.

3   Disconnect the radiator hoses (see Section 3).
4   Remove the radiator mounting bolts **(see illustration 4.2)**.
5   Lift the radiator away from the frame. Inspect the mounting bolt grommets and replace them if they're worn or deteriorated.
6   Installation is the reverse of the removal steps, with the following additions:

a) Tighten the mounting bolts securely, but don't overtighten them and distort the grommets.
b) Fill the cooling system (see Chapter 1).
c) Run the engine and check for coolant leaks.

5 Water pump - removal, inspection and installation

⚠ *Warning: The engine must be completely cool before beginning this procedure.*

Removal

1   Drain the cooling system (see Chapter 1).

2   Disconnect the hoses from the water pump.
3   Remove the pump cover bolts **(see illustration)**. The bolts are different lengths, so tag them for reinstallation.
4   Take off the pump cover and O-ring **(see illustration)**.
5   To remove the impeller and shaft, remove the right crankcase cover together with the shaft (see Chapter 2).
6   Unscrew the impeller from its shaft and remove the washer **(see illustrations)**.
7   Pull the impeller shaft out of the pump body, twisting it as you remove it to prevent damage to the seal.

Inspection

8   Check the impeller seal for wear or damage. This seal separates the coolant from the engine oil. If the oil is milky or foamy, coolant may have been leaking into it past the seal. Refer to Section 6 and replace the seal.
9   To inspect the bearing and impeller driven gear, wiggle the impeller shaft and check for play. If it can be wiggled from side to side, the bearing needs to be replaced. Lift the impeller shaft out of the bearing in the cover as described above. Spin the bearing and check it for roughness, looseness or noise and replace it as described in Section 6 if any problems are found.

5.4 Unbolt the housing and remove the O-ring - note the location of the dowel pin (arrow)

5.6a **Place a wrench on the flats of the impeller shaft to keep it from turning . . .**

5.6b **. . . unscrew the impeller . . .**

5.6c **. . . take the impeller off, then remove the washer (arrow) . . .**

5.6d **. . . and take the impeller shaft out of the crankcase cover**

Installation

10   Installation is the reverse of the removal steps, with the following additions:

   a) *Use a new O-ring.*
   b) *Engage the water pump gear with the primary drive gear.*
   c) *Tighten the water pump bolts to the torque listed in this Chapter's Specifications.*
   d) *Fill the cooling system (see Chapter 1).*
   e) *Run the engine and check for coolant leaks.*

6   Water pump seals and bearing - replacement

1   If coolant has been leaking from the weep hole (if equipped, see Chapter 1), the water pump seal needs to be replaced.
2   Remove the right crankcase cover (see Chapter 2) and the water pump impeller shaft (see Section 5).
3   Carefully pry the seals out of their bore with a screwdriver, being careful not to gouge the crankcase cover **(see illustration)**.

4   The ball bearing is mounted in the right crankcase cover. If it needs to be replaced, drive the bearing out with a socket or bearing driver **(see illustration)**.

6.3 **Pry the seals out of the cover, noting which way the open sides face**

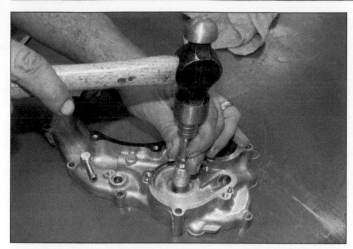

6.4 Inspect the bearing and, if it needs to be replaced, drive it out with a socket or bearing driver

7.1 WR models are equipped with a catch tank or reservoir tank

5   Drive in a new bearing with a bearing driver or socket that bears against the bearing outer race.

6   Tap in the new seals with a socket the same diameter as the seal. Install the seals with the open sides facing away from each other.

7   Reservoir or catch tank (WR models) - removal and installation

1   The reservoir or catch tank is located behind the left side cover (see illustration). Remove the seat and side cover for access (see Chapter 8).

2   Disconnect the hoses from the tank. Remove the tank mounting bolts and take it off.

3   Installation is the reverse of the removal steps.

Notes

Chapter 4
Fuel and exhaust systems

Contents

Degrees of difficulty

Easy, suitable for novice with little experience	**Fairly easy,** suitable for beginner with some experience	**Fairly difficult,** suitable for competent DIY mechanic	**Difficult,** suitable for experienced DIY mechanic	**Very difficult,** suitable for expert DIY or professional

Specifications

General
Fuel type .... See Chapter 1

Carburetor (YZ250F)
2001
  Type .... Keihin FCR-MX37
  ID mark
    All except Europe 5NL100
    Europe 5NL210
  Main jet 175
  Main air jet 200
  Needle/clip position
    All except Europe OBELP-4
    Europe OBEJP-4
  Pilot jet 42
  Pilot air jet 100
  Standard pilot screw setting .... 1-3/4 turns out
  Starter jet 72
  Leak jet 105
  Float height .... 8 mm (0.31 inch)

Carburetor (YZ250F) (continued)

2002

Type	Keihin FCR-MX37
ID mark	
All except Europe	5SG1 00
Europe	5SG2 10
Main jet	178
Main air jet	200
Needle/clip position	OBEPK-4
Pilot jet	40
Pilot air jet	100
Standard pilot screw setting	
All except Europe	1-5/8 turns out
Europe	1-1/2 turns out
Starter jet	72
Leak jet	
All except Europe	100
Europe	95
Float height	8 mm (0.31 inch)

2003

Type	Keihin FCR-MX37
ID mark	
All except Europe	5UL1 00
Europe	5UL2 10
Main jet	178
Main air jet	200
Needle/clip position	
All except Europe	OBELP-4
Europe	OBEKP-4
Pilot jet	40
Pilot air jet	100
Standard pilot screw setting	1-5/8 turns out
Starter jet	72
Leak jet	100
Float height	8 mm (0.31 inch)

2004

Type	Keihin FCR-MX37
ID mark	5XC1 00
Main jet	180
Main air jet	200
Needle/clip position	OBEKR-4
Pilot jet	42
Pilot air jet	100
Standard pilot screw setting	1-5/8 turns out
Starter jet	72
Leak jet	90
Float height	8 mm (0.31 inch)

2005

Type	Keihin FCR-MX37
ID mark	
US	5XC5 50
All others	5XC6 60
Main jet	182
Main air jet	200
Needle/clip position	
US	OBELR-4
All others	OBEKS-4
Pilot jet	42
Pilot air jet	100
Standard pilot screw setting	
US	1-7/8 turns out
All others	2-1/8 turns out
Starter jet	72
Leak jet	
US	90
All others	95
Float height	8 mm (0.31 inch)

2006

Type	Keihin FCR-MX37

ID mark	
US and Canada	5XC9 90
All others	5XCA A0
Main jet	
US and Canada	178
All others	175
Main air jet	200
Needle/clip position	
US and Canada	OBELQ-3
All others	OBEKP-3
Pilot jet	42
Pilot air jet	105
Standard pilot screw setting	
US	2-1/2 turns out
All others	2-1/4 turns out
Starter jet	72
Leak jet	
US	80
All others	85
Float height	8 mm (0.31 inch)
2007	
Type	Keihin FCR-MX37
ID mark	
US and Canada	5XCG G0
All others	5XCH H0
Main jet	
US and Canada	178
All others	175
Main air jet	200
Needle/clip position	OBEPQ-4
Pilot jot	42
Pilot air jet	105
Standard pilot screw setting	
US	1-7/8 turns out
All others	2-1/8 turns out
Starter jet	72
Leak Jet	
US	80
All others	70
Float height	8 mm (0.31 inch)
2008	
Type	Keihin FCR-MX37
ID mark	5XCL L0
Main jet	178
Main air jet	200
Needle/clip position	NFPR-5
Pilot jet	42
Pilot air jet	105
Standard pilot screw setting	2-3/8 turns out
Starter jet	72
Leak jet	70
Float height	8 mm (0.31 inch)

Carburetor (WR250F)

2001	
Type	Keihin FCR-MX37
ID mark	
US	5PH1 00
Europe	5PH2 10
All others	5PH4 20
Main jet	170
Main air jet	200
Needle/clip position	
US	OBELP-3
Europe	OBEGP-4
All others	OBEHP-4
Pilot jet	42
Pilot air jet	75

Carburetor (WR250F) (continued)

Standard pilot screw setting	1-3/4 turns out
Starter jet	72
Leak jet	
US	60
All others	105
Float height	8 mm (0.31 inch)

2002

Type	Keihin FCR-MX37
ID mark	
US	5HP5 50
Europe	5PH6 60
All others	5PH8 80
Main jet	175
Main air jet	200
Needle/clip position	
US	OBELQ-3
Europe	OBELN-4
All others	OBEKP-3
Pilot jet	40
Pilot air jet	75
Standard pilot screw setting	
US	2 turns out
Europe	1 turn out
All others	1-1/8 turns out
Starter jet	72
Leak jet	
US	60
Europe	90
All others	105
Float height	8 mm (0.31 inch)

2003

Type	Keihin FCR-MX37
ID mark	
US	5UM1 00
All others	5UM2 10
Main jet	
US	175
All others	182
Main air jet	200
Needle/clip position	
US	OBDVS-4
All others	OBELP-4
Pilot jet	40
Pilot air jet	75
Standard pilot screw setting	
US	2 turns out
All others	1-1/8 turns out
Starter jet	72
Leak jet	
US	90
All others	105
Float height	8 mm (0.31 inch)

2004

Type	Keihin FCR-MX37
ID mark	
US	5UM5 50
All others	5UM6 60
Main jet	
US	175
All others	170
Main air jet	200
Needle/clip position	
US	OBDUS-4
All others	OBEKQ-4
Pilot jet	40
Pilot air jet	
US	70
All others	60

Standard pilot screw setting

US	2 turns out
All others	1-1/2 turns out
Starter jet	72

Leak jet

US	70
All others	90
Float height	8 mm (0.31 inch)

2005

Type	Keihin FCR-MX37
ID mark	5UM8 80
Main jet	168
Main air jet	200
Needle/clip position	OBERP-5
Pilot jet	40
Pilot air jet	70
Standard pilot screw setting	1-3/4 turns out
Starter jet	72
Leak jet	95
Float height	8 mm (0.31 inch)

2006

Type	Keihin FCR-MX37
ID mark	5UMB B0
Main jet	168
Main air jet	115
Needle	NGSU
Pilot jet	42
Pilot air jet	70
Standard pilot screw setting	Not specified
Starter jet	68
Leak jet	70
Float height	8 mm (0.31 inch)

2007

Type	Keihin FCR-MX37
ID mark	5UME E0
Main jet	170
Main air jet	115
Needle	NJRU
Pilot jet	42
Pilot air jet	70
Standard pilot screw setting	Not specified
Starter jet	68
Leak jet	70
Float height	8 mm (0.31 inch)

2008

Type	Keihin FCR-MX37
ID mark	5UME E0
Main jet	170
Main air jet	115
Needle	NJRU
Pilot jet	42
Pilot air jet	70
Standard pilot screw setting	Not specified
Starter jet	68
Leak jet	70
Float height	8 mm (0.31 inch)

Carburetor (YZ400F)

1998

Type	Keihin FCR 39H
ID mark	5BE1 01
Main jet	175
Main air jet	200
Needle/clip position	OBDVR-4
Pilot jet	45
Pilot air jet	100
Standard pilot screw setting	1-3/4 turns out
Starter jet	65
Float height	9 mm (0.35 inch)

Carburetor (YZ400F) (continued)
1999

Type	Keihin FCR-39H
ID mark	5BE1 00
Main jet	175
Main air jet	200
Needle/clip position	OBDVR-4
Pilot jet	45
Pilot air jet	100
Standard pilot screw setting	1-3/8 turns out
Starter jet	65
Float height	9 mm (0.35 inch)

Carburetor (WR400F)
1998

Type	Keihin FCR-39H
ID mark	5BF1 01
Main jet	168
Main air jet	200
Needle/clip position	DTM-3
Pilot jet	45
Pilot air jets	
No. 1	75
No. 2	90
Standard pilot screw setting	1-3/8 turns out
Starter jet	65
Float height	9 mm (0.35 inch)

1999

Type	Keihin FCR-39H
ID mark	
US	5GS1 00
All others	5GS2 10
Main jet	168
Main air jet	200
Needle/clip position	
US	OBDTM-3
All others	OBDXM-4
Pilot jet	45
Pilot air jet	
No. 1	75
No. 2	90
Standard pilot screw setting	1-1/4 turns out
Starter jet	
US	60
All others	62
Float height	9 mm (0.35 inch)

2000

Type	Keihin FCR-39H
ID mark	
US	5GS5 50
All others	5GS6 60
Main jet	165
Main air jet	200
Needle/clip position	
US	OBDRS-3
All others	OBDRQ-4
Pilot jet	42
Pilot air jet	
No. 1	75
No. 2	90
Standard pilot screw setting	
US	1-5/8 turns out
All others	1-1/2 turns out
Starter jet	65
Float height	8 mm (0.35 inch)

Carburetor (YZ426F)

2000

Type	Keihin FCR-MX39
ID mark	
US	5JG1 00
All others	5JG2 10
Main jet	162
Main air jet	200
Needle/clip position	
US	OBEKR-4
All others	OBEKQ-4
Pilot jet	42
Pilot air jet	100
Standard pilot screw setting	
US	1-3/4 turns out
All others	1-1/2 turns out
Starter jet	72
Float height	8 mm (0.31 inch)

2001

Type	Keihin FCR-MX39
ID mark	5JG5 50
Main jet	162
Main air jet	200
Needle/clip position	OBEJP-4
Pilot jet	42
Pilot air jet	100
Standard pilot screw setting	1-1/4 turns out
Starter jet	72
Float height	8 mm (0.31 inch)

2002

Type	Keihin FCR-MX39
ID mark	5SF1 00
Main jet	162
Main air jet	200
Needle/clip position	OBEJP-4
Pilot jet	42
Pilot air jet	100
Standard pilot screw setting	1-3/8 turns out
Starter jet	72
Float height	8 mm (0.31 inch)

Carburetor (WR426F)

2001

Type	Keihin FCR-MX39
ID mark	
US	5NG1 00
All others	5NG2 10
Main jet	165
Main air jet	200
Needle/clip position	
US	OBDRR-4
All others	OBDQR-4
Pilot jet	42
Pilot air jet	75
Standard pilot screw setting	
US	1-5/8 turns out
All others	7/8 turn out
Starter jet	65
Float height	0 mm (0.01 inch)

2002

Type	Keihin FCR-MX39
ID mark	
US	5NG5 50
All others	5NG6 60
Main jet	165
Main air jet	200

Carburetor (WR426F) (continued)

Needle/clip position
US OBDRR-4
All others OBDQR-4
Pilot jet 42
Pilot air jet 75
Standard pilot screw setting
US .... 1-5/8 turns out
All others .... 1-1/8 turns out
Starter jet 65
Float height .... 8 mm (0.31 inch)

Carburetor (YZ450F)

2003
Type .... Keihin FCR-MX39
ID mark 5TA1 00
Main jet 165
Main air jet 200
Needle/clip position NCVQ-4
Pilot jet 42
Pilot air jet 100
Standard pilot screw setting .... 2 turns out
Starter jet 72
Float height .... 8 mm (0.31 inch)

2004
Type .... Keihin FCR-MX39
ID mark
US 5TA1 00
All others 5XD2 10
Main jet
US 165
All others 172
Main air jet 200
Needle/clip position
US NCVQ-4
All others OBEMR-3
Pilot jet
US 42
All others 45
Pilot air jet 100
Standard pilot screw setting .... 2 turns out
Starter jet 72
Float height .... 8 mm (0.31 inch)

2005
Type .... Keihin FCR-MX39
ID mark 5XD5 50
Main jet 165
Main air jet 200
Needle/clip position NFLR-4
Pilot jet 45
Pilot air jet 100
Standard pilot screw setting .... 1-3/4 turns out
Starter jet 72
Leak jet 40
Float height .... 8 mm (0.31 inch)

2006
Type .... Keihin FCR-MX39
ID mark 2S21 00
Main jet 165
Main air jet 200
Needle/clip position NFPR-4
Pilot jet 42
Pilot air jet 100
Standard pilot screw setting .... 2-1/8 turns out
Starter jet 72
Leak jet 55
Float height .... 8 mm (0.31 inch)

2007

Type	Keihin FCR-MX39
ID mark	2S27 70
Main jet	160
Main air jet	200
Needle/clip position	NFLR-4
Pilot jet	45
Pilot air jet	100
Standard pilot screw setting	1-1/4 turns out
Starter jet	72
Leak jet	55
Float height	8 mm (0.31 inch)

2008

Type	Keihin FCR-MX39
ID mark	2S2B B0
Main jet	160
Main air jet	200
Needle/clip position	NFLR-3
Pilot jet	45
Pilot air jet	100
Standard pilot screw setting	Not specified
Starter jet	72
Leak jet	55
Float height	8 mm (0.31 inch)

Carburetor (WR450F)

2003

Type	Keihin FCR-MX39
ID mark	
US	5TJ1 00
All others	5TJ2 00
Main jet	
US	150
All others	160
Main air jet	200
Needle/clip position	
US	OBDUT-4
All others	OBDUQ-4
Pilot jet	
US	45
All others	48
Pilot air jet	
US	70
All others	80
Standard pilot screw setting	
US	1-3/4 turns out
All others	1-1/2 turns out
Starter jet	65
Float height	8 mm (0.31 inch)

2004

Type	Keihin FCR-MX39
ID mark	
US	5TJ5 50
All others	5TJ2 10
Main jet	
US	165
All others	160
Main air jet	200
Needle/clip position	
US	OBDWR-4
All others	OBDUQ-4
Pilot jet	
US	45
All others	48
Pilot air jet	
US	70
All others	80

Carburetor (WR450F) (continued)

Standard pilot screw setting
  US .... 1-3/4 turns out
    All others .... 1-1/2 turns out
Starter jet 65
Leak jet
  US 60
    All others 50
Float height .... 8 mm (0.31 inch)

2005
  Type.. Keihin FCR-MX39
  ID mark... 5TJ8 80
  Main jet 165
  Main air jet 200
  Needle/clip position .... NFNT
  Pilot jet 45
  Pilot air jet 80
  Standard pilot screw setting .... Not specified
  Starter jet 65
  Leak jet 50
  Float height .... 8 mm (0.31 inch)

2006
  Type.. Keihin FCR-MX37
  ID mark... 5TJ8 80
  Main jet 165
  Main air jet 200
  Needle/clip position .... NFNT
  Pilot jet 45
  Pilot air jet 80
  Standard pilot screw setting .... Not specified
  Starter jet 65
  Leak jet 50
  Float height .... 8 mm (0.31 inch)

2007
  Type.. Keihin FCR-MX39
  ID mark... 5TJE E0
  Main jet 162
  Main air jet 200
  Needle/clip position .... NFNT
  Pilot jet 45
  Pilot air jet 70
  Standard pilot screw setting .... Not specified
  Starter jet 65
  Leak jet 60
  Float height .... 8 mm (0.31 inch)

2008
  Type.. Keihin FCR-MX39
  ID mark... 5TJE E0
  Main jet 162
  Main air jet 200
  Needle/clip position .... NFNT
  Pilot jet 45
  Pilot air jet 70
  Standard pilot screw setting .... Not specified
  Starter jet 65
  Leak jet 60
  Float height .... 8 mm (0.31 inch)

Accelerator pump adjusting rod diameter (to be inserted under throttle valve when adjusting the accelerator pump)

1998 through 2001 .... Not specified
2002 through 2004
  YZ250F.. 1.25 mm (0.049 inch)
  WR250F .... 1.5 mm (0.059 inch)
  YZ426F, YZ450F.. 3.4 mm (0.134 inch)
  WR426F, WR450F
    2002 US models, all 2003 and 2004 models............................. 3.10 mm (0.122 inch)
    2002 except US .... 2.25 mm (0.089 inch)

2005
YZ250F	0.9 mm (0.035 inch)
WR250F	1.5 mm (0.059 inch)
YZ450F	1.3 mm (0.051 inch)
WR450F	3.1 mm (0.122 inch)

2006
YZ250F
US and Canada	0.8 mm (0.031 inch)
All others	0.7 mm (0.028 inch)
WR250F	0.8 mm (0.031 inch)
YZ450F	1.25 mm (0.049 inch)
WR450F	3.1 mm (0.122 inch)

2007 and later
YZ250F, WR250F	0.8 mm (0.031 inch)
YZ450F	1.25 mm (0.049 inch)
WR450F	3.1 mm (0.122 inch)

1 General information

All models use a flat-slide carburetor with an accelerator pump. The slide acts as the throttle valve. The accelerator pump injects extra gasoline into the fuel mixture during acceleration. For cold starting on all models, a choke plunger is actuated by a knob. On some models, an additional knob opens a hot start valve; on others, the hot start valve is controlled by a handlebar lever.

The exhaust system consists of a pipe and muffler with a replaceable core.

Later WR models have an air induction system to reduce exhaust emissions.

2 Fuel tank - removal and Installation

⚠️ *Warning: Gasoline is extremely flammable, so take extra precautions when you work on any part of the fuel system.* **Don't smoke or allow open flames or bare light bulbs near the work area, and don't work in a garage where a gas-type appliance (such as a water heater or clothes dryer) is present. Since gasoline is carcinogenic, wear nitrile gloves when there's a possibility of being exposed to fuel, and, if you spill any fuel on your skin, rinse it off immediately with soap and water. Mop up any spills immediately and do not store fuel-soaked rags where they could ignite. When you perform any kind of work on the fuel system, wear safety glasses and have an extinguisher suitable for a class B type fire (flammable liquids) on hand.**

2.2a If the fuel tap is easy to get to, squeeze the ends of the clamp together, slide it down the hose, then disconnect the fuel line from the tap . . .

Removal

1 Remove the seat and side covers (see Chapter 8).
2 Turn the fuel tap to Off and disconnect

2.2c . . . and cap the fitting to keep out dirt . . .

2.2b . . . if not, slide the clamp up off the carburetor fitting and disconnect the hose . . .

the fuel line. Disconnect it at the tap if there's room (see illustration). If not, disconnect it at the carburetor fitting (see illustration). If you disconnect the fuel line at the carburetor, cap the fitting or at least turn it downward to keep out dirt (see illustrations).

2.2d . . . or turn it downward if there's nothing to cap it with

2.3 Pull the vent hose off of the tank fitting or out of the steering stem nut

2.4a On 2005 and earlier models, note how the ends of the tank strap are shaped (one is designed for the button on the tank and the other for the hook on the frame), then unhook the strap . . .

2.4b . . . and remove the mounting bolt(s) at the front of the tank

3   Pull the fuel tank vent hose off the fitting on the filler cap **(see illustration)**.

4   If you're working on a 2005 or earlier model, unhook the strap from the rear of the tank and unbolt the tank at the front **(see illustrations)**.

5   If you're working on a 2006 or later model, unbolt the tank at the front, then lift the cover material off of the rear bolt and unbolt the tank at the rear **(see illustrations)**.

6   Lift the fuel tank off the bike together with the fuel tap.

Installation

7   Before installing the tank, check the condition of the rubber mounting bushings, the isolators on the frame and the rubber strap at the rear (2005 and earlier models) - if they're hardened, cracked, or show any other signs of deterioration, replace them **(see illustrations)**.

8   When installing the tank, reverse the removal procedure. On 2005 and earlier models, the rubber tab on the retaining strap

2.5a On 2006 and later models, remove the mounting bolt at the front of the tank . . .

goes under the tank **(see illustration)**. On all models, the arrow on the one-way valve in the vent hose points toward the tank **(see illustration)**. Make sure the tank does not pinch any wires. Tighten the tank mounting bolts securely, but don't overtighten them and strip the threads.

3 Carburetor overhaul - general information

1   Poor engine performance, hesitation, hard starting, stalling, flooding and backfiring are all signs that major carburetor main-

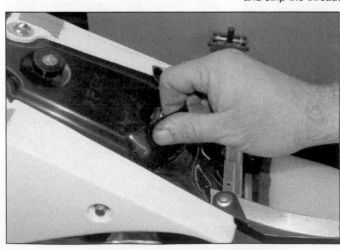

2.5b . . . lift the cover off the rear mounting bolt . . .

2.5c . . . unscrew the bolt and lift the tank off

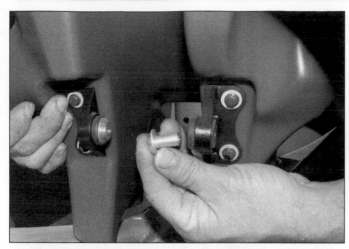

2.7a On 2005 and earlier models, inspect the tank bushings

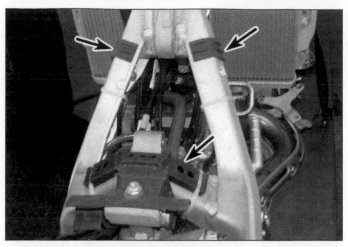

2.7b Check for deteriorated or damaged isolators on the frame (arrows) (2006 and later shown)

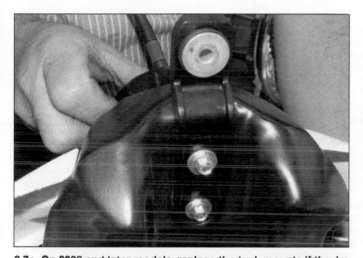

2.7c On 2006 and later models, replace the tank mounts if they're cracked or deteriorated - this is the front mount . . .

2.7d . . . and this is the rear mount

tenance may be required.

2   Keep in mind that many so-called carburetor problems are really not carburetor problems at all, but mechanical problems within the engine or ignition system malfunctions. Try to establish for certain that the carburetor is in need of maintenance before beginning a major overhaul.

3   Check the fuel tap and its strainer screen, the fuel lines, the intake manifold clamps, the O-ring between the intake manifold and cylinder head, the vacuum hoses, the air filter element, the cylinder compression, the spark plug and the ignition timing before assuming that a carburetor overhaul is required. If the bike has been unused for more than a couple of weeks, drain the float chamber and refill the tank with fresh fuel.

4   Most carburetor problems are caused by dirt particles, varnish and other deposits which build up in, and block, the fuel and air passages. Also, in time, gaskets and O-rings shrink or deteriorate and cause fuel and air leaks which lead to poor performance.

5   When the carburetor is overhauled, it is

generally disassembled completely and the parts are cleaned thoroughly with a carburetor cleaning solvent and dried with filtered, unlubricated compressed air. The fuel and air passages are also blown through with compressed air to force out any dirt that may

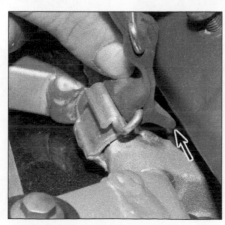

2.8a The tab on the retaining strap goes between the tank and frame (arrow)

have been loosened but not removed by the solvent. Once the cleaning process is complete, the carburetor is reassembled using a new top gasket, O-rings and, generally, a new inlet needle valve and seat.

6   Before disassembling the carburetor,

2.8b The arrow on the one-way valve points toward the tank

4.3 Loosen the clamps (arrows)

4.8a Note how the hoses are routed over the top . . .

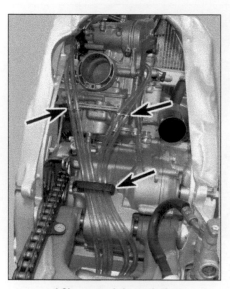

4.8b . . . and through the retainers (arrows)

make sure you have the necessary gasket, O-rings and other parts, some carburetor cleaner, a supply of rags, some means of blowing out the carburetor passages and a clean place to work.

4 Carburetor - removal and installation

1   Remove the seat and both side covers (see Chapter 8).
2   Where necessary for access, remove the fuel tank. If there's room to remove the carburetor with the fuel tank installed, disconnect the fuel line from the tap (see Section 2).
3   Loosen the clamping bands at the front and rear of the carburetor (see illustration).
4   On 2006 and later models, remove the subframe and rear shock absorber (see Chapters 8 and 6).
5   On models equipped with a throttle

position sensor, disconnect the electrical connector.
6   Remove the throttle cable cover from the side of the carburetor and disconnect the throttle cable(s) (see Section 7).
7   On bikes equipped with a hot start lever on the handlebar, slide back the boot and unscrew the hot start plunger from the carburetor (see Section 7).
8   Free the carburetor from the intake tube (all models) and the air cleaner tube (2005 and earlier models). Note how the hoses are routed and free them from the retainer (see illustrations). Remove the carburetor to the side (2005 and earlier) or to the rear (2006 and later (see illustrations).

Installation
9   Installation is the reverse of the removal steps, with the following additions:
a) Adjust the throttle freeplay (see Chapter 1).
b) Adjust the idle speed and fuel/air mixture (see Chapter 1).

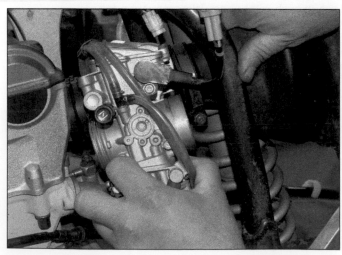

4.8c Remove the carburetor from the left side (2005 and earlier) . . .

4.8d . . . or out the back (2006 and later)

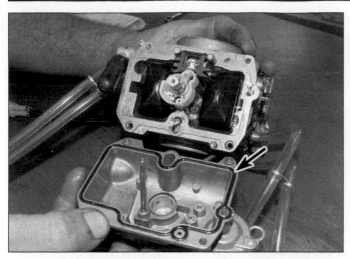

5.3a Remove the float chamber screws, take off the float chamber and remove the O-ring (arrow)

5.3b Remove the accelerator pump cover screws while holding the cover down against the spring pressure . . .

5 Carburetor - disassembly, cleaning and inspection

⚠️ **Warning: Gasoline is extremely flammable, so take extra precautions when you work on any part of the fuel system. See the Warning in Section 2.**

Disassembly

1   Remove the carburetor from the machine as described in Section 4.
2   Set the carburetor on a clean working surface. Take note of how the vent hoses are routed, including locations of hose retainers.
3   To disassemble the carburetor, refer to the accompanying illustrations (see illustrations).

Cleaning

Caution: Use only a carburetor cleaning

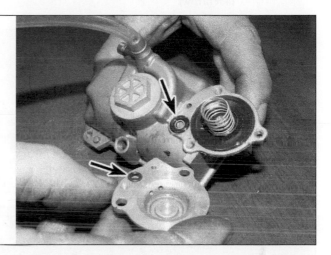

5.3c . . . then lift off the cover and remove the O-rings (arrows) . . .

solution that is safe for use with plastic parts (be sure to read the label on the container).

4   Submerge the metal components in the carburetor cleaner for approximately thirty

minutes (or longer, if the directions recommend it).
5   After the carburetor has soaked long enough for the cleaner to loosen and dissolve most of the varnish and other deposits,

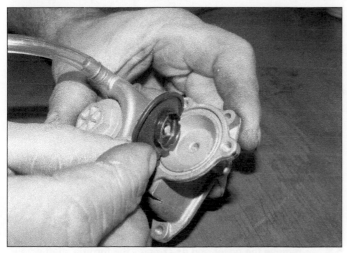

5.3d . . . and remove the accelerator pump diaphragm

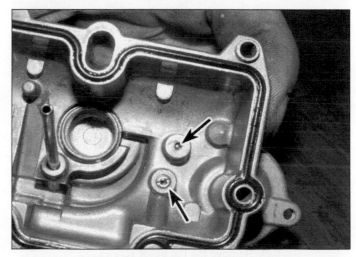

5.3e Unscrew the leak jet (lower arrow) and check the spring-loaded ball (upper arrow) for free movement

5.3f Bottom the pilot screw lightly, counting the number of turns, then unscrew it all the way and remove the spring, washer (right arrow) and O-ring (left arrow)

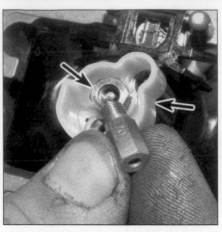

5.3g Unscrew the main jet, then unscrew the needle jet (left arrow) and take off the baffle (right arrow)

5.3h Unscrew the starter jet (upper arrow) and pilot jet (lower arrow)

5.3i Push out the float pivot pin with a piece of wire, then lift off the floats, together with the needle valve (arrow)

5.3j Unscrew the choke valve from the carburetor

5.3k Remove the top cover screws . . .

5.3l . . . remove the cover and its O-ring - the throttle shaft screw (lower arrow) is not normally removed (unless the throttle valve is going to be removed) - the needle holder (upper arrow) . . .

5.3m . . . secures the spring and clip

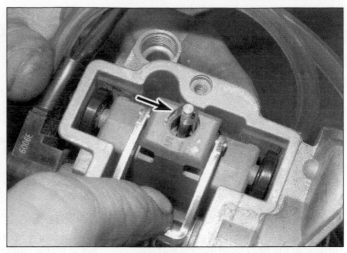

5.3n Lift the throttle piston for access to the jet needle (arrow) . . .

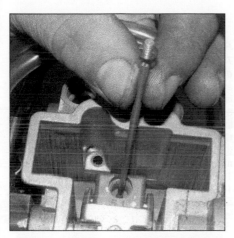

5.3o . . . and lift out the jet needle with its clip

5.3p Remove the throttle shaft screw (see illustration 5.3l) **and pull out the throttle valve assembly. Caution:** *Don't remove the throttle shaft from the carburetor body.* **Note:** *Before reinstalling the throttle shaft screw, apply a drop of non-hardening thread-locking agent to the threads*

5.6 Check the choke plunger seat for wear or deterioration

use a brush to remove the stubborn deposits. Rinse it again, then dry it with compressed air. Blow out all of the fuel and air passages in the carburetor body.

Caution: Never clean the jets or passages with a piece of wire or a drill bit, as they will be enlarged, causing the fuel and air metering rates to be upset.

Inspection

6   Check the operation of the choke plunger. If it doesn't move smoothly, replace it. Check the plunger seat for wear or damage **(see illustration)** and replace it if problems are found. Inspect the hot start valve in the same manner.

7   Check the tapered portion of the pilot screw for wear or damage. Replace the screw if necessary.

8   Check the carburetor body, float chamber and carburetor top for cracks, distorted sealing surfaces and other damage. If any defects are found, replace the faulty component, although replacement of the entire

carburetor will probably be necessary (check with your parts supplier for the availability of separate components).

9   Check the jet needle for straightness by rolling it on a flat surface (such as a piece of glass). Replace it if it's bent or if the tip is worn.

10   Check the tip of the fuel inlet valve needle. If it has grooves or scratches in it, it must be replaced. Push in on the rod in the other end of the needle, then release it - if it doesn't spring back, replace the valve needle.

11   Check the O-rings on the float chamber and the main jet access plug (in the float chamber). Replace them if they're damaged.

12   Check the floats for damage. This will usually be apparent by the presence of fuel inside one of the floats. If the floats are damaged, they must be replaced.

13   Insert the throttle valve in the carbure-

tor body and see that it moves up-and-down smoothly. Check the throttle valve and rollers for wear **(see illustration)**. If it's worn excessively or doesn't move smoothly in the bore, replace it.

5.13 Check the throttle valve plate and rollers for damage and wear (one of the rollers comes off - the others don't)

6.5 Measure the float height from the surface of the float bowl mating surface. The tang on the float should just push the needle down far enough to seat it, but not enough to compress the spring-loaded pin in the needle

6.8 Carefully tighten the accelerator pump screw (upper arrow) all the way, then back it off until there's no play in the lever (lower arrow)

7.5 Slide the cover back from the throttle housing

6 Carburetor - reassembly, float height check and accelerator pump check

Reassembly

Caution: When installing the jets, be careful not to over-tighten them - they're made of soft material and can strip or shear easily.

Note: *When reassembling the carburetor, be sure to use new O-rings.*

1   Install the clip on the jet needle if it was removed. Place it in the needle groove listed in this Chapter's Specifications. Install the needle and clip in the throttle valve.

2   Install the pilot screw along with its spring, washer and O-ring, turning it in until it seats lightly. Now, turn the screw out the number of turns listed in this Chapter's Specifications.

3   Reverse the disassembly steps to install the jets.

4   Attach the fuel inlet valve needle to the float. Set the float into position in the carburetor, making sure the valve needle seats correctly. Install the float pivot pin.

Float height check

5   To check, hold the carburetor so the float hangs down, then tilt it back until the valve needle is just seated. Measure the distance from the float chamber gasket surface to the top of the float and compare your measurement to the float height listed in this Chapter's Specifications **(see illustration)**. Bend the float tang as necessary to change the adjustment.

6   Install the float chamber gasket or O-ring. Place the float chamber on the carburetor and install the screws, tightening them securely. Install the main jet access plug in the bottom of the float chamber, using a new O-ring, and tighten it securely.

Accelerator pump adjustment

7   Raise the throttle valve by hand. Slip a drill bit or similar rod under the throttle valve and let it down against the rod. The diameter is listed in this Chapter's Specifications.

8   Tighten the accelerator pump adjusting screw as far as it will go (but don't overtighten it) **(see illustration)**.

9   Wiggle the link lever with your fingers and check for freeplay **(see illustration 6.8)**. There should be some.

10   Back out the adjusting screw just far enough to remove the freeplay in the link lever.

7 Throttle and hot start cables - removal and installation

1   These motorcycles are equipped with an accelerator (opening) cable and a decelerator (closing) cable. Later models have a lever on the left handlebar that operates the hot start valve through a cable.

Throttle cables

2   Remove the fuel tank (see Section 2).

3   At the handlebar, loosen the throttle cable adjuster all the way (see Chapter 1).

4   Look for a punch mark on the handlebar next to the split in the throttle housing. If you don't see a mark, make one so the throttle housing can be installed in the correct position.

5   Slide the cable boots back **(see illustration)**.

6   Pull back the rubber boot and remove the throttle housing screws **(see illustration)**. Separate the throttle housing halves, noting how the alignment pins and insert fit **(see illustration)**.

7   Rotate the throttle grip to create slack

7.6a Slide the rubber boot out of the way and remove the housing screws (arrow)

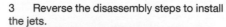

7.6b Separate the housing halves and note the locations of the guide and pins (arrows)

7.7 Rotate one cable to align with the slot, slip the cable end out of the pulley, then do the same with the other cable

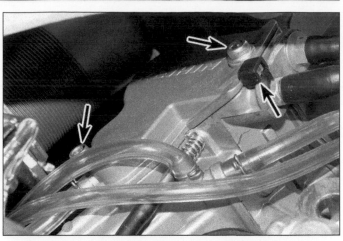

7.8a Remove the throttle pulley cover screws and note the location of the rubber piece (arrows)

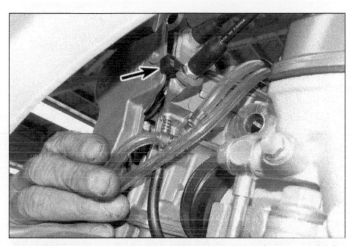

7.8b Take the cover off - the rubber piece (arrow) comes with it

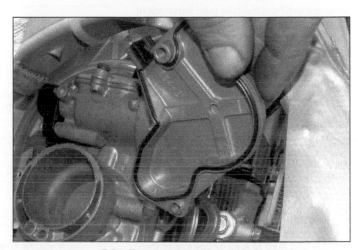

7.8c Remove the cover O-ring

in one cable. Lift the cable out of the groove, turn it to align with the removal slot and slip the cable end plug out of the pulley (see illustration). Do the same thing with the other cable.

8 At the carburetor, remove the throttle pulley cover (see illustrations).
9 Loosen the upper cable locknut to create slack in the cable (see illustration). Lift the cable out of the groove, turn

it to align with the removal slot and slip the cable end plug out of the pulley (see illustration). Do the same thing with the other cable.
10 Note how the cables are routed and

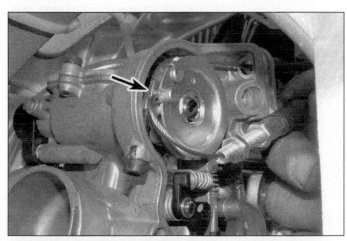

7.9a With the cable housing detached from the carburetor, slip the cable out of the pulley, turn it to align with the slot and slip the end plug out . . .

7.9b . . . then do the same thing with the other cable

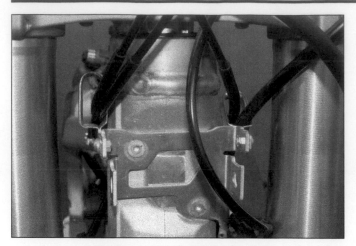

7.10 Note how the cables are routed

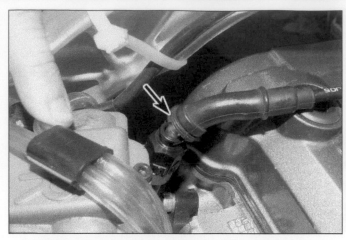

7.20a Slip the boot back from the hot start valve . . .

remove them from the bike **(see illustration)**.
11 Route the cables into place. Make sure they don't interfere with any other components and that they aren't kinked or bent sharply.
12 Lubricate the carburetor ends of the cables with multi-purpose grease. Reverse the disconnection steps to connect the throttle cables to the carburetor throttle pulley.
13 Reverse the disconnection steps to connect the cables to the throttle grip.
14 Operate the throttle and make sure it returns to the idle position by itself under spring pressure.

 Warning: If the throttle doesn't return by itself, find and solve the problem before continuing with installation. A stuck throttle can lead to loss of control of the motorcycle.

15 Follow the procedure outlined in Chapter 1 ("Throttle and choke operation/grip freeplay - check and adjustment") to adjust the cable.
16 Turn the handlebars back and forth to make sure the cables do not cause the steering to bind.

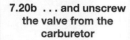

7.20b . . . and unscrew the valve from the carburetor

17 Once you're sure the cable operates properly, install the fuel tank.
18 With the engine idling, turn the handlebars through their full travel (full left lock to full right lock) and note whether idle speed increases. If it does, the cables are routed incorrectly. Correct this dangerous condition before riding the bike.

Hot start cable

19 This cable, used on models with a hot start lever on the left handlebar, connects the hot start lever to the hot start valve in the carburetor.
20 Slip the cable boot off the hot start valve at the carburetor **(see illustration)**. Unscrew the valve with a wrench and take it out **(see illustration)**.
21 If necessary, compress the spring and detach the valve from the cable **(see illustration)**.
22 Remove the lever mounting bolts, take the lever off the handlebar and disconnect the cable from the lever **(see illustration)**.

7.21 To separate the valve from the cable, compress the spring and slip the cable end out of the valve

7.22 Remove the upper screw to separate the hot start lever from the handlebar; remove the lower screw and nut to detach the lever from the bracket

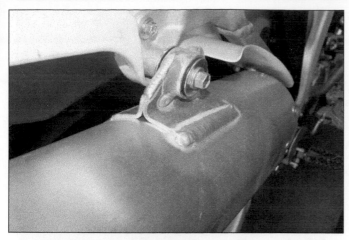

8.2a Unbolt the muffler from the frame . . .

8.2b . . . unbolt the pipe . . .

8 Exhaust system - removal and installation

1 Remove the right side cover (see Chapter 8).
2 Loosen the muffler clamp bolt and unscrew the mounting bolts **(see illustrations)**. Work the muffler free of the pipe and remove the gasket.
3 Detach the front pipe from the cylinder head and remove it from the machine **(see illustration)**.
4 To replace the glass wool muffler, refer to Chapter 1.
5 Installation is the reverse of removal. Use a new gasket in the exhaust port **(see illustration)** and at the joint between the pipe and muffler.

9 Throttle position sensor - check, replacement and adjustment

Note: *The throttle position sensor supplies the CDI unit with throttle angle data, which the CDI unit uses to determine the optimum ignition timing for any given engine rpm and rider demand. Normally, the throttle position sensor won't have to be removed, but if the bike exhibits sluggish performance, a defective or out-of-adjustment sensor is a possibility.*

Check

1 Follow the wiring harness from the throttle position sensor to the connector and unplug it (see Section 4).
2 Measure the resistance of the sensor coil. To do this, connect an ohmmeter between the blue and black wire terminals of the sensor electrical connector (the sensor side, not the wiring harness side). Connect ohmmeter positive to the blue wire and ohmmeter negative to the black wire). The ohmmeter should indicate 4000 to 6000 ohms.
Note: *On some models the black wire might have a blue tracer (stripe).*
3 If you're working on a 2002 or later model, back the throttle stop screw out all the way so the throttle will close completely (count and write down the number of turns so you can return the screw to its original position after this step). See the idle adjustment procedure in Chapter 1 if you're not sure where the throttle stop screw is.

4 If you're working on a 1998 through 2001 model, let the throttle close, but leave the throttle stop screw in the normal position.
5 Measure the variable resistance of the sensor. To do this, move the ohmmeter positive lead to the yellow wire's terminal in the sensor side of the connector (leave the ohmmeter negative lead connected to the black wire's terminal). Raise and lower the throttle valve. With the throttle all the way closed, the ohmmeter should indicate between zero and 2000 ohms (2001 and earlier models) or zero and 3000 ohms (2002 and later models). With the throttle all the way open, it should indicate 4000 to 6000 ohms. The resistance doesn't need to be exactly as specified, but it does need to vary smoothly. If the readings are incorrect, if the resistance doesn't change at all, or if resistance changes abruptly as the throttle is moved, replace and adjust the sensor as described below. Do not remove the sensor unless it needs to be replaced, or performance will be impaired.
6 If you're working on a 2000 or later model, check the sensor input voltage. To do this, disconnect the sensor wiring connector. Connect the thin wires into the wire terminals for the blue and blue/black wires

8.3 . . and detach the pipe from the cylinder head - some models have an Allen bolt in location A

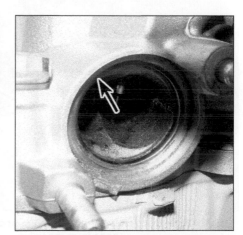

8.5 Use a new gasket (arrow)

9.8 Throttle position sensor mounting screw (some models have two)

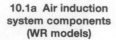

10.1a Air induction system components (WR models)

A Bracket
B Air induction valve
C Vacuum hose
D Air induction tube

in the harness side of the connector. Connect the voltmeter to the thin wires (positive to blue and negative to blue/black). With the engine idling, the voltmeter should indicate 4 to 6 volts. If the voltage is outside this range, check the wiring back to the CDI unit. If the wiring is good, the CDI unit may be defective. Since this is an expensive component that can't be returned once purchased, it's a good idea to substitute a known good CDI unit or have it checked by a dealer service department before buying a new one.

Removal

7   On models with a single screw securing the sensor, this procedure requires a T25 Torx bit. These are often available as part of inexpensive screwdriver sets. On models with two screws, take a look at the screws; if there's a raised post in the center of the screws, you'll need a T20 security (tamper resistant) Torx bit. These are less commonly available, so be sure you can get the proper tool before starting the procedure.

8   Remove the mounting screw(s) **(see illustration)**. Take the sensor off.

9   Check the sensor slot for wear or damage and replace the sensor with a new one if problems are found.

Installation and adjustment
1998 through 2001 models
10   Place the sensor on the carburetor, aligning the sensor slot with the carburetor tab, and tighten the mounting screws slightly (leave them loose enough so the sensor can be rotated). Leave the electrical connector unplugged for now.

11   Measure the resistance of the sensor with the throttle wide open (engine off) as described in Step 5. Multiply this number by 0.13, then by 0.15, and write the results down (for example, of the wide open throttle resistance is 5000 ohms, the two numbers will 650 and 750). These numbers are the high and low end of the resistance range when the throttle position sensor is in the closed throttle position.

12   Connect an ohmmeter between the yellow and black wire terminals in the sensor side of the connector, ohmmeter positive to

yellow and ohmmeter negative to black.

13   Place the throttle in the closed position. Rotate the sensor so the reading on the ohmmeter is within the range written down in Step 11, then tighten the mounting screws and plug in the connector. Refer to Chapter 1 and reset idle speed and mixture.

2002 and later models
14   Install the carburetor on the engine (if it was removed) and install the sensor on the carburetor, aligning the sensor slot with the carburetor tab. Tighten the mounting screw(s) slightly (leave them loose enough so the sensor can be rotated). Connect the sensor wiring connector.

15   Backprobe the connector with a pair of thin wires, one in the yellow wire's terminal and one in the black wire's terminal. **Caution:** *Don't let the wires touch each other or the resulting short circuit may damage the sensor or CDI unit.*

16   Connect a digital voltmeter to the thin wires, positive to yellow and negative to black.

10.1b Slide the clamp (arrow) down the hose and detach the hose from the valve

10.1c On some models, the air tube is attached to the front of the cylinder and head and secured by a bolt (left arrow) and clamp (right arrow)

17   Start the engine and let it idle. Adjust the idle speed if necessary, referring to Chapter 1.

18   Rotate the throttle position sensor until the voltmeter indicates 0.58 to 0.78 volts.

19   Once the correct reading is obtained, mark the sensor position on the carburetor with a felt pen. Shut off the engine.

20   Tighten the mounting screw(s), making sure that the felt pen marks stay aligned.

22   Remove the voltmeter and thin wires.

22   Refer to Chapter 1 and reset idle speed and mixture.

10 Air induction system (WR models) - removal, inspection and installation

Removal

1   Unbolt the bracket, remove the air cut-off valve and disconnect the hoses (see illustrations).

Inspection

2   Check the hoses and air induction tube for wear, cracks or deterioration. Replace them as needed.

3   Try to blow air through the two large hose fittings. Air should flow from the side fitting through the lower fitting, but not the other way. If it flows the wrong way or both ways, or doesn't flow either way, replace the valve.

4   Connect a hand vacuum pump to the small fitting and pump the pressure to 47 to 87 kPa (14 to 25 inches Hg).

Caution: Don't exceed the maximum pressure or the valve may be damaged.

Blow air through the large fittings again. Air should now flow from the lower fitting through the side fitting, but not the other way.

Notes

Chapter 5
Ignition and electrical systems

Contents

Degrees of difficulty

Easy, suitable for novice with little experience	Fairly easy, suitable for beginner with some experience	Fairly difficult, suitable for competent DIY mechanic	Difficult, suitable for experienced DIY mechanic	Very difficult, suitable for expert DIY or professional 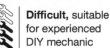

Specifications

YZ250F

Ignition coil resistance
2001 and 2002
    Primary.. 0.2 to 0.3 ohms
    Secondary.. 9500 to 14,300 ohms
2003 through 2005
    Primary.. 0.08 to 0.10 ohms
    Secondary.. 4600 to 6800 ohms
2006
    Primary.. 0.2 to 0.3 ohms
    Secondary.. 9500 to 14,300 ohms
2007
    Primary.. 0.08 to 0.10 ohms
    Secondary.. 4600 to 6800 ohms

Spark plug cap resistance
2001 through 2005	Not specified
2006	4000 to 6000 ohms
2007 and later	Not specified

Alternator stator coil resistance
2001 through 2005	
No. 1 charging coil (green-to-brown)	720 to 1080 ohms
No. 2 charging coil (black-to-pink)	44 to 66 ohms
2006	
No. 1 charging coil (black-to-black/red)	720 to 1080 ohms
No. 2 charging coil (green/blue-to-green/white)	44 to 66 ohms
2007 and later	
No. 1 charging coil (green-to-brown)	720 to 1080 ohms
No. 2 charging coil (black-to-pink)	44 to 66 ohms

Alternator pulse generator resistance
2001 through 2005 (white-to-red)	248 to 372 ohms
2006 (white/blue-to-white/red)	248 to 372 ohms
2007 and later (white-to-red)	248 to 372 ohms

Torque specifications
Alternator rotor nut	
2001 and 2002	48 Nm (35 ft-lbs)
2003 and later	56 Nm (40 ft-lbs)
Stator coil screws	
2001 through 2005	10 Nm (86 inch-lbs)
2006	7 Nm (61 inch-lbs)
2007 and later	10 Nm (86 inch-lbs)

YZ400F, YZ426F, YZ450F

Ignition coil resistance
1998 through 2002	
Primary	0.2 to 0.3 ohms
Secondary	9500 to 14,300 ohms
2003 and later	
Primary	0.08 to 0.10 ohms
Secondary	4600 to 6800 ohms

Spark plug cap resistance
Spark plug cap resistance	Not specified

Alternator stator coil resistance
No. 1 charging coil (green-to-brown)	720 to 1080 ohms
No. 2 charging coil (black-to-pink)	44 to 66 ohms

Alternator pulse generator resistance (white-to-red)
Alternator pulse generator resistance (white-to-red)	248 to 372 ohms

Torque specifications
Alternator rotor nut	
1998 through 2003	48 Nm (35 ft-lbs)
2004 and later	56 Nm (40 ft-lbs)
Stator coil screws	10 Nm (86 inch-lbs)

WR250F

Fuses
Main and spare	10 amps

Ignition coil resistance
2001 through 2003	
Primary	0.2 to 0.3 ohms
Secondary	9500 to 14,300 ohms
2004 and later	
Primary	0.08 to 0.10 ohms
Secondary	4600 to 6800 ohms

Spark plug cap resistance .... Not specified

Alternator stator coil resistance
2001 through 2002
    No. 1 charging coil (green-to-brown) .... 640 to 960 ohms
    No. 2 charging coil (green-to-pink) .... 464 to 696 ohms
    Lighting coil (black-to-yellow) .... 0.16 to 0.24 ohms
2003 and later
    Charging coil (white-to-ground) .... 0.288 to 0.432 ohms
    Lighting coil (yellow-to-ground) .... 0.224 to 0.336 ohms

Alternator pulse generator resistance (white-to-red) .... 248 to 372 ohms

Torque specifications
Alternator rotor nut
    2001 through 2002 .... 48 Nm (35 ft-lbs)
    2003
        Step 1 .... 70 Nm (50 ft-lbs)
        Step 2 .... Loosen
        Step 3 .... 70 Nm (50 ft-lbs)
    2004
        Step 1 .... 65 Nm (47 ft-lbs)
        Step 2 .... Loosen
        Step 3 .... 65 Nm (47 ft-lbs)
Stator coil screws
    2001 and 2002 .... 10 Nm (86 inch-lbs)
    2003 and later .... 7 Nm (51 inch-lbs)
Starter motor mounting bolts .... 10 Nm (86 inch-lbs)

WR400F, W426F, WR450F

Fuses
Main and spare .... 10 amps

Ignition coil resistance
1998 through 2002
    Primary .... 0.2 to 0.3 ohms
    Secondary .... 9500 to 14,300 ohms
2003 and later
    Primary .... 0.08 to 0.10 ohms
    Secondary .... 4600 to 6800 ohms

Spark plug cap resistance .... Not specified

Alternator stator coil resistance
1998 through 2002
    No. 1 charging coil (green-to-brown) .... 640 to 960 ohms
    No. 2 charging coil (green-to-pink) .... 464 to 696 ohms
    Lighting coil (black-to-yellow) .... 0.24 to 0.36 ohms
2003 and later
    Charging coil (white-to-ground) .... 0.288 to 0.432 ohms
    Lighting coil (yellow-to-ground) .... 0.224 to 0.336 ohms

Alternator pulse generator resistance (white-to-red) .... 248 to 372 ohms

Torque specifications
Alternator rotor nut
    1998 through 2002 .... 48 Nm (35 ft-lbs)
    2003 and later
        Step 1 .... 65 Nm (47 ft-lbs)
        Step 2 .... Loosen
        Step 3 .... 65 Nm (47 ft-lbs)
Stator coil screws
    1998 through 2002 .... 10 Nm (86 inch-lbs)
    2003 and later .... 7 Nm (51 inch-lbs)
Starter motor mounting bolts .... 10 Nm (86 inch-lbs)

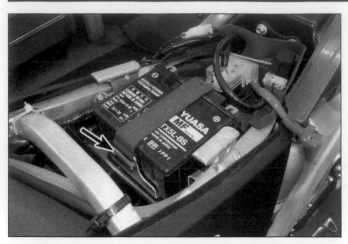

3.3a Unhook the retaining strap and lift the battery out

3.3b Make sure the terminal nut is in position in the terminal before installing the cable bolt

1 General information

The only electrical circuit on YZ models is the ignition system. Because YZ models are intended for motocross competition, they do not have a battery, ignition switch, fuses, turn signals or lights. The engine is started with a kickstarter and turned off with a kill button on the left handlebar.

All WR models have a lighting system (headlight and taillight) in addition to the ignition system. 2003 and later WR models have an electric starter, battery and charging system. 2006 and later WR models have an electronic display that includes a speedometer (which can display current speed or average speed), two tripmeters and a clock (which can also measure elapsed time).

The ignition system consists of an alternator that generates the current, a capacitive discharge ignition (CDI) unit that receives and stores it, and a pulse generator that triggers the CDI unit to discharge its current into the ignition coil, where it is stepped up to a voltage high enough to jump the spark plug gap. The coil on later models is built into the spark plug boot. To aid in locating a problem in the ignition circuit, wiring diagrams are included at the end of this manual.

The CDI ignition system functions on the same principle as a breaker point ignition system with the pulse generator and CDI unit performing the tasks previously associated with the breaker points and mechanical advance system. As a result, adjustment and maintenance of breakerless ignition components is eliminated (with the exception of spark plug replacement).

Note: *Keep in mind that electrical parts, once purchased, can't be returned. To avoid unnecessary expense, make very sure the faulty component has been positively identified before buying a replacement part.*

2 Electrical troubleshooting

Electrical problems often stem from simple causes, such as loose or corroded connections. Prior to any electrical troubleshooting, always visually check the condition of the wires and connections in the circuit.

If testing instruments are going to be utilized, use the diagrams to plan where you will make the necessary connections in order to accurately pinpoint the trouble spot.

The basic tools needed for electrical troubleshooting include a test light or voltmeter, an ohmmeter or a continuity tester (which includes a bulb, battery and set of test leads) and a jumper wire, preferably with a circuit breaker incorporated, which can be used to bypass electrical components.

A continuity check is performed to see if a circuit, section of a circuit or individual component is capable of passing electricity through it. Connect one lead of a self-powered test light or ohmmeter to one end of the circuit being tested and the other lead to the other end of the circuit. If the bulb lights (or the ohmmeter indicates little or no resistance), there is continuity, which means the circuit is passing electricity through it properly. The kill switch can be checked in the same way.

Remember that the electrical circuit on these motorcycles is designed to conduct electricity through the wires, kill switch, etc. to the electrical component (CDI unit, etc.). From there it is directed to the frame (ground) where it is passed back to the alternator. Electrical problems are basically an interruption in the flow of electricity.

Because of their nature, the individual ignition system components can be checked but not repaired. If ignition system troubles occur, and the faulty component can be isolated, the only cure for the problem is to replace the part with a new one. Keep in mind that most electrical parts, once purchased, can't be returned. To avoid unnec-

essary expense, make very sure the faulty component has been positively identified before buying a replacement part.

Most battery damage is caused by heat, vibration, and/or low electrolyte levels, so keep the battery securely mounted, check the electrolyte level frequently and make sure the charging system is functioning properly.

3 Battery - inspection and maintenance

1   The battery used on 2003 and later WR models is a maintenance-free type.
2   Remove the seat (see Chapter 8). Check around the base inside of the battery for sediment, which is the result of sulfation caused by low electrolyte levels. These deposits will cause internal short circuits, which can quickly discharge the battery. Look for cracks in the case and replace the battery if either of these conditions is found.

 Warning: Always disconnect the negative cable first and reconnect it last to prevent sparks that could cause the battery to explode.

3   Check the battery terminals and cable ends for tightness and corrosion. If corrosion is evident, remove the cables from the battery and clean the terminals and cable ends with a wire brush or knife and emery paper. If you need to remove the battery, unhook the retaining strap and lift it out of the carrier **(see illustration)**. Make sure the terminal nuts are in place **(see illustration)**. Reconnect the cables and apply a thin coat of petroleum jelly to the connections to slow further corrosion.
4   The battery case should be kept clean to prevent current leakage, which can discharge the battery over a period of time (especially when it sits unused). Wash the outside of the case with a solution of baking soda and water. Rinse the battery thoroughly, then dry it.
5   If acid has been spilled on the frame

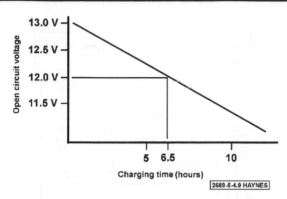

4.10 Draw a line straight across from the open circuit voltage to the bar, then straight down to find the charging time. *Note: This is based on a temperature of 68-degrees F (20-degrees C)*

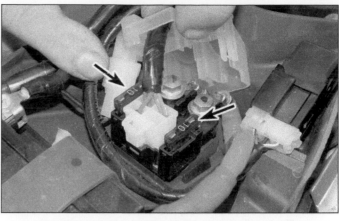

5.1 The main fuse and a spare are located on the starter relay

or battery box, neutralize it with the baking soda and water solution, dry it thoroughly, then touch up any damaged paint.

6　If the motorcycle sits unused for long periods of time, disconnect the cables from the battery terminals. Refer to Section 4 and charge the battery approximately once every month.

4　Battery - charging

1　If the machine sits idle for extended periods or if the charging system malfunctions, the battery can be charged from an external source.
2　Charging the maintenance-free battery used on 2003 and later WR models requires a variable-voltage charger, digital voltmeter and ammeter. If the charger doesn't have an ammeter built in, you can hook up an external ammeter between the positive terminal on the battery charger and the positive terminal on the battery.

Caution: Never connect an ammeter between the battery terminals. The ammeter's fuse will blow, or if doesn't have a fuse, the ammeter will be ruined.

3　When charging the battery, always remove it from the machine. If the battery case is translucent, check the electrolyte level by looking through the case before hooking up the charger. If the electrolyte level is low, the battery must be discarded; never remove the sealing plug to add water.
4　Disconnect the battery cables (negative cable first), then connect a digital voltmeter between the battery terminals and measure the voltage (open circuit voltage).
5　If open circuit voltage is 12.8 volts or higher, the battery is fully charged. If it's lower, recharge the battery.
6　A quick charge can be used in an emergency, provided the maximum charge rates and times are not exceeded (exceeding the maximum rate or time may ruin the battery). A quick charge should always be followed as

soon as possible by a charge at the standard rate and time.
7　Hook up the battery charger leads (positive lead to battery positive terminal and negative lead to battery negative terminal), then, and only then, plug in the battery charger.

 Warning: The gas escaping from a charging battery is explosive, so keep open flames and sparks well away from the area. Also, the electrolyte is extremely corrosive and will damage anything it comes in contact with.

8　Start charging at a high voltage setting (20 to 25 volts, but no more than 25 volts) and watch the ammeter for about 5 minutes. The charging amperage should exceed the maximum charging amperage listed on the battery. If it doesn't, replace the battery with a new one.
9　When the charging current increases beyond the specified maximum, reduce the charging voltage to reduce the charging current to the rate listed on the battery. Do this periodically as the battery charges (at least every five hours).
10　Allow the battery to charge for the specified time **(see illustration)**. If the battery overheats or gases excessively, the charging rate is too high. Either disconnect the charger or lower the charging rate to prevent damage to the battery.
11　After the specified time, unplug the charger first, then disconnect the leads from the battery.
12　Wait 30 minutes, then measure voltage between the battery terminals. If it's 12.8 volts or higher, the battery is fully charged. If it's between 12.0 and 12.7 volts, charge the battery again.

5　Fuse (2003 and later WR models) - check and replacement

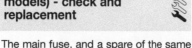

1　The main fuse, and a spare of the same capacity, are mounted under the seat forward of the battery **(see illustration)**.
2　The fuse can be removed and checked

visually. A blown fuse is easily identified by a break in the element.
3　If the fuse blows, be sure to check the wiring harnesses very carefully for evidence of a short circuit. Look for bare wires and chafed, melted or burned insulation. If a fuse is replaced before the cause is located, the new fuse will blow immediately.
4　Never, under any circumstances, use a higher rated fuse or bridge the fuse holder terminals, as damage to the electrical system or a fire could result.
5　Occasionally a fuse will blow or cause an open circuit for no obvious reason. Corrosion of the fuse ends and fuse holder terminals may occur and cause poor fuse contact. If this happens, remove the corrosion with a wire brush or emery paper, then spray the fuse end and terminals with electrical contact cleaner.

6　Ignition system - check

 Warning: Because of the very high voltage generated by the ignition system, extreme care should be taken when these checks are performed.

1　If the ignition system is the suspected cause of poor engine performance or failure to start, a number of checks can be made to isolate the problem.

Engine will not start

3　If no spark occurs, the following checks should be made:
4　Make sure all electrical connectors are clean and tight. Check all wires for shorts, opens and correct installation.
5　Refer to Section 7 and check the ignition coil primary and secondary resistance.
6　Check the pulse generator (and exciter coil if equipped) (see Section 10).
7　If the preceding checks produce positive results but there is still no spark at the plug, refer to Section 8 and check the CDI unit.

6.9 A simple spark gap testing fixture can be made from a block of wood, two nails, a large alligator clip, a screw and a piece of wire

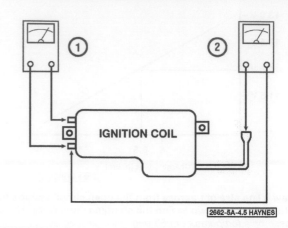

7.6 Ignition coil test

1 *Measure primary winding resistance*
2 *Measure secondary winding resistance*

Engine starts but misfires

8   If the engine starts but misfires, make the following checks before deciding that the ignition system is at fault.

9   The ignition system must be able to produce a spark across a six-millimeter (1/4-inch) gap (minimum). A simple test fixture (see illustration) can be constructed to make sure the minimum spark gap can be jumped. Make sure the fixture electrodes are positioned seven millimeters apart.

10   Connect the spark plug wire to the protruding test fixture electrode, then attach the fixture's alligator clip to a good engine ground.

11   Crank the engine over with the kickstarter and see if well-defined, blue sparks occur between the test fixture electrodes. If the minimum spark gap test is positive, the ignition coil is functioning properly. If the spark will not jump the gap, or if it is weak (orange colored), refer to Steps 4 through 7 of this Section and perform the component checks described.

12   If you're working on a WR model with electric start, make sure the battery is fully charged. These models are sensitive to the state of battery charge and may misfire with a partially charged battery, even if the battery is charged enough to start the bike.

---

7   Ignition coil - check, removal and installation

---

1   The ignition coil on 2002 and earlier models is mounted on the frame behind the engine. The ignition coil on 2003 and later models is built into the spark plug cap and fits directly onto the spark plug. This eliminates the need for a spark plug wire.

2   In order to determine conclusively that the ignition coil is defective, it should be tested by an authorized Yamaha dealer service department which is equipped with the special electrical tester required for this check.

3   However, the coil can be checked visually (for cracks and other damage) and the primary and secondary coil resistances can be measured with an ohmmeter. If the coil is undamaged, and if the resistances are as specified, it is probably capable of proper operation.

2002 and earlier models

Check

4   To check the coil for physical damage, it must be removed (see Steps 9 and 10). To check the resistance, remove the fuel tank (see Chapter 4), unplug the primary circuit electrical connector(s) from the coil and remove the spark plug wire from the spark plug. Mark the locations of all wires before disconnecting them.

5   Label the primary wires, then disconnect them. Unscrew the spark plug cap from the plug wire.

6   Connect an ohmmeter between the primary (small) terminals (see illustration). Set the ohmmeter selector switch in the Rx1 position and compare the measured resistance to the primary resistance values listed in this Chapter's Specifications.

7   Connect the ohmmeter between the coil primary terminal and the spark plug wire. Place the ohmmeter selector switch in the Rx100 position and compare the measured resistance to the secondary resistance val-

ues listed in this Chapter's Specifications.

8   If the resistances are not as specified, unscrew the spark plug cap from the plug wire and check the resistance between the ground wire's primary terminal and the end of the spark plug wire (see illustration). If it's now within specifications, the spark plug cap is bad. If it's still not as specified, the coil is probably defective and should be replaced with a new one.

Removal and installation

9   To remove the coil, refer to Chapter 4 and remove the fuel tank, then disconnect the spark plug wire from the plug. Unplug the coil primary circuit electrical connector(s). Some models have a single primary circuit connector and another wire that connects to one of the coil mounting bolts to provide ground (see illustration).

10   Remove the coil mounting bolts and lift the coil out.

11   Installation is the reverse of the removal steps.

2003 and later models

Check

12   To check the coil for physical damage, it must be removed from the spark plug (see the spark plug removal procedure in Chap-

7.8 Unscrew the spark plug cap from the plug wire and measure its resistance with an ohmmeter

7.9 Disconnect the primary terminals (left arrows) - if a mounting bolt secures a ground wire (right arrow) be sure to reconnect it on installation

8.5a The CDI unit is mounted under the frame tube near the steering head or behind the front number plate . . .

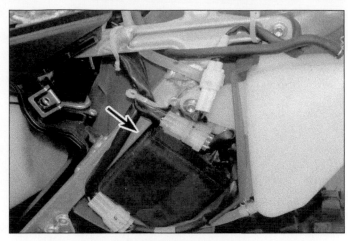

8.5b . . . or to the rear of the bike under the side cover

ter 1). Unplug the primary circuit electrical connector from the coil.

13 Connect an ohmmeter between the primary (small) terminals. Set the ohmmeter selector switch in the Rx1 position and compare the measured resistance to the primary resistance values listed in this Chapter's Specifications.

14 Connect the ohmmeter between the coil primary terminal and the spark plug terminal in the bottom end of the coil. Place the ohmmeter selector switch in the Rx100 position and compare the measured resistance to the secondary resistance values listed in this Chapter's Specifications.

15 If the resistances are not as specified, the coil is probably defective and should be replaced with a new one.

Removal and installation

16 The coil is removed as part of the spark plug removal procedure (see Chapter 1). Do not use a screwdriver or pliers to pry it out, or it may be damaged. Once the coil is out of the spark plug well, disconnect the primary wire connector from its upper end.

17 Installation is the reverse of removal. Lubricate the rubber seal at the upper end with silicone spray.

---

8 CDI unit - check, removal and installation

Check

1 The CDI unit is tested by process of elimination (when all other possible causes of ignition problems have been checked and eliminated, the CDI unit is at fault).

2 Check the ignition coil, alternator coils, pulse generator and kill switch as described elsewhere in this Chapter.

3 Carefully check the wiring harnesses for breaks or bad connections.

4 If the harness and all other system

components tested good, the CDI unit may be defective. Before buying a new one, it's a good idea to substitute a known good CDI unit.

Removal and installation

5 Locate the CDI unit. On 2006 and earlier models it's located near the steering head (see illustration); on these models, remove the fuel tank (see Chapter 4). On 2007 and later WR models it's located behind the left side cover (see illustration). On 2007 and later YZ models it's located behind the front number plate.

6 Unplug the connector from the CDI unit and work the unit out of its mounting band or remove the mounting screws.

7 Installation is the reverse of the removal steps.

---

9 Charging system testing (battery-equipped models)

General information and precautions

1 If the performance of the charging system is suspect, the system as a whole should be checked first, followed by testing of the individual components (the alternator and the regulator/rectifier). **Note:** *Before beginning the checks, make sure the battery is fully charged and that all system connections are clean and tight.*

2 Checking the output of the charging system and the performance of the various components within the charging system requires the use of a voltmeter and ohmmeter or the equivalent multimeter.

3 When making the checks, follow the procedures carefully to prevent incorrect connections or short circuits, as irreparable damage to electrical system components may result if short circuits occur.

4 If the necessary test equipment is not available, it is recommended that charg-

ing system tests be left to a dealer service department or a reputable motorcycle repair shop.

Leakage and output test

5 If a charging system problem is suspected, perform the following checks. Start by removing the left or right side cover for access to the battery (see Chapter 7).

Leakage test

6 Turn the ignition switch Off and disconnect the cable from the battery negative terminal.

7 Set the multimeter to the mA (milliamps) function and connect its negative probe to the battery negative terminal, and the positive probe to the disconnected negative cable. Although Yamaha doesn't specify a leakage limit for these models, it should be very low, about 0.1 milliamp or less.

8 If the reading is too high there is probably a short circuit in the wiring. Thoroughly check the wiring between the various components (see the wiring diagrams at the end of the book).

9 If the reading is satisfactory, disconnect the meter and connect the negative cable to the battery, tightening it securely. Check the alternator output as described below.

Output test

10 Start the engine and let it warm up to normal operating temperature.

11 With the engine idling, attach the positive (red) voltmeter lead to the positive (+) battery terminal and the negative (black) lead to the battery negative (-) terminal. The voltmeter selector switch (if equipped) must be in the 0-20 DC volt range.

12 Slowly increase the engine speed until voltage reaches its maximum (don't over-rev the engine) and compare the voltmeter reading to the value listed in this Chapter's Specifications.

13 If the output is as specified, the alternator is functioning properly.

14 Low voltage output may be the result of damaged windings in the alternator charging

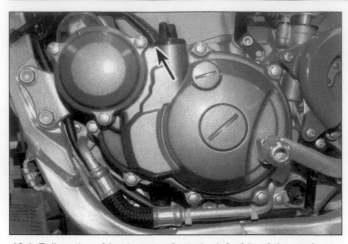

10.1 Follow the wiring harness from the left side of the engine to the connector and unplug it

10.5a Rotors like this are used on models without a battery - remove the nut . . .

coil or wiring problems between the alternator and battery. Make sure all electrical connections are clean and tight, then refer to the following Section to check the alternator coils. If the wiring and the charging coil are good, the problem may be a defective regulator/rectifier.

15   High voltage output (above the specified range) indicates a defective voltage regulator/rectifier.

10 Alternator and ignition pulse generator - check and replacement

Check

1   Follow the alternator wiring harness from the left side of the engine to the connector on the left side of the vehicle frame, then unplug it (see illustration).

2   To check the stator coil(s), connect an ohmmeter between the specified terminals in the side of the connector that runs back to the coil on the left side of the engine (refer to this Chapter's Specifications for wire col-

ors and resistance values). If the readings are much outside the value listed, replace the coil as described below. The stator, which the coil is part of, is available only as a complete unit.

3   To check the pulse generator, connect the ohmmeter between the specified terminals in the side of the connector that runs back to the ignition pulse generator (refer to this Chapter's Specifications for wire colors and resistance values). If the readings are much outside the value listed in this Chapter's Specifications, replace the pulse generator as described below.

Rotor replacement

Removal

Note: *To remove the alternator rotor, the special Yamaha puller or an aftermarket equivalent will be required. Don't try to remove the rotor without the proper puller, as it's almost sure to be damaged. Pullers are readily available from motorcycle dealers and aftermarket tool suppliers.*

4   Remove the left crankcase cover (see Chapter 2).

10.5b . . . and the washer behind it

5   Hold the alternator rotor with a universal holder. You can also use a strap wrench. If you don't have one of these tools and the engine is in the frame, the rotor can be locked by placing the transmission in gear and holding the rear brake on. Unscrew the rotor nut or bolt and remove the washer (see illustrations).

10.5c Battery-equipped models use rotors like this one - remove the nut . . .

10.5d . . . and the washer behind it

10.6a Thread the rotor puller onto the rotor, hold its flats with a wrench and tighten the puller bolt to push the rotor off the crankshaft (non-battery models)

10.6b Here's the rotor removal tool in use on a battery-equipped model

6   Thread an alternator puller into the center of the rotor and use it to remove the rotor **(see illustrations)**. If the rotor doesn't come off easily, tap sharply on the end of the puller to release the rotor's grip on the tapered crankshaft end.

7   Pull the rotor off **(see illustration)**. If you're working on a WR model, remove the thrust washer from behind it. Check the Woodruff key; if it's not secure in its slot, pull it out and set it aside for safekeeping **(see Illustration)**. On WR models, you'll need to remove it to remove the starter clutch drive gear. A convenient method is to stick the Woodruff key to the magnets inside the rotor **(see illustration)**, but be certain not to forget it's there, as serious damage to the rotor and stator coils will occur if the engine is run with anything stuck to the magnets.

Installation

8   Take a look to make sure there isn't anything stuck to the inside of the rotor **(see illustration 10.7c)**.

9   Degrease the center of the rotor and the end of the crankshaft.

10   Make sure the Woodruff key is positioned securely in its slot **(see illustration 10.7b)**.

11   If you're working on a WR model, make sure the starter clutch drive gear is in place on the crankshaft, then install the thrust washer **(see illustration 10.7a)**. Align the rotor slot with the Woodruff key. Place the rotor on the crankshaft.

12   Install the rotor washer and nut. Hold the rotor from turning with one of the methods described in Step 5 and tighten the nut to the torque listed in this Chapter's Specifications.

13   The remainder of installation is the reverse of the removal steps.

Stator coil and pulse generator replacement

14   The stator coils and pulse generator are replaced as a unit.

15   Remove the left crankcase cover (see

10.7a Take the rotor off the crankshaft and remove the thrust washer behind it (if equipped)

Chapter 2). On models without a battery, remove the alternator rotor as described above.

10.7b Remove the Woodruff key from its slot . . .

10.7c . . . you can store it by sticking it to the rotor magnets, but be sure to reinstall it before installing the rotor

10.16a On models without a battery, unbolt the stator plate from the crankcase

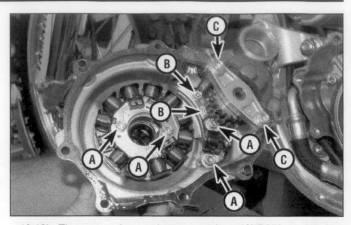

10.16b The stator plate on battery equipped WR250 models is inside the left crankcase cover

A   Stator plate bolts
B   Harness retainer bolts
C   Starter reduction gear retaining plate bolts

16   Remove the screws that secure the stator and take it off the engine, together with the pulse generator (see illustrations).
17   Installation is the reverse of the removal steps. Tighten the screws securely, but don't overtighten them and strip the threads.

11 Regulator/rectifier (2003 and later WR models) - check and replacement

1   On early models, the regulator/rectifier is mounted on the right-hand side of the bike above the engine. On later models, it's mounted on the frame near the steering head (see illustration).

Check

2   Start by checking the charging system voltage. To do this, connect a voltmeter across the battery terminals (positive to positive, negative to negative). Start the engine and run it at varying speeds up to 5000 rpm. The voltmeter should indicate 14 to 15 volts. If it's too high, the regulator/rectifier is probably at fault. If it's too low, the problem could be with the stator charging coil, rotor (weak or damaged magnets) or regulator/rectifier. Check the stator coils as described in Section 10. Check the rotor for physical damage, such as cracked or chipped magnets.
3   The regulator/rectifier is tested by a process of elimination. When all other possible causes of charging system failure, including wiring problems, have been checked and eliminated, the rectifier/regulator is at fault. Since this is difficult to detect and since the new part can't be returned if it doesn't solve the problem, it's best to substitute a known good unit or have the system tested by a dealer service or other qualified motorcycle electrical shop before replacing the rectifier/regulator.
4   Where necessary for access, remove the seat and fuel tank (see Chapters 8 and 4).
5   Follow the wiring harnesses from the regulator/rectifier to the connector and disconnect it.
6   Remove the regulator/rectifier mounting screws and lift it off the bike.
7   Installation is the reverse of the removal steps.

12 Starter circuit (2003 and later WR models) - component check and replacement

1   The electric starter used on these models uses a starter relay, starting circuit cut-off relay and diode.

Starter relay

2   Remove the seat and fuse box cover (see Chapter 8 and Section 5).

10.16c On battery-equipped WR450 models, remove the stator plate screws (left arrows) and pulse generator screws (right arrows)

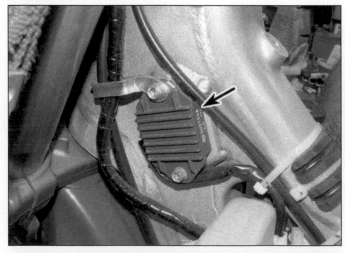

11.1 Here's the regulator/rectifier on an aluminum-frame model

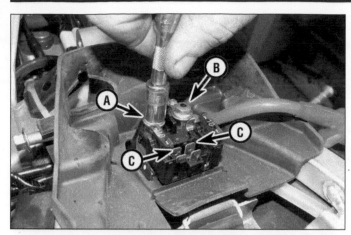

12.3 Disconnect the negative battery cable first, as shown here, then the positive cable before removing the relay from its rubber mount

A   Battery negative terminal   C   Relay terminals
B   Battery positive terminal

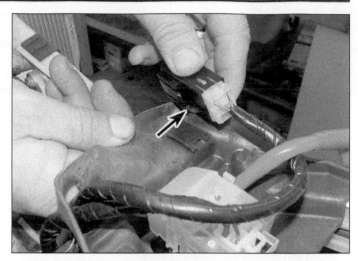

12.8 Slip the cut-off relay off its mount, then disconnect the wiring connector

3   Disconnect the battery cables (negative cable first) and wiring connector from the relay **(see illustration)**.
4   Connect an ohmmeter between the battery terminals on the relay **(see illustration 12.3)**. The ohmmeter should indicate Infinite resistance (no continuity).
5   Connect a fully charged battery (the motorcycle's battery will work) to the relay terminals, using lengths of wire. The ohmmeter between the relay's battery terminals should now indicate continuity (little or no resistance).
6   If the relay doesn't perform as described, remove it from its mount and install a new one.

Starting circuit cut-off relay

7   Remove the seat (see Chapter 8).
8   Remove the relay from its mount and disconnect it from its wiring connector **(see illustration)**.
9   Connect an ohmmeter between the battery terminals on the relay **(see illustration**

12.3). The ohmmeter should indicate infinite resistance (no continuity).
10   Connect a fully charged battery (the motorcycle's battery will work) to the relay terminals, using lengths of wire. The ohmmeter between the relay's battery terminals should now indicate continuity (little or no resistance).
11   If the relay doesn't perform as described, replace it.

Diode

12   Locate the diode in the wiring harness on the left side of the motorcycle **(see illustration)**. It's a square metal unit that looks like a turn signal flasher relay. It can be identified by its wiring connectors. Unwrap the tape from the diode and disconnect it from the wiring harness.
13   Connect an ohmmeter between the terminals for the blue/red wire and the light blue wire (ohmmeter positive terminal to blue/red; ohmmeter negative terminal to light blue). There should be little or no resistance. With the connections reversed, there should be

infinite resistance (no continuity).
14   Connect the ohmmeter between the terminals for the blue/red wire and the blue/yellow wire (ohmmeter positive terminal to blue/red; ohmmeter negative terminal to blue/yellow). There should be little or no resistance. With the connections reversed, there should be infinite resistance (no continuity).
15   If the relay doesn't perform as described, replace it.

13   Starter motor (2003 and later WR models) - check and replacement

Check

 Warning 1: This check may cause sparks. Make sure there is no leaking gasoline or anything else flammable in the vicinity.

 Warning 2: Make sure the transmission is in Neutral or the bike will jump forward during Step 2.

Caution: The jumper cable used for this test must be of a gauge at least as heavy as the battery cables or it may melt.

1   Locate the starter motor and disconnect its cable.
2   Connect a jumper cable from the battery positive terminal directly to the starter motor terminal. The starter should crank the engine.
3   If the starter doesn't crank at all, replace it. If it turns but doesn't crank the engine, remove and inspect the starter reduction gears and slipper clutch (see Sections 14 and 15).

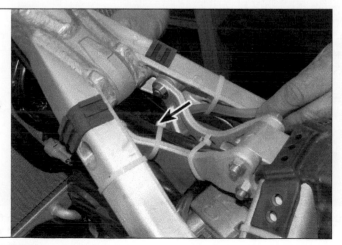

12.12 The starting circuit diode is inside a wiring harness in this area

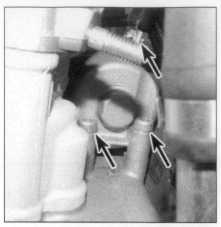

13.5 Disconnect the starter cable (upper arrow) and remove the mounting bolts (lower arrows)

Replacement

4   Disconnect the negative cable from the battery.
5   Disconnect the starter cable from the motor **(see illustration)**.

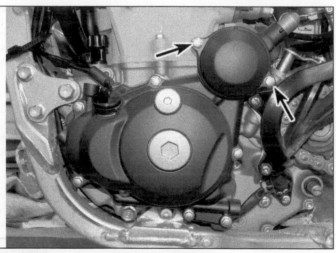

14.2b On WR450 models, the slipper clutch is behind the cylinder - its cover is secured by two bolts

6   Remove the motor mounting bolts and take it out **(see illustration 13.5)**. Check the O-ring for damage and replace it if necessary.
7   Installation is the reverse of the removal steps. Tighten the mounting bolts to the torque listed in this Chapter's Specifications.

14.2a On WR250 models, the slipper clutch is forward of the engine - its cover is secured by three bolts

14 Starter slipper clutch or idler gears (2003 and later WR models) - removal and installation

1   2003 models with electric start use a set of idler gears, mounted inside the left crankcase cover. 2004 and later models use a torque limiting slipper clutch. The idler gears or slipper clutch connect the starter pinion to the starter reduction gears.

Removal
2   Remove the cover mounting bolts **(see illustrations)**.
3   Take the idler gears or slipper clutch off the engine. On 2004 and later models, remove the thrust washer between the slipper clutch and crankcase cover **(see illustration)**. Remove the slipper clutch from its cover and remove the cover thrust washer **(see illustration)**.

Inspection
4   On 2003 models, inspect the idler gears and replace them if there are any problems,

14.3a Remove the cover and take out the thrust washer . . .

14.3b . . . then remove the slipper clutch from the cover and take out the remaining thrust washer

15.4a WR450 models use a set of starter reduction gears . . .

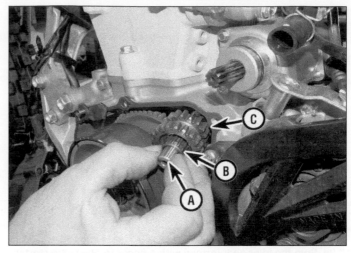

15.4b . . . to remove them, pull out the shaft and bearing

A	Shaft	C Gears
B	Bearing	

such as chipped teeth.

5   On 2004 and later models, the slipper clutch components aren't available separately, so if there's visible wear or damage, replace the slipper clutch as a unit.

Installation

6   Installation is the reverse of the removal steps. Use a new O-ring, lubricated with clean engine oil (see illustration 14.0a).

---

15 Starter reduction gears (2003 and later WR models) - removal and installation

1   WR models with electric start use two reduction gears, an idler gear and the starter clutch drive gear. The idler gear (small gear) transfers torque from the starter

slipper clutch to the starter clutch drive gear (large gear). The starter clutch turns the crankshaft.

Idler gear removal

2   Remove the left crankcase cover (see Chapter 2).

WR250 models

3   The idler gear on WR250 models is mounted inside the left crankcase cover. Unbolt the idler plate from the cover and lift the gear off, noting which direction it faces (see illustration 10.16b).

WR450 models

4   The idler gear on WR450 models is mounted in the outside of the crankcase (see illustration). Pull the gear shaft and bearing out and remove the gear, noting which direction it faces (see illustration).

Starter clutch gear removal

5   Remove the left crankcase cover and alternator rotor (see Chapter 2 and Section 10). Remove the thrust washer behind the rotor.

6   Slip the drive gear off of the crankshaft, noting which direction it faces (see illustration).

Inspection

7   Check the gears for wear and damage such as chipped teeth. Check the shaft (built into the gear on 250 models) for scoring or heat damage that might indicate lack of lubrication. Replace the gear if problems are found.

8   Check the bushing on the inside of the starter clutch gear for wear or damage. Replace the gear if problems are found.

9   Since needle roller bearing wear is difficult to see, the idler gear bearing on WR450 models should be replaced if its condition is in doubt.

Installation

10   Installation is the reverse of the removal procedure.

---

16 Starter clutch (2003 and later WR models) - removal and installation

Check

1   If the starter motor spins but doesn't crank the engine, you can perform a quick check to see if the starter clutch could be causing the problem.

2   Remove the slipper clutch (see Section 14). Reach into the slipper clutch recess and spin the starter idler gear with a fin-

15.6 Take the reduction gear off the crankshaft and note which way it faces

16.2a Spin the idler gear (arrow) with a finger to check the starter clutch - this is a WR250 . . .

16.2b . . . and this is a WR450

16.4a Carefully pry the starter clutch retainer loose, working around the edge . . .

16.4b . . . then lift it off and take the starter clutch out of the rotor

ger **(see illustrations)**. It should spin freely and smoothly in one direction and not at all in the other direction. If it spins both ways, the starter clutch is probably damaged. If movement of the gear is rough or uneven, the starter clutch or one of the reduction gears may be damaged. Remove the starter clutch for further inspection as described below.

Removal

3   Remove the alternator rotor (see Section 10). The starter clutch is mounted on the back of the rotor.

4   Carefully pry up the clip that secures the starter clutch to the back of the rotor, working gradually around the clip **(see illustration)**. Lift the clip off, then carefully pry the starter clutch out in the same manner **(see illustration)**.

Inspection

5   Check all parts for visible wear and damage and replace any parts that show problems.

6   Test the starter clutch. Place the clutch into the back of the alternator rotor. Hold the rotor steady and try to twist the clutch. It should turn in one direction only. If it turns both ways or neither way, replace it.

Installation

7   Installation is the reverse of the removal steps. Be sure the starter clutch is installed with its arrow facing the correct direction **(see illustration)**.

16.7 Be sure the directional arrow is visible when you put the starter clutch back in the rotor

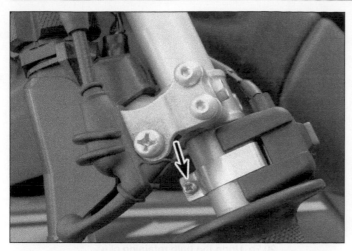

17.5 Remove the clamp screw (arrow) and take the kill switch of the handlebar

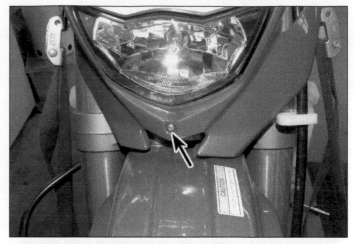

19.2 Turn the adjusting screw (arrow) to change the vertical adjustment of the headlight

17 Kill switch - check, removal and installation

1   The kill switch, mounted on the left handlebar, shorts the ignition circuit to ground when its button is pressed.

Check

2   Follow the wires from the switch to their connectors and unplug them.
3   Connect an ohmmeter between the wire terminals in the switch side of the connectors (not the side that leads back to the wiring harness). With the switch in the released position, the ohmmeter should show no continuity (infinite resistance); with the button pushed, the ohmmeter should show continuity (little or no resistance).
4   Repeat the test several times. The ohmmeter should move from continuity to no continuity each time the button is released. If it continues to show continuity after it's released, the ignition system is being shorted out constantly and won't produce a spark.

Removal and installation

5   To remove the switch, remove its mounting screw, separate the clamp and take it off the handlebar **(see illustration)**. Remove the wiring harness retainers and unplug the switch electrical connector.
6   Installation is the reverse of removal. Note that the clamp screw secures the switch ground wire.

18 Lighting circuit (WR models) - check

1   If the headlight or taillight doesn't work, check the bulb. If it's good, check the socket for corrosion and the wiring for breaks or bad connections. The taillight on later models is a light-emitting diode (LED) assembly that's replaced as a unit.
2   If neither light works, check the switch. Disconnect its wiring connector and connect an ohmmeter to the switch terminals. The ohmmeter should show little or no resistance when the switch is On, and infinite resistance when it's Off. If not, replace it.
3   If neither light works and the switch is good, check the alternator lighting coil (see Section 10).

19 Headlight (WR models) - adjustment

1   The headlight can be adjusted vertically with an adjustment screw below the headlight assembly.
2   To adjust the beam, turn the adjustment screw in or out **(see illustration)**.

20 Light bulbs (WR models) - replacement

Headlight

1   Remove the headlight assembly from the motorcycle **(see illustrations)**.

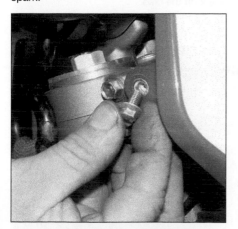

20.1a Remove the headlight assembly mounting bolt from each side . . .

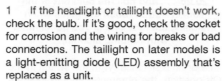

20.1b . . . and lift the assembly off the mounting posts (arrows)

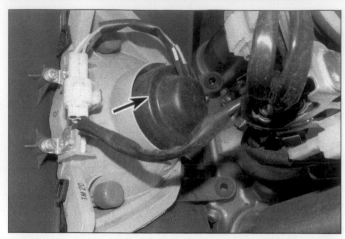

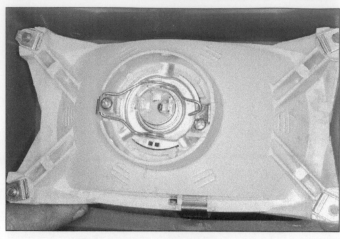

20.2a Pull off the rubber cover - on installation, make sure the TOP mark is up

20.2b Undo the bulb retaining clip . . .

2   Pull the rubber cover back from the bulb **(see illustration)**. Release the bulb clip and take the bulb out of the socket **(see illustrations)**.

3   Installation is the reverse of the removal steps. Be sure the TOP mark on the bulb cover is upward.

Taillight

4   On early models with a taillight bulb, remove the lens securing screws and take the lens off the top of the rear fender. Press the bulb into the socket against the spring pressure and turn it counterclockwise to remove it. Press the new bulb into the socket, turn it clockwise and release it. Replace the lens gasket if it's damaged or deteriorated. Reinstall the lens and tighten the screws securely, but don't overtighten them and crack the lens.

5   On later models, the taillight is a light emitting diode (LED) assembly that's replaced as a unit. Follow the wiring harness from the light to the connector and unplug it. Remove the mounting screws, working beneath the rear fender, and take the lamp out **(see illustration)**. Installation is the reverse of the removal steps.

20.2c . . . and pull the bulb out of the socket - don't touch the glass when installing the new one (if you do, clean it with rubbing alcohol)

20.5 Remove the three mounting screws and take the taillight assembly off the fender

21 Main switch (WR models) - check and replacement

Check

1   If the main switch doesn't light up when turned on, make sure the battery is fully charged (see Section 4).

2   If charging the battery doesn't cause the switch to light up, or if the switch doesn't work at all, remove the headlight assembly for access to the main switch, then unplug its wiring connector **(see illustration)**.

3   Connect the positive lead of an ohmmeter to the red wire's terminal in the connector (the switch side, not the harness side). Connect the ohmmeter negative lead to the brown wire's terminal. With the switch in the On position, the ohmmeter should indicate

little or no resistance. With the switch in the Off position, the ohmmeter should indicate infinite resistance. If the switch doesn't perform as described, replace it.

4   To check the switch illumination, connect the positive terminal of a 12-volt battery to the connector's red/black wire terminal, using a length of wire. Connect the battery negative terminal to the switch's black wire

terminal using another length of wire. If the switch doesn't light up, replace it.

Replacement

5   Remove the headlight assembly for access and unplug its wiring connector.

6   Remove the switch mounting bolts and take it off the bike.

7   Installation is the reverse of the removal steps.

21.2 The main switch and multi-function display are mounted in front of the upper triple clamp - connector locations vary

A   Main switch connector
B   Main switch mounting bolts
C   Multi-function display input connector
D   Multi-function display output connector

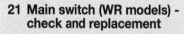

22.8 The speed sensor for the multi-function display is mounted on the front axle

23.1 Unplug the connector and remove the screws to remove the neutral switch

22 Multi-function display (2006 and later WR models) - check, removal and installation

Check

1   Remove the number plate and headlight assembly for access to the wiring connectors (see Chapter 8 and Section 20).
2   Locate the input connector to the display **(see illustration 21.2)**. Location varies with model, but the connector can be identified by its wire colors.
3   Without disconnecting the connector, slip two probes (pins or thin, stiff wire) into the backs of the terminals for the brown and black wires.

Caution: Do not let the probes touch each other, or the resulting short circuit will damage the display.

4   Set a voltmeter to 20 volts DC and connect its positive terminal to the brown wire's probe and its negative terminal to the black wire's probe.
5   Turn the main switch to On. The voltmeter should indicate at least 10 volts. If not, check the wiring to the display for breaks or bad connections.
6   Turn the main switch to Off and disconnect the voltmeter and probes. Move the probes to the display's output connector (red and black/white wires) **(see illustration 21.2)**. Connect the voltmeter positive lead to the red wire's probe and the negative lead to the black/white wire's probe. The voltmeter should indicate at least 4.5 volts. If not, the display may be defective and in need of replacement. Since this is an expensive part that can't be returned once purchased, it's a good idea to have the diagnosis confirmed by a Yamaha dealer before replacing the display.
7   Support the motorcycle so the front tire is off the grounds.
8   Follow the wiring harness from the

speed sensor **(see illustration)** to the connector (white, black and red wires). Insert the probes into the terminals for the black and white wires, then connect the voltmeter positive lead to the white wire's probe and the negative lead to the black wire's probe.
9   Turn the main switch to On. Slowly rotate the front tire by hand and watch the voltmeter. The indication should vary from 0.6 volts to 4.8 volts with each full turn of the wheel. If not, replace the speed sensor.

Replacement
Multi-function display

10   If you haven't already done so, remove the number plate and headlight assembly for access to the wiring connectors (see Chapter 8 and Section 20).
11   Disconnect the electrical connectors **(see illustration 21.2)**. Unscrew the display mounting bolts and take the display off the bracket.
12   Installation is the reverse of the removal steps.

Speed sensor

13   Follow the wiring harness from the speed sensor to the connector and unplug it.
14   Remove the front wheel fro the motorcycle (see Chapter 7).
15   Slip the speed sensor off of the hub **(see illustration 21.8)**.
16   Installation is the reverse of the removal steps. Be sure to align the two slots in the sensor with the tabs on the hub.

23 Neutral switch - check and replacement

Check

1   Locate the neutral switch on the left side of the engine **(see illustration)**. Disconnect its wire.

2   Set an ohmmeter to RX-1 and connect its positive terminal to the electrical terminal on the switch. Connect the ohmmeter to a good ground (bare metal on the engine).
3   Move the shift lever through the gear positions. The ohmmeter should show continuity when the transmission is in Neutral and no continuity in any other gear position.
4   If the switch doesn't perform properly, replace it.

Replacement

5   If you haven't already done so, disconnect the switch wire.
6   Remove the mounting screws and take the switch off the crankcase.
7   Installation is the reverse of the removal steps. Use a new sealing washer.

24 Clutch switch - check and replacement

Check

1   Follow the wiring harness from the switch at the clutch lever to the connector, then disconnect the connector.
2   Connect the leads of an ohmmeter to the terminals in the switch side of the wiring harness. With the clutch lever pulled in, there should be continuity through the switch (0 ohms).
3   With the clutch lever released, the meter should indicate infinite resistance.
4   If the switch fails either of these tests, replace it.

Replacement

5   Detach the switch from the clutch lever bracket, then disconnect the switch electrical connector.
6   Installation is the reverse of the removal steps.

Notes

Chapter 6
Steering, suspension and final drive

Contents

Degrees of difficulty

Easy, suitable for novice with little experience	**Fairly easy,** suitable for beginner with some experience	**Fairly difficult,** suitable for competent DIY mechanic	**Difficult,** suitable for experienced DIY mechanic	**Very difficult,** suitable for expert DIY or professional

Specifications

Front forks

Fork type
  1998 through 2004 (all models).. Bottom-mount cartridge
  2005 and later
    YZ models .... Top-mount cartridge (twin chamber)
    WR models .... Bottom-mount cartridge
Oil type
  1998 through 2004.. Yamaha suspension oil "01" or equivalent
  2005 and later .... Yamaha suspension oil "S1" or equivalent

Front forks (continued)
Oil capacity
- 1998
  - YZ400 .... 575 cc (19.45 US fl oz, 20.24 Imp oz)
  - WR400 .... 560 cc (18.9 US fl oz, 19.7 Imp oz)
- 1999, 2000
  - YZ400, YZ426 .... 578 cc (19.5 US fl oz, 20.3 Imp oz)
  - WR400
    - All except Europe .... 563 cc (19.0 US fl oz, 19.8 Imp oz)
    - Europe .... 583 cc (19.7 US fl oz, 20.5 Imp oz)
- 2001
  - YZ250 .... 568 cc (19.2 US fl oz, 20.0 Imp oz)
  - WR250
    - US and Canada .... 568 cc (19.2 US fl oz, 20.0 Imp oz)
    - Europe .... 578 cc (19.5 US fl oz, 20.3 Imp oz)
    - Australia, New Zealand, South Africa .... 568 cc (19.2 US fl oz, 20.0 Imp oz)
  - YZ426 .... 578 cc (19.5 US fl oz, 20.3 Imp oz)
  - WR426
    - All except Europe .... 573 cc (19.4 US fl oz, 20.2 Imp oz)
    - Europe .... 583 cc (19.7 US fl oz, 20.5 Imp oz)
- 2002
  - YZ250 .... 563 cc (19.0 US fl oz, 19.8 Imp oz)
  - WR250
    - US and Canada .... 568 cc (19.2 US fl oz, 20.0 Imp oz)
    - Europe .... 578 cc (19.5 US fl oz, 20.3 Imp oz)
    - Australia, New Zealand, South Africa .... 568 cc (19.2 US fl oz, 20.0 Imp oz)
  - YZ426 .... 568 cc (19.2 US fl oz, 20.0 Imp oz)
  - WR426
    - All except Europe .... 568 cc (19.2 US fl oz, 20.0 Imp oz)
    - Europe .... 578 cc (19.5 US fl oz, 20.3 Imp oz)
- 2003
  - YZ250
    - All except Europe .... 563 cc (19.0 US fl oz, 19.8 Imp oz)
    - Europe .... 573 cc (19.4 US fl oz, 20.2 Imp oz)
  - WR250
    - US and Canada .... 568 cc (19.2 US fl oz, 20.0 Imp oz)
    - Europe .... 573 cc (19.4 US fl oz, 20.2 Imp oz)
    - Australia, New Zealand, South Africa .... 578 cc (19.5 US fl oz, 20.3 Imp oz)
  - YZ450 .... 568 cc (19.2 US fl oz, 20.0 Imp oz)
  - WR450 .... 573 cc (19.4 US fl oz, 20.2 Imp oz)
- 2004
  - YZ250 .... 662 cc (22.4 US fl oz, 23.3 Imp oz)
  - WR250
    - US and Canada .... 578 cc (19.5 US fl oz, 20.3 Imp oz)
    - All except US and Canada .... 573 cc (19.4 US fl oz, 20.2 Imp oz)
  - YZ450
    - All except Europe .... 662 cc (22.4 US fl oz, 23.3 Imp oz)
    - Europe .... 668 cc (22.6 US fl oz, 23.5 Imp oz)
  - WR450
    - US and Canada .... 578 cc (19.5 US fl oz, 20.3 Imp oz)
    - All except US and Canada .... 573 cc (19.4 US fl oz, 20.2 Imp oz)
- 2005
  - YZ250 .... 430 cc (14.5 US fl oz, 15.1 Imp oz)
  - WR250
    - All except Europe .... 715 cc (24.2 US fl oz, 25.2 Imp oz)
    - Europe .... 725 cc (24.5 US fl oz, 25.5 Imp oz)
  - YZ450
    - All except Europe .... 445 cc (15.0 US fl oz, 15.7 Imp oz)
    - Europe .... 455 cc (15.4 US fl oz, 16.0 Imp oz)
  - WR450
    - US and Canada .... 718 cc (24.3 US fl oz, 25.3 Imp oz)
    - All except US and Canada .... 715 cc (24.2 US fl oz, 25.2 Imp oz)
- 2006
  - YZ250
    - All except Europe .... 522 cc (17.6 US fl oz, 18.4 Imp oz)
    - Europe .... 527 cc (17.8 US fl oz, 18.6 Imp oz)
  - WR250 .... 650 cc (22.0 US fl oz, 22.9 Imp oz)

YZ450
    All except Europe .... 542 cc (18.3 US fl oz, 19.1 Imp oz)
    Europe .... 532 cc (18.0 US fl oz, 10.7 Imp oz)
WR450
    US and Canada .... 650 cc (21.9 US fl oz, 22.8 Imp oz)
    Europe .... 653 cc (22.1 US fl oz, 23.0 Imp oz)
    Australia, New Zealand, South Africa .... 655 cc (22.1 US fl oz, 23.1 Imp oz)
2007
    YZ250 .... 527 cc (17.8 US fl oz, 18.6 Imp oz)
    WR250 .... 648 cc (21.9 US fl oz, 22.8 Imp oz)
    YZ450
        All except Europe .... 527 cc (17.8 US fl oz, 18.6 Imp oz)
        Europe .... 537 cc (18.2 US fl oz, 18.9 Imp oz)
    WR450
        US and Canada .... 648 cc (21.9 US fl oz, 22.8 Imp oz)
        All except US and Canada .... 655 cc (22.1 US fl oz, 23.1 Imp oz)
2008
    YZ250 .... 521 cc (17.6 US fl oz, 18.3 Imp oz)
    WR250 .... 648 cc (21.9 US fl oz, 22.8 Imp oz)
    YZ450 .... 541 cc (18.3 US fl oz, 19.0 Imp oz)
    WR450 .... 648 cc (21.9 US fl oz, 22.8 Imp oz)

Oil level (standard)*
1998
    YZ400 .... 130 mm (5.12 inches)
    WR400 .... 145 mm (5.71 inches)
1999
    YZ400 .... 130 mm (5.12 inches)
    WR400
        All except Europe .... 145 mm (5.71 inches)
        Europe .... 125 mm (4.92 inches)
2000
    YZ246 .... 130 mm (5.12 inches)
    WR400
        All except Europe .... 135 mm (5.31 inches)
        Europe .... 125 mm (4.92 inches)
2001
    YZ250 .... 140 mm (6.61 inches)
    WR250
        All except Europe .... 140 mm (5.51 inches)
        Europe .... 130 mm (5.12 inches)
    YZ426 .... 130 mm (5.12 inches)
    WR426
        All except Europe .... 135 mm (5.31 inches)
        Europe .... 125 mm (4.92 inches)
2002
    YZ250 .... 140 mm (5.51 inches)
    WR250
        All except Europe .... 140 mm (5.51 inches)
        Europe .... 130 mm (5.12 inches)
    YZ426 .... 135 mm (5.31 inches)
    WR426
        All except Europe .... 135 mm (5.31 inches)
        Europe .... 125 mm (4.92 inches)
2003
    YZ250 .... 140 mm (5.51 inches)
    WR250
        US and Canada .... 135 mm (5.31 inches)
        Europe .... 130 mm (5.12 inches)
        Australia, New Zealand, South Africa .... 125 mm (4.92 inches)
    YZ450 .... 135 mm (5.31 inches)
    WR450 .... 130 mm (5.12 inches)
2004
    YZ250 .... 125 mm (4.92 inches)
    WR250
        US and Canada .... 125 mm (4.92 inches)
        All except US and Canada .... 130 mm (5.12 inches)
    YZ450
        All except Europe .... 125 mm (4.92 inches)
        Europe .... 120 mm (4.72 inches)

Front forks
Oil level (standard)* (continued)

WR450

US and Canada..	125 mm (4.92 inches)
All except US and Canada..................................	130 mm (5.12 inches)

2005

YZ250..	See Section 5

WR250

All except Europe..	130 mm (5.12 inches)
Europe..	123 mm (4.84 inches)
YZ450..	See Section 5

WR450

US and Canada..	128 mm (5.04 inches)
All except US and Canada..................................	130 mm (5.12 inches)

2006

YZ250..	See Section 5
WR250...	130 mm (5.12 inches)
YZ450..	See Section 5

WR450

US and Canada..	130 mm (5.12 inches)
Europe..	127 mm (5.00 inches)
Australia, New Zealand, South Africa	125 mm (4.92 inches)

2007 and later

YZ250..	See Section 5
WR250...	132 mm (5.20 inches)
YZ450..	See Section 5
WR450...	132 mm (5.20 inches)

*From top of fork tube, with spring removed and damper rod compressed all the way.

Oil level limits

1998 through 2003 (all models)

Minimum..	80 mm (3.15 inches)
Maximum...	150 mm (5.91 inches)

2004

YZ models

Minimum	105 mm (4.13 inches)
Maximum	135 mm (5.31 inches)

WR models

Minimum	80 mm (3.15 inches)
Maximum	150 mm (5.91 inches)

2005

YZ models	Not specified

WR models

Minimum	80 mm (3.15 inches)
Maximum	150 mm (5.91 inches)

2006 and later

YZ models	Not specified

WR models

Minimum	95 mm (3.74 inches)
Maximum	150 mm (5.91 inches)

Fork spring free length

1998 through 2003 (all models)

Standard...	460 mm (18.1 inches)
Limit..	455 mm (17.9 inches)

2004

YZ models

Standard...	479 mm (18.9 inches)
Limit..	474 mm (18.7 inches)

WR models

Standard...	460 mm (18.1 inches)
Limit..	455 mm (17.9 inches)

2005

YZ models

Standard...	465 mm (18.3 inches)
Limit..	460 mm (18.1 inches)

WR models

Standard...	460 mm (18.1 inches)
Limit..	455 mm (17.9 inches)

2006 and later
    YZ models
        Standard.. 454 mm (17.9 inches)
        Limit... 449 mm (17.7 inches)
    WR models
        Standard.. 460 mm (18.1 inches)
        Limit... 455 mm (17.9 inches)
Protrusion from upper triple clamp
    1998 .... Zero
    1999, 2000
        YZ models.. Zero
        WR models
            All except Europe) .... 5 mm (0.20 inch)
            (Europe) .... Zero
    2001, 2002
        YZ250 .... 5 mm (0.20 inch)
        WR250
            US and Canada .... 7 mm (0.28 inch)
            Europe .... 8 mm (0.31 inch)
            Australia, New Zealand and South Africa .... 10.5 mm (0.41 inch)
        YZ426 .... Zero
        WR426 .... Zero
    2003, 2004
        YZ250 .... 5 mm (0.20 inch)
        WR250
            All except Australia, New Zealand and South Africa.............. 5 mm (0.20 inch)
            Australia, New Zealand and South Africa .... 10.0 mm (0.39 inch)
        YZ450 .... Zero
        WR450 .... Zero
    2005 and later
        YZ250 .... 5 mm (0.20 inch)
        WR250 .... 5 mm (0.20 inch)
        YZ450 .... Zero
        WR450 .... Zero

Fork tube bend limit.. 0.2 mm (0.008 inch)

Drive chain

Length limit (10 links)
    1998 .... 152.5 mm (6.004 inches)
    1999
        YZ400 .... 152.5 mm (6.004 inches)
        WR400 .... 150.1 mm (5.91 inches)
    2000 through 2004
        YZ models .... 152.5 mm (6.004 inches)
        WR models .... 150.1 mm (5.91 inches)
Length limit (15 links)
    2005 and later
        YZ models .... 242.9 mm (9.563 inches)
        WR models .... 239.3 mm (9.42 inches)

Torque specifications

Handlebar bracket bolts
    1998 through 2000.. 23 Nm (16 ft-lbs)
    2001 and later .... 28 Nm (20 ft-lbs)
Handlebar bracket nuts (2006 and later models)
    2006 .... 40 Nm (29 ft-lbs)
    2007 and later .... 34 Nm (24 ft-lbs)
Triple clamp bolts
    1998 through 2006
        Lower... 20 Nm (14 ft-lbs)
        Upper... 23 Nm (17 ft-lbs)
    2007 and later (upper and lower) .... 21 Nm (15 ft-lbs)
Fork cap to fork (bottom mount cartridge type) .... 30 Nm (22 ft-lbs)
Fork cartridge to fork (top mount cartridge type)........................ 30 Nm (22 ft-lbs)
Fork base valve
    1998 through 2004 (all models)... 55 Nm (40 ft-lbs)
    2005 and later YZ models (top mount base valve) .... 29 Nm (21 ft-lbs)
    2005 and later WR models.. 55 Nm (40 ft-lbs)

Torque specifications (continued)

Fork damping adjuster (2005 and later) .... 55 Nm (40 ft-lbs)
Steering stem bearing adjusting nut .... See Chapter 1
Steering stem nut .... 145 Nm (105 ft-lbs)
Rear shock absorber mounting bolts
    Upper .... 56 Nm (40 ft-lbs)
    Lower .... 53 Nm (38 ft-lbs)
Shock linkage to swingarm and frame
    1998 through 2005 .... 80 Nm (58 ft-lbs)
    2006 and later
        Relay arm to swingarm .... 70 Nm (50 ft-lbs)
        Relay arm to connecting rod .... 80 Nm (58 ft-lbs)
        Relay arm to frame .... 80 Nm (58 ft-lbs)
Swingarm pivot bolt nut
    1998 through 2005 .... 85 Nm (61 ft-lbs)
    2006
        YZ models .... 85 Nm (61 ft-lbs)
        WR models .... 80 Nm (58 ft-lbs)
    2007 .... 85 Nm (61 ft-lbs)
Engine sprocket nut .... 75 Nm (54 ft-lbs)
Rear sprocket bolts/nuts
    1998 through 2002 .... 42 Nm (30 ft-lbs)
    2003 and later
        YZ models .... 42 Nm (30 ft-lbs)
        WR models .... 50 Nm (36 ft-lbs)

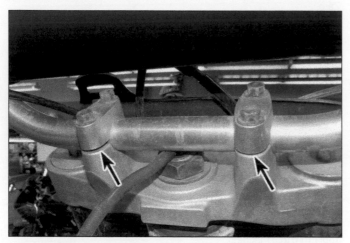

2.3a Remove the bolts and lift off the upper brackets; on installation, tighten the front bolts first, then the rear bolts, and don't try to close up the gap by overtightening

2.3b On models with removable handlebar brackets, unscrew the nuts from below (arrows)

1 General information

The steering system on these models consists of a one-piece braced handlebar and a steering head attached to the front portion of the frame. The steering stem rides in tapered roller bearings.

The front suspension uses one of two basic fork designs; bottom cartridge or top cartridge (refer to this Chapter's Specifications for the type used on your bike).

The rear suspension consists of a single shock absorber with a concentric coil spring, a swingarm and progressive rising rate suspension linkage. The suspension linkage causes the suspension to stiffen as its travel increases. This allows a softer ride over small bumps in the terrain, together with firmer suspension control over large irregularities.

2 Handlebars - removal, inspection and installation

1 The handlebars rest in brackets on top of the upper triple clamp. If the handlebars must be removed for access to other components, such as the steering head bearings, simply remove the bolts and take the handlebars off the bracket. It's not necessary to disconnect the throttle or clutch cables, brake hose or the kill switch wires, but it is a good idea to support the assembly with a piece of wire or rope, to avoid unnecessary strain on the cables.

2 If the handlebars are to be removed completely, refer to Chapter 2 for the clutch lever removal procedure, Chapter 4 for the throttle housing removal procedure, Chapter 5 for the kill switch removal procedure and Chapter 7 for the brake master cylinder removal procedure.

3 Remove the upper bracket bolts, lift off the brackets and remove the handlebars **(see illustration)**. On later models, the lower brackets can be removed from the triple clamp if necessary. To do this, remove the bracket nuts from below and lift the brackets out **(see illustration)**.

4 Check the handlebars and brackets for cracks and distortion and replace them if any problems are found.

5 Place the handlebars in the lower brack-

3.4a Note the position of the fork in the upper triple clamp, then loosen the pinch bolts

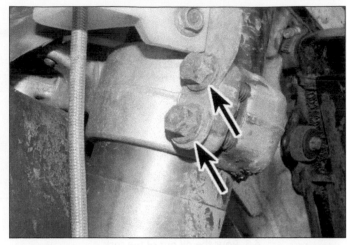

3.4b Loosen the lower triple clamp bolts and remove the fork

ets. Line up the punch mark on the handlebar with the parting line of the upper and lower brackets.

6   Install the upper brackets with their punch marks facing forward. Tighten the front bolts, then the rear bolts, to the torque listed in this Chapter's Specifications.

Caution: If there's a gap between the upper and lower brackets at the rear after tightening the bolts, don't try to close it by tightening beyond the recommended torque. You'll only crack the brackets.

3   Front forks - removal and installation

Removal

1   Support the bike securely upright with its front wheel off the ground so it can't fall over during this procedure.

2   Remove the front wheel, unbolt the brake caliper and detach the brake hose retainer from the left fork leg (see Chapter 7).

3   If you plan to disassemble the forks, loosen the upper triple clamp bolts, then loosen the fork cap bolts (see Section 4 or Section 5). This can be done later, but it will be easier while the forks are securely held in the triple clamps. Also, set the damping adjuster (if equipped) to its softest setting to prevent damage to the adjuster when the fork is reassembled. Count the number of turns and write this down so you can return it to its original setting after the fork is installed.

4   Loosen the upper and lower triple clamp bolts **(see illustrations)**.

5   Lower the fork leg out of the triple clamps, twisting it if necessary.

Installation

6   Slide each fork leg into the lower triple clamp.

7   Slide the fork legs up, installing the tops of the tubes into the upper triple clamp. Posi-

tion the upper end of each fork tube at the distance from the upper triple clamp listed in this Chapter's Specifications. Make sure the forks protrude an equal amount above each triple clamp.

8   Tighten the triple clamp bolts to the torque listed in this Chapter's Specifications.

9   Make sure the damping adjusters are at the same setting for both forks.

10   The remainder of installation is the reverse of the removal steps.

4   Bottom-mount cartridge forks - disassembly, inspection and reassembly

Disassembly

⚠️ *Warning: Do not disassemble the cartridge. This is a highly specialized job that even professional mechanics will send out to suspension modification shops. Incorrect assembly can cause erratic handling.*

4.2a Counting the turns, turn the damping adjuster (arrow) in until it bottoms, then back it out until it's loose

Note: *Overhaul of the forks on these models requires special tools for which there are no good substitutes. An air compressor and air tools are also very helpful. The tools can be ordered from your local Yamaha dealer, or you may be able to buy equivalent tools from aftermarket suppliers. Read through the procedure and arrange to get the special tools or substitutes before starting. If you don't disassemble the forks on a regular basis, it may be more practical to have the job done by a Yamaha dealer or other motorcycle repair shop.*

1   This fork design is used on 2004 and earlier YZ models, as well as all WR models.

2   Turn the damping adjuster in until it bottoms, counting the turns, and write this number down so the damping adjuster can be returned to its original setting **(see illustration)**. Now, back it out until it's loose. Unscrew the cap bolt from the fork tube (it was loosened while the fork was still in the triple clamps) **(see illustration)**. If you forget to loosen the cap bolt, temporarily put the fork back into the upper triple clamp and tighten it, then loosen the clamp bolt. Don't use a vise or the inner fork tube may be distorted.

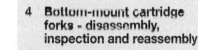

4.2b Unscrew the fork cap from the tube

4.3 Pour the oil out of the fork

4.4 Hold the locknut with a wrench and unscrew the fork cap from the damper rod

4.5a Remove the rod cap, noting that its oil holes are downward . . .

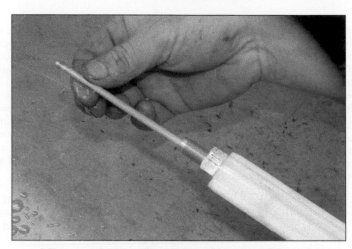

4.5b . . . and remove the pushrod (if equipped) from the damper rod

3   Turn the fork upside down and pour the oil into a container (see illustration).
4   Compress the spring far enough to expose the locknut below the fork cap. Hold the locknut with an open-end wrench and unscrew the fork cap from the damper rod (see illustration).
5   Remove the pushrod from the damper rod (see illustrations).
6   Remove the spring seat and spring from the inner fork tube (see illustration).
7   Pry the dust seal away from the outer fork tube and slide it off the end of the inner fork tube (see illustration).
8   Pry the oil seal retainer out of its groove,

4.6 Remove the spring seat from the spring and pull the spring out of the tube

4.7 Pry the dust seal out of the fork tube . . .

4.8 . . . and pry out the oil seal retainer, taking care not to scratch the fork tube

4.9a Yank the tubes sharply apart several times to separate them, then remove the inner tube from the outer tube . . .

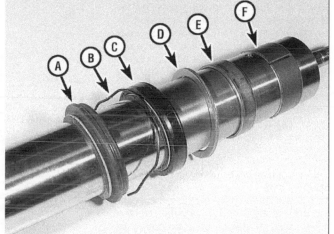

4.9b . . . which will provide access to the bushings and seals

A   Dust seal
B   Retainer
C   Oil seal
D   Back-up ring
E   Outer tube bushing
F   Inner tube bushing

4.9c Note which way the back-up ring faces

taking care not to scratch the inner fork tube (see illustration).

9   Grasp one fork tube in each hand, compress them together, then yank them apart sharply as far as they'll go. Do this several times until the tubes separate; the slide hammer-like motion is necessary to pull the slider bushing out of its bore in the outer fork tube (see illustration). Remove the inner tube, bushings, back-up ring and oil seal (see illustrations).

10   Check for built-up dirt or corrosion in the hex of the base valve at the bottom of the fork (see illustration). Clean as needed. Hold the cartridge from turning and unscrew the base valve from the bottom of the fork (see illustrations). If you don't have the special tool, you can spin the base valve loose

4.10a Clean all dirt from the base valve . . .

4.10b . . . and unscrew it from the fork leg . . .

4.10c . . . an air wrench with a hex bit is a convenient way to do this

4.10d If you don't have a compressor, use a damper rod tool . . .

with an air wrench. However, you'll still need a way to hold the damper rod during assembly so you can tighten the center bolt to the correct torque. The ideal way to do this is with a damper rod holder (see illustrations). You can order the Yamaha tool or an aftermarket equivalent from a Yamaha dealer. Another alternative is to install the fork spring and screw on the cap. The spring tension may be enough to keep the cartridge from turning as you loosen the base valve.

11   Pull the cartridge out of the fork (see illustration).

Inspection

12   Check the bushings for wear or damage (see illustration 4.9b). Replace them if problems are found.
13   Check the O-rings and bushings on the base valve for wear or damage. Replace them if problems are found. It's a good idea to replace the O-rings whenever the fork is disassembled.
14   Check the damper O-rings for wear or damage. Replace it if problems are found. It's a good idea to replace the O-rings when-

ever the fork is disassembled.
15   Check the bushing on the collar for wear and damage and replace as needed.
16   Check the adjuster O-ring for wear or damage. It's a good idea to replace the O-ring whenever the fork is disassembled. Always replace the copper washer when the adjuster is removed.
17   Check the plated surface of the inner fork tube for scratches or other damage. Check it for bending (ideally with a dial indicator, but you can also roll it on a flat surface such as a piece of glass). If it's bent more than the limit listed in this Chapter's Specifications, replace it. Do not try to straighten a bent fork tube.
18   Check the damper rod (and the pushrod that fits inside it) for bending. Replace either component if it's bent; do not try to straighten them.
19   Check the base valve for wear or damage and replace it if problems are found. Replace the base valve O-rings with new ones whenever the base valve is removed.
20   Check the spring guides, O-rings and other components on the cartridge rod for

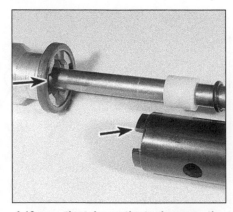

4.10e . . . the tabs on the tool engage the slots in the damper rod (arrows) (damper rod removed for clarity)

wear or damage (see illustration). If necessary, remove them from the rod and install new ones. Count the number of turns as you remove the locknut from the top of the pushrod so it can be returned to its original position.

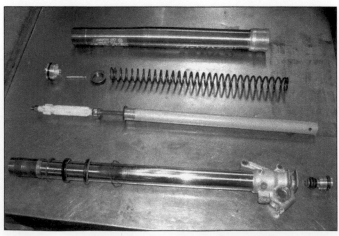

4.11 Bottom-mount cartridge fork details

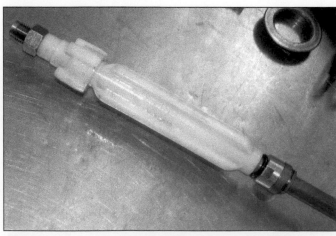

4.20 Don't remove the spring guides and other components from the damper rod unless they need to be replaced

4.21 Tape the end of the fork tube so the sharp edges won't cut the seals

4.22 This is the factory tool used to install bushings and seals

Assembly

21   Wrap the end of the inner fork tube with electrical tape to protect the new oil seal on installation (**see illustration**). Coat the lip of the new oil seal with the recommended fork oil. Install the bushings (if removed), backup ring and oil seal on the inner fork tube (**see illustration 4.9b**).

22   Install the inner tube in the outer tube. Slide the outer tube's bushing all the way down the inner fork tube until it rests against its bore in the outer fork tube. Install the back-up ring on top of the bushing. With a seal driver or equivalent tool (**see illustration**), tap against the backup ring to drive the new bushing into its bore. Once the bushing is seated, use the same tool to seat the oil seal in the case just below the retainer ring groove.

23   Install the retainer ring and make sure it seats securely in its groove, then install the dust seal.

24   Install the cartridge in the outer fork tube.

25   Place new O-ring(s) on the base valve.

26   Coat the threads of the base valve with a non-permanent thread locking agent. Hold the cartridge with the tool mentioned in Step 10, then tighten the base valve to the torque listed in this Chapter's Specifications.

27   Compress the fork all the way, then slowly add the recommended fork oil until it's even with the top of the outer fork tube.

28   Pump the damper rod slowly up-and-down at least ten times, then push it all the way down.

29   Add fork oil to bring the level up to the top of the fork tube again.

30   Slowly pump the outer fork tube another ten times. **Note:** *Do not pump the fork more than eight inches, or air will be sucked into the oil and the procedure will have to be done over.*

31   Wait ten minutes for any air bubbles to flow out. **Note:** *The oil level won't be accurate if it's checked too soon.* With the fork vertical, measure the oil level in the fork (**see**

4.31 Measure the oil level with the fork fully compressed and the spring removed

illustration) and compare it to the value listed in this Chapter's Specifications. Add or drain oil as necessary to correct the level. **Note:** *Minimum oil level will make the suspension slightly softer near full compression; maximum oil level will make the suspension slightly stiffer near full compression.*

> ⚠️ **Warning: To prevent unstable handling, make sure the oil level is exactly the same in both forks.**

32   Check the size of the hole in the spring seat. If it's too small for the locknut to fit through it, install the spring and spring seat in the fork now. If it's large enough for the locknut to fit through it, don't install it yet.

33   If the locknut was removed from the damper rod, thread it back on to the distance listed in this Chapter's Specifications.

34   If the spring seat will fit over the locknut, install it at this point.

35   Install the pushrod (and pushrod cap if equipped) in the damper rod (**see illustrations 4.5a and 4.5b**).

36   Thread the fork cap onto the damper rod until it reaches the locknut, then hold the locknut with a wrench and tighten the cap

4.36 Hold the locknut with a wrench and tighten the cap

against the locknut to the torque listed in this Chapter's Specifications (**see illustration**).

37   Extend the outer fork tube and thread the fork cap into it (tighten it to the specified torque later, when the fork leg is held in the triple clamps).

---

5   Top-mount (twin chamber) cartridge forks - disassembly, inspection and reassembly

1   This fork design is used on 2005 and later YZ models. **Note:** *This procedure requires special tools. Read through the procedure before starting to see what's needed. The Yamaha special tools can be ordered from a Yamaha dealer, and aftermarket substitutes may be available.*

Disassembly

2   Remove the plastic protectors from the fork legs if you haven't already done so.

3   Thoroughly clean the outside of the fork, paying special attention to the surface

5.4 Unscrew the damping adjuster from the bottom of the fork leg (the hex portion, not the adjusting screw in the center)

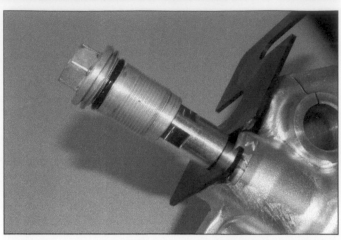

5.5 Pull the damping adjuster out to expose the locknut and slip a support tool under it

5.6a Hold the locknut with a wrench and unscrew the damping adjuster . . .

5.6b . . . and take it off the damper rod

of the inner fork tube and the cavity around the center bolt on the bottom of the fork.

4   Unscrew the adjuster from the bottom of the fork leg (see illustration).

5   Pull the adjuster out and slip a support tool under the locknut to keep it from sliding back into the fork leg (see illustration).

6   Place a wrench on the flats of the locknut and unscrew the adjuster (see illustrations).

Caution: DO NOT unscrew the locknut from the damper rod. If you do, the damper rod may fall into the cartridge and *be impossible to remove; if this happens, the cartridge will have to be replaced.*

7   Lift the pushrod out of the damper rod (see illustration).

8   Pour the oil out of the fork into a container (see illustration).

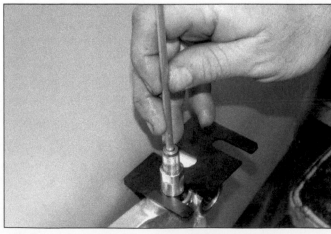

5.7 Pull the pushrod out of the damper rod - inspect the O-ring and replace the copper washer

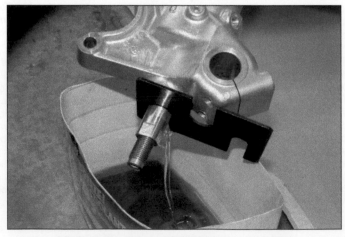

5.8 Pour the oil out of the fork into a container

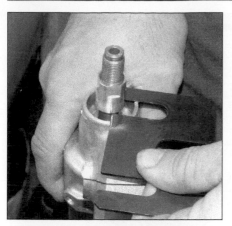

5.9 Remove the support tool; DO NOT remove the locknut or the damper rod may fall into the cartridge and be impossible to get out

5.10 Pry the dust seal out of its bore

5.11 Pry the oil seal retainer out, taking care not to scratch the fork tube

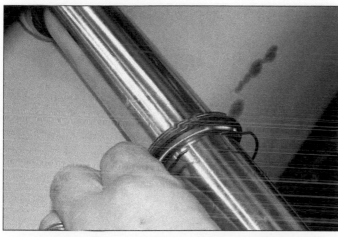

5.12a Slide the retainer and dust seal down the fork tube, taking care not to let the retainer ends scratch the tube . . .

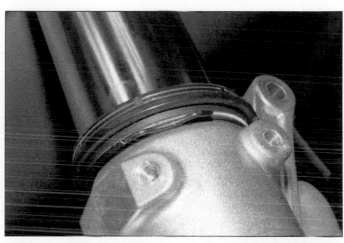

5.12b . . . position the retainer and seal against the bracket at the bottom of the tube

9   Remove the support tool (see illustration).
10   Pry the dust seal away from the outer fork tube, taking care not to scratch the inner fork tube (see illustration).
11   Pry the oil seal retainer out of its groove (see illustration).

12   Slide the dust seal and retainer down to the bottom of the fork leg, taking care not to scratch the tube (see illustrations). This is necessary to get them out of the way for the next step.
13   Compress the fork tubes together, then yank them sharply apart. Do this several

times to separate the fork tubes and expose the bushings (see illustrations). Remove the spring from the damper rod, then take the collar out of the fork tube (see illustration).
14   Spread the bushings just enough to remove them from the fork leg, then slide them off, followed by the back-up ring, oil

5.13a Yank the tubes apart to expose the bushings and oil seal . . .

5.13b . . . then separate the inner and outer tubes . . .

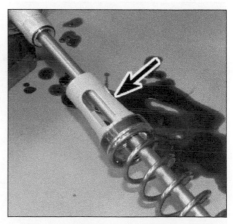

5.13c . . . and remove the fork spring and collar (arrow) from the damper rod

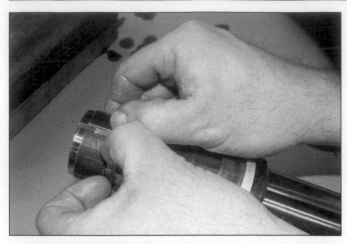

5.14a If you need to remove the bushings, spread them just enough to slip off the fork tube

5.14b The oil back-up ring is directional - note which way the tabs face

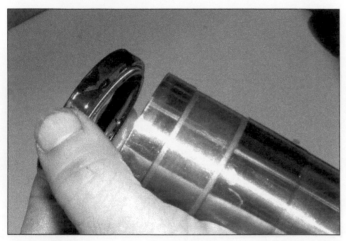

5.14c Slide off the oil seal, noting which side the coil spring is on

5.15 Unscrew the cartridge and lift it out of the fork tube, together with the damper rod

seal, dust seal and retaining ring (see illustrations).

15 Unscrew the damper rod from the fork tube, using a fork cap wrench (tool no. YM-01501-90890-0501, shown and described in the following step). Lift the damper rod out of the fork tube (see illustration).

16 Secure the damper in a padded vise (see illustration). Hold the damper from turning with a cap bolt ring wrench (tool no. YM-01501-90890-0501) and unscrew the base valve with a cap bolt wrench (tool no. YM-01500-90890-0500 (see illustration). These are both eight-sided (not hex). The ring wrench is 49 mm (1.928 inches) across the flats. The cap bolt wrench is 35.6 mm (1.405 inches) across the flats.

5.16a Secure the cartridge in a padded vise . . .

5.16b . . . and unscrew the base valve, using special tools

5.17a Lift the base valve out of the cartridge . . .

5.17b . . . and pour the oil out of the cartridge

17 Once the base valve is unscrewed, lift it out of the cartridge (see illustration). Pour the oil out of the cartridge (see illustration).

Inspection

18 Check the bushings for wear or damage. Replace them if problems are found.

19 Check the O-rings and bushings on the base valve for wear or damage (see illustration). Replace them if problems are found. It's a good idea to replace the O-rings whenever the fork is disassembled.

20 Check the damper O-rings for wear or damage (see illustrations). Replace it if problems are found. It's a good idea to replace the O-rings whenever the fork is disassembled.

21 Check the bushing on the collar for wear and damage and replace as needed (see illustration 5.20b).

22 Check the adjuster O-ring for wear or damage (see illustration 5.6b). It's a good idea to replace the O-ring whenever the fork is disassembled. Always replace the copper washer when the adjuster is removed.

23 Check the plated surface of the inner

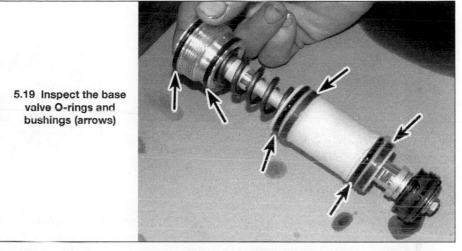

5.19 Inspect the base valve O-rings and bushings (arrows)

fork tube for scratches or other damage. Check it for bending (ideally with a dial indicator, but you can also roll it on a flat surface such as a piece of glass). If it's bent more than the limit listed in this Chapter's Specifications, replace it. Do not try to straighten a bent fork tube.

24 Make sure all parts are spotlessly clean before beginning assembly. Take care not to get any dirt in the fork when assembling it.

Assembly

25 Pull the damper rod all the way out of the cartridge (see illustration).

5.20a Inspect the O-ring at the top of the cartridge . . .

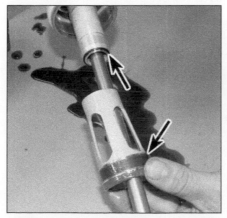

5.20b . . . and at the bottom (upper arrow) - also check the bushing on the collar (lower arrow)

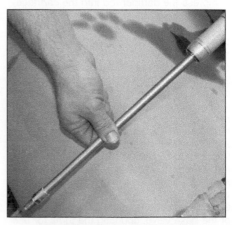

5.25 Extend the damper rod all the way . . .

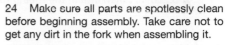

5.26 . . . and pour the specified amount of oil into the top of the cartridge

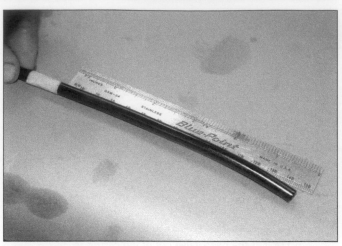

5.28a To make an oil level gauge, measure the specified length on a piece of rubber tubing and wrap it with tape . . .

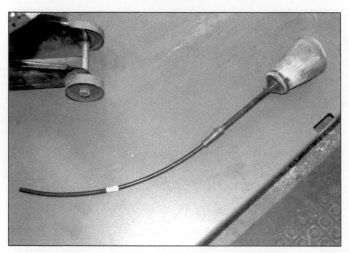

5.28b . . . then attach the tubing to a baster . . .

5.28c . . . and insert the marked tube into the cartridge - if oil level is too high, the baster can be used to suck some out

26 Hold the cartridge vertical and pour the specified amount of oil into the top **(see illustration)**.

27 With the cartridge held vertical, slowly compress the damper rod into the cartridge, then pull it back down, several times to bleed air out of the oil. **Note:** *Do not extend the damper rod more than 200 mm (8 inches) or air will be pulled into the oil and the bleeding procedure will have to be done over.*

28 Measure the oil level in the cartridge and add or remove oil to bring to the correct level. This can be done with a measuring tool made of a baster, some hose and a piece of tape **(see illustrations)**.

29 Once you've got the oil level correct in the cartridge, check the position of the locknut. Thread it all the way onto the damper rod, using fingers only (don't tighten it with a wrench).

30 Loosen the adjuster screw in the top of the base valve, counting the turns for use later.

31 Mount the cartridge in a vise with padded jaws, taking care not to damage the cartridge **(see illustration 6.16a)**.

32 Push up on the damper rod, then place the base valve into position, push it into the cartridge and tighten it **(see illustrations)**. As you do this, let the damper rod extend.

5.32a Push the base valve into the cartridge . . .

33 Check to make sure the damper rod is extended. If it isn't, repeat Step 32.

34 Finish tightening the base valve in the cartridge to the specified torque, using the tools described in Step 16.

5.32b . . . and tighten it temporarily

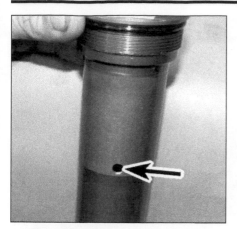

5.36 Excess oil will run out this hole in the cartridge

5.38a Lubricate the oil seal with suspension grease or fork oil applied to the seal lip so it will slip onto the fork tube easily

5.38b Cover the end of the fork tube with a plastic bag to protect the seal

35 With the cartridge held vertical, slowly extend and compress the damper rod at least 10 times to distribute the oil in the cartridge.

36 Place the end of the damper rod on a pile of rags on the floor (to protect the damper rod from damage). Push the cartridge down to compress the damper all the way, allowing any excess oil to flow out of the oil hole in the cartridge (see illustration).

37 Once this is done, extend and compress the damper rod by hand, checking for smooth movement. If movement is tight or uneven, repeat the oil filling procedure.

38 Note how the oil seals and bushings are installed (see illustration 5.13a). Lubricate the oil seal with suspension grease or fork oil (see illustration), then install the seals and bushings on the inner fork tube. Protect the oil seal by covering the end of the tube with a plastic bag (see illustration). Note the direction of the oil seal and the back-up ring (see illustration 5.14b).

39 Assemble the fork tubes to each other. Drive the outer bushing and oil seal into position using a seal driver (see illustration

5.17). Once the bushings and oil seals are in place, install the retaining ring and make sure it seats securely in its groove.

40 Lubricate the inner fork tube with multi-purpose grease where the dust seal fits into it, then install the dust seal.

41 Extend and compress the fork tubes, checking for rough or uneven movement. If this is found, find out what the problem is before continuing.

42 Measure the length of the exposed threads between the damper rod locknut and the end of the damper rod. There should be at least 19 mm (0.75 inch) of exposed threads.

43 Install the collar and fork spring on the damper. The larger end of the collar faces the spring (see illustration 5.13c).

44 Hold the fork leg at an angle, with its open end (top end) facing down. Install the damper rod in the fork tube. Pull the damper rod end out of the bottom of the fork tube and secure it with a holding tool (see illustration).

45 Install the pushrod in the damper rod (see illustration 5.7).

46 Loosen the screw in the center of the rebound damping adjuster until it's finger-tight, counting the turns. Thread the damping adjuster onto the damper rod, using fingers only (see illustration 5.6b). Check the gap between the adjuster and the locknut on the damper rod - it should be 0.5 to 1.0 mm (0.02 to 0.04 inch).

47 Hold the locknut with a wrench and tighten the damping adjuster against it to the torque listed in this Chapter's Specifications.

48 Thread the damping adjuster into the bottom of the fork, then tighten it to the torque listed in this Chapter's Specifications.

49 Pour the recommended amount of fork oil into the top of the fork (see illustration).

50 Thread the cartridge into the upper end of the fork leg and tighten it temporarily. Tighten it to the torque listed in this Chapter's Specifications after securing it in the triple clamps.

51 If you removed the guide for the fork leg protector, slip it onto the fork tube with its wider ring downward.

5.44 Slip the support tool under the locknut

5.49 Pour the specified amount of oil into the outer fork tube

6.3a Loosen the steering stem nut . . .

6.3b . . . and unscrew it from the steering stem . . .

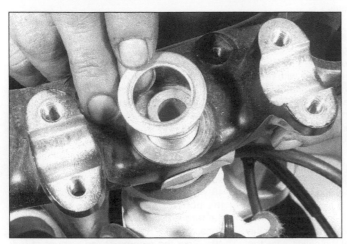

6.3c . . . then lift off the washer and upper triple clamp

6.4a Remove the bearing adjusting nut(s) . . .

6 Steering head bearings - replacement

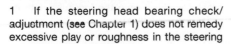

1   If the steering head bearing check/adjustment (see Chapter 1) does not remedy excessive play or roughness in the steering head bearings, the entire front end must be disassembled and the bearings and races replaced with new ones.

2   Remove the handlebars (see Section 2), the front wheel and brake caliper (see Chapter 7), the front fender (see Chapter 8) and the forks (see Section 3).

3   Loosen the steering stem nut with a socket (see illustration). Remove the nut, washer and upper triple clamp (see illustrations).

4   Using a spanner wrench of the type described in Chapter 1, remove the stem locknut(s) and bearing cover (see illustrations) while supporting the steering head from the bottom.

5   Remove the steering stem and lower triple clamp assembly (see illustration). If it's

6.4b . . . and lift off the bearing cover

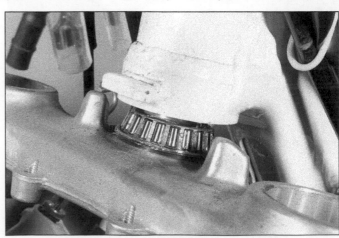

6.5 Lower the steering stem out of the steering head

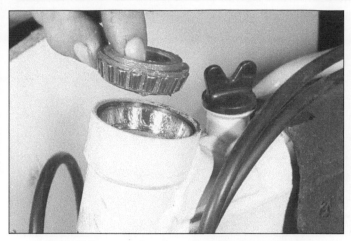

6.6 Lift the upper bearing out of the steering head

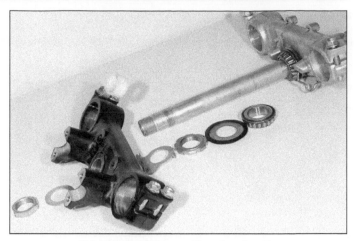

6.7 Steering stem and bearing details

6.9 Place the drift against the edge of the lower bearing race (arrow) and tap evenly around it to drive the bearing out

6.11 Leave the lower bearing and grease seal (arrow) on the steering stem unless you plan to replace them

stuck, gently tap on the top of the steering stem with a plastic mallet or a hammer and a wood block.

6   Remove the upper bearing (see illustration).

7   Clean all the parts with solvent and dry them thoroughly, using compressed air, if available (see illustration). If you do use compressed air, don't let the bearings spin as they're dried - it could ruin them. Wipe the old grease out of the frame steering head and bearing races.

8   Examine the races in the steering head for cracks, dents, and pits. If even the slightest amount of wear or damage is evident, the races should be replaced with new ones.

9   To remove the races, drive them out of the steering head with a bearing driver or a hammer and drift punch (see illustration). A slide hammer with the proper internal-jaw puller will also work. Since the races are an interference fit in the frame, installation will be easier if the new races are left overnight in a refrigerator. This will cause them to contract and slip into place in the frame with very little effort. When installing the races,

use a bearing driver the same diameter as the outer race, or tap them gently into place with a hammer and punch or a large socket. Do not strike the bearing surface or the race will be damaged.

10   Check the bearings for wear. Look for cracks, dents, and pits in the races and flat spots on the bearings. Replace any defective parts with new ones. If a new bearing is required, replace both of them as a set.

11   Check the grease seal under the lower bearing and replace it with a new one if necessary (see illustration).

12   To remove the lower bearing and grease seal from the steering stem, you may need to use a bearing puller, which can be rented. Don't remove this bearing unless it, or the grease seal underneath, must be replaced. Removal will damage the grease seal, so replace it whenever the bearing is removed.

13   Inspect the steering stem/lower triple clamp for cracks and other damage. Do not attempt to repair any steering components. Replace them with new parts if defects are found.

14   Pack the bearings with high-quality grease (preferably a moly-based grease) (see illustration). Coat the outer races with grease also.

15   Install the grease seal and lower bearing onto the steering stem. Drive the lower

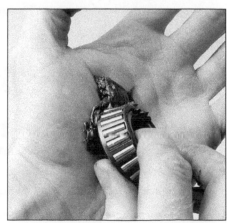

6.14 Work the grease completely into the bearing

7.3 Pad the frame with tape or a rag so the shock reservoir won't scratch it during removal

7.4a Remove the nut, washer and bolt at the top of the shock, then unbolt the lower end from the shock linkage

bearing onto the steering stem using a pipe the same diameter as the bearing inner race. Drive the bearing on until it's fully seated.

16 Insert the steering stem/lower triple clamp into the frame head. Install the upper bearing, bearing cover and adjusting nut. Refer to the adjustment procedure in Chapter 1 and tighten the adjusting nut to the torque listed in the Chapter 1 Specifications.

17 Make sure the steering head turns smoothly and that there's no play in the bearings.

18 Install the upper triple clamp, then install the washer and nut, but don't tighten the nut yet.

19 Slide the forks through the lower triple clamp and into the upper triple clamp, then install the front wheel (this is to align the upper triple clamp). Tighten the lower triple clamp bolts securely, then install the steering stem washer and nut or bolt. Tighten the nut or bolt to approximately half the torque listed in this Chapter's Specifications to seat the

upper triple clamp on the steering stem.

20 Loosen the lower triple clamp bolts and install the forks to their proper height (see Section 3 and this Chapter's Specifications), then tighten the upper and lower triple clamp bolts to the torque listed in this Chapter's Specifications.

21 Tighten the steering stem nut or bolt to its full torque.

7 Rear shock absorber - removal, inspection and installation

Removal

1 Support the bike securely so it can't be knocked over during this procedure. Support the swingarm with a jack so the suspension can be raised or lowered, and to prevent the bike from sagging when the shock absorber is removed.

2 Remove the muffler (see Chapter 4).

3 Remove the seat and sub-frame (see Chapter 8).

HAYNES HiNT *To keep from scratching the frame and shock reservoir, cover the right side of the frame with duct tape or a rag* (see illustration).

4 Remove the upper and lower mounting bolts (see illustrations). **Note:** *On early models, the bolts have hex heads; on later models, they have round heads with a flat on one side, which fits into a matching flat in the component it attaches to.*

5 Remove the shock from the bike.

Inspection

6 The shock absorber can be overhauled, but it's a complicated procedure that requires special tools not readily available to the typical owner. If inspection reveals problems, have the shock rebuilt by a dealer or motorcycle repair shop.

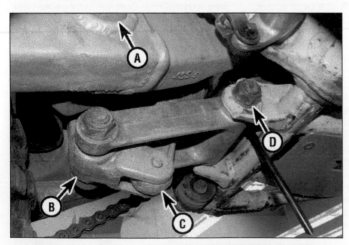

7.4b Shock linkage details (1998 through 2001 models)

A Access hole plug - relay arm-to-swingarm bolt
B Relay arm
C Shock absorber lower end
D Connecting rod-to-frame attaching point

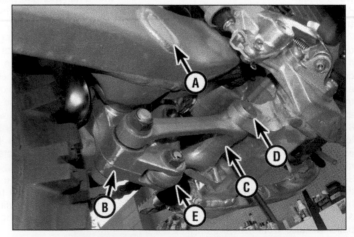

7.4c Shock linkage details (2002 and later)

A Access hole plug - relay arm-to-swingarm bolt
B Relay arm
C Connecting rod
D Connecting rod-to-frame attaching point
E Shock absorber lower end

7.8 Check the shock absorber bearings or bushings for wear or damage

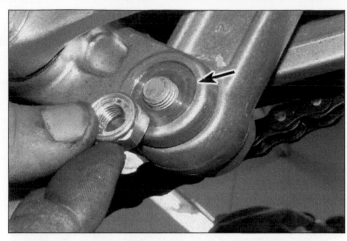

8.4 Remove the nuts - pry the washers out if necessary

7   Check the shock absorber for damage and oil leaks. If these can be seen, have the shock overhauled.

8   Check the spherical bearing at the upper end of the shock for leaking grease, looseness or signs of damage (see illustration). If any of these problems can be seen, have the bearing pressed out and a new one pressed in by a dealer or motorcycle repair shop.

9   Clean all parts thoroughly with solvent and dry them with compressed air, if available. Check all parts for scoring, damage or heavy corrosion and replace them as necessary.

Installation

10   Installation is the reverse of the removal steps. Tighten the bolts to the torques listed in this Chapter's Specifications.

8   Shock linkage - removal, inspection and installation

1   The shock linkage consists of two parts, a relay arm and a connecting rod. The shock

absorber lower end is attached to the relay arm, which in turn is attached to the connecting rod. The connecting rod is connected to the swingarm and frame.

2   The connecting rod on all models is made in one piece.

Removal

3   Support the bike securely so it can't be knocked over during this procedure. Support the swingarm with a jack so the suspension can be raised or lowered as needed for access.

4   Remove the shock absorber lower mounting bolt. Unbolt the linkage from the swingarm and frame (see illustrations 7.4b, 7.4c and the accompanying illustration).

5   Take the linkage out of the bike, then separate the relay arm from the connecting rod.

Inspection

6   Slip the collars out of the needle bearings (see illustrations). Check the bearings for wear or damage. If the bearings are okay, pack them with molybdenum disulfide grease and reinstall the collars.

7   If the dust seals are worn or appear to have been leaking, pry them out and press in new ones with a seal driver or socket the same diameter as the seals.

8   If the bearings need to be replaced, press them out and press new ones in. To prevent damage to the new bearings, you'll need a shouldered drift with a narrow diameter the same size as the inside diameter of the bearings. If you don't have the proper tool, have the bearings replaced by a Yamaha dealer or other motorcycle repair shop. A well-equipped automotive machine shop should also be able to do the job.

Installation

9   Installation is the reverse of the removal steps. Lubricate the bearings and thrust washers with molybdenum disulfide grease. Tighten the nuts and bolts to the torques listed in this Chapter's Specifications. Connections are as follows:

a) Relay arm rear hole to swingarm (rear hole upward) (see illustration 7.4a or 7.4b)

b) Relay arm center hole to connecting rod rear ends (separate ends)

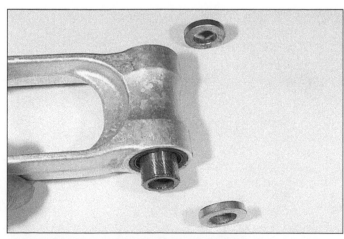

8.6a Pull the covers off the shock link pivots and inspect the seals and needle bearings

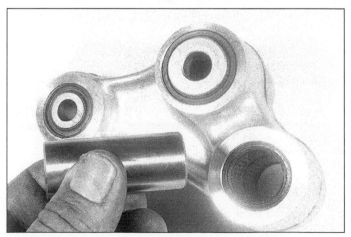

8.6b Pull out the collar to inspect the needle roller bearings

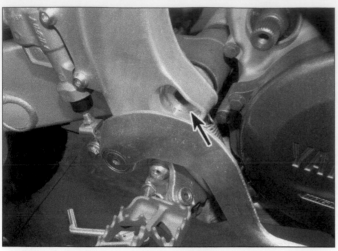

10.5a The swingarm pivot bolt head fits into a shaped recess in the frame (arrow)

10.5b With the engine supported, remove the pivot bolt nut and washer and pull out the bolt

c) Relay arm front hole to shock absorber
d) Connecting rod front end to frame

9 Swingarm bearings - check

1   Refer to Chapter 7 and remove the rear wheel, then refer to Section 7 and remove the rear shock absorber.
2   Grasp the rear of the swingarm with one hand and place your other hand at the junction of the swingarm and frame. Try to move the rear of the swingarm from side-to-side. Any wear (play) in the bushings should be felt as movement between the swingarm and the frame at the front. The swingarm will actually be felt to move forward and backward at the front (not from side-to-side). If any play is noted, the bearings should be replaced with new ones (see Section 11).
3   Next, move the swingarm up and down through its full travel. It should move freely, without any binding or rough spots. If it doesn't move freely, refer to Section 11 for servicing procedures.

10 Swingarm - removal and installation

1   Refer to Section 12 and disconnect the drive chain.
2   Remove the rear wheel and the brake pedal (see Chapter 7). Detach the brake hose from the retainer on the swingarm and support the caliper so it doesn't hang by the hose.
3   Unbolt the shock absorber and shock linkage from the swingarm (see Sections 7 and 8).

4   Check the chain guards on the swingarm for wear or damage and replace them if necessary (see Chapter 1).
5   Support the swingarm from below, then remove its pivot bolt nut and washer and pull the bolt out (see illustrations).
6   Check the chain slider, chain adjuster plates and brake disc guard for wear or damage. Replace them as necessary.
7   Installation is the reverse of the removal steps, with the following additions:

a) Lubricate the swingarm bearings and thrust washers with molybdenum disulfide grease (see Section 11). Make sure the bushings are installed all the way into the swingarm (see illustration 11.2).
b) Tighten the swingarm pivot bolt and nut to the torque listed in this Chapter's Specifications.
c) Refer to Chapter 1 and adjust the drive chain and rear brake pedal (drum brake models).

11 Swingarm bearings - replacement

1   Refer to Section 10 and remove the swingarm.
2   Pull the pivot collars out of the swingarm bearings, then inspect the bearings and seals (see illustration).
3   If the dust seals are worn or appear to have been leaking, pry them out and press in new ones with a seal driver or socket the same diameter as the seals.
4   Check the needle bearings for wear or damage. Needle bearing replacement requires a special puller, a press and a shouldered drift the same diameter as the inside of the bearings. If you don't have these, have the bearings replaced by a Yamaha dealer or other motorcycle repair shop.
5   Coat the bearings and pivot collars with moly-based grease and slip the collar into the swingarm. Install the dust covers.

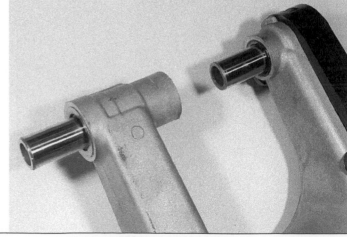

11.2 Slide the pivot collars out

12.1 Remove the clip from the master link; its open end (arrow) faces rearward when the chain is on the top run and forward when the chain is on the bottom run

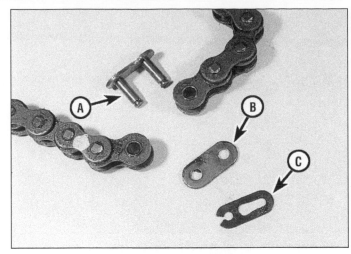

12.2 Master link details

A   Link
B   Plate
C   Clip

12 Drive chain - removal, cleaning, inspection and installation

Removal

1   Turn the rear wheel to place the drive chain master link where it's easily accessible (see illustration).
2   Remove the clip and plate and pull the master link out of the chain (see illustration).
3   Remove the engine sprocket cover (see illustration).
4   Lift the chain off the sprockets and remove it from the bike.

 HAYNES HiNT *If you're removing the chain for access to other components, secure the chain to the top of the swingarm so you can roll the bike without jamming the chain (see illustration).*

5   Check the chain guards and rollers on the swingarm and frame for wear or damage and replace them as necessary.

Cleaning and inspection

6   Soak the chain in a high flash point solvent for approximately five or six minutes. Use a brush to work the solvent into the spaces between the links and plates.
7   Wipe the chain dry, then check it carefully for worn or damaged links. Replace the chain if wear or damage is found at any point.
8   Stretch the chain taut and measure its length between the number of pins listed in this Chapter's Specifications. Compare the measured length to the specified value and replace the chain if it's beyond the limit. If the chain needs to be replaced, refer to Section 13 and check the sprockets. If they're worn, replace them also. If a new chain is installed on worn sprockets, it will wear out quickly.

9   Lubricate the chain with the type of lubricant listed in the Chapter 1 Specifications.

Installation

10   Installation is the reverse of the removal steps, with the following additions:
   a) *Install the master link clip so its opening faces the back of the motorcycle when the master link is in the upper chain run (see illustration 12.1).*
   b) *Refer to Chapter 1 and adjust the chain.*

13 Sprockets - check and replacement

1   Support the bike securely so it can't be knocked over during this procedure.
2   Whenever the sprockets are inspected, the chain should be inspected also and

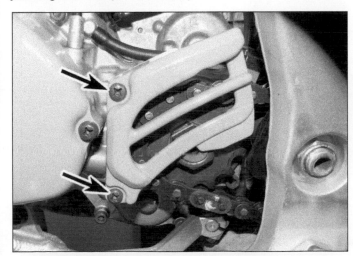

12.3 The engine sprocket is visible through the cover slots - remove the screws (arrows) to take off the cover

12.4 Use a wire-tie to secure the chain to the swingarm

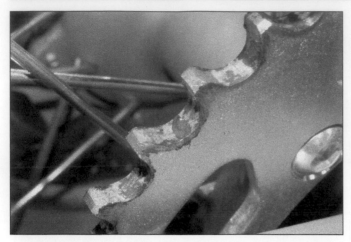

13.3 Inspect the sprocket teeth for excessive or uneven wear

13.6a Bend back the lockwasher tabs . . .

replaced if it's worn. Installing a worn chain on new sprockets will cause them to wear quickly.

3   Check the teeth on the engine sprocket and rear sprocket for wear (see illustration). The engine sprocket is visible through the cover slots.

4   If the sprockets are worn, remove the chain (see Section 12) and the rear wheel (see Chapter 7).

5   Remove the sprocket from the rear wheel hub.

6   Bend back the tabs on the engine sprocket lockwasher (see illustration). Unscrew the nut, remove the lockwasher and take the sprocket off the transmission shaft (see illustrations).

7   Inspect the seal behind the engine sprocket. If it has been leaking, remove the collar and O-ring (see Chapter 2). Pry the seal out (taking care not to scratch the seal bore) and tap in a new seal with a socket the same diameter as the seal.

8   Installation is the reverse of the removal steps, with the following additions:

a) Install the sprocket with its dished side toward the engine.

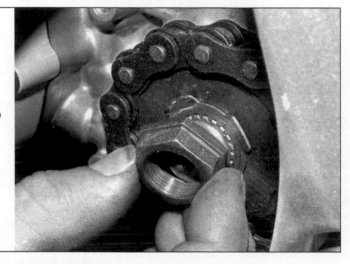

13.6b . . . unscrew the nut and remove the lockwasher . . .

b) Tighten the engine sprocket nut to the torque listed in this Chapter's Specifications.

c) Tighten the driven sprocket bolts to the torque listed in this Chapter's Specifications.

d) Install the master link clip so its opening faces the back of the motorcycle when the master link is in the upper chain run (see illustration 12.1).

e) Refer to Chapter 1 and adjust the chain.

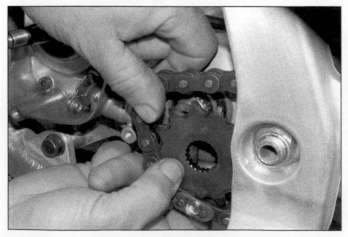

13.6c . . . and slip the sprocket off the transmission shaft

13.6d The dished side of the sprocket faces the engine

Chapter 7
Brakes, wheels and tires

Contents

Degrees of difficulty

Easy, suitable for novice with little experience	**Fairly easy,** suitable for beginner with some experience	**Fairly difficult,** suitable for competent DIY mechanic	**Difficult,** suitable for experienced DIY mechanic	**Very difficult,** suitable for expert DIY or professional

Specifications

Disc brakes

Brake fluid type	See Chapter 1
Brake pad minimum thickness...	See Chapter 1
Front disc thickness	
Standard	3.0 mm (0.12 inch)
Limit*	2.5 mm (0.10 inch)
Rear disc thickness	
1998	
Standard	4.5 mm (0.18 inch)
Limit*...	4.0 mm (0.16 inch)
1999 and later	
Standard	4.0 mm (0.16 inch)
Limit	3.5 mm (0.14 inch)
Disc runout limit	
Front	
1998 through 2000	0.15 mm (0.006 inch)
2001 and later...	Not specified
Rear...	0.15 mm (0.006 inch)

Refer to marks stamped into the disc (they supersede information printed here)

Wheels and tires

Tire pressures..	See Chapter 1
Tire tread depth..	see Chapter 1
Axle runout limit (front and rear)	0.5 mm (0.020 inch)
Wheel out-of-round and lateral runout limit (front and rear)..................	2.0 mm (0.08 inch)

Torque specifications

Front axle	105 Nm (75 ft-lbs)
Front axle pinch bolts..	23 Nm (17 ft-lbs)
Rear axle nut	
1998	115 Nm (85 ft-lbs)
1999 and later	125 Nm (90 ft-lbs)
Front caliper	
Caliper bracket bolts	
All except 2008 YZ250F and YZ450F.........................	23 Nm (17 ft-lbs)
2008 YZ250F and YZ450F................................	28 Nm (20 ft-lbs)
Pad retaining pins	18 Nm (13 ft-lbs)
Rear caliper	
Pad retaining pins	18 Nm (13 ft-lbs)
Pad pin plugs	3 Nm (26 inch-lbs)
Brake hose union bolts	
1998 and 1999	26 Nm (19 ft-lbs)
2000 and later	30 Nm (22 ft-lbs)
Brake disc-to-wheel bolts*	
Front	12 Nm (108 inch-lbs)
Rear	14 Nm (120 inch-lbs)
Front master cylinder	
Mounting bolts	9 Nm (78 inch-lbs)
Brake lever pivot bolt	
1998 through 2000	7 Nm (61 inch-lbs)
2001 and later..	6 Nm (52 inch-lbs)
Brake lever pivot bolt locknut	
1998 through 2000	7 Nm (61 inch-lbs)
2001 and later..	6 Nm (52 inch-lbs)
Rear master cylinder mounting bolts	10 Nm (86 in-lbs)
Pedal pivot bolt	
1998 through 2000......................................	19 Nm (13 ft-lbs)
2001 and later	26 Nm (19 ft-lbs)

*Apply non-permanent thread locking agent to the threads.

1 General information

The front wheel on all motorcycles covered by this manual is equipped with a hydraulic disc brake using a pin slider caliper. The front caliper has dual pistons. The rear wheel on all models is equipped with a hydraulic disc brake using a pin slider caliper and a single piston.

Caution: Disc brake components rarely require disassembly. Do not disassemble components unless absolutely necessary. If any hydraulic brake line connection in the system is loosened, the entire system should be disassembled drained and cleaned, then properly filled and bled upon reassembly. Do not use solvents on internal brake components. Solvents will cause seals to swell and distort. Use only clean brake fluid or alcohol for cleaning.

Use care when working with brake fluid as it can injure your eyes and it will damage painted surfaces and plastic parts.

2 Front brake pads - replacement

⚠ *Warning: The dust created by the brake system is harmful to your health. Yamaha hasn't used asbestos in brake parts for a number of years, but aftermarket parts may contain it. Even so, all brake dust should be considered harmful. Never blow it out with compressed air and don't inhale any of it. An approved filtering mask should be worn when working on the brakes. Use only clean brake fluid, brake system cleaner or denatured (rubbing) alcohol for cleaning brake parts.*

Removal

1 Support the bike securely upright.
2 Unscrew the plug from the pad pin **(see illustration)**. Loosen the pad pin while the caliper is still bolted to the bracket **(see illustration)**.
3 Remove the caliper mounting bolts and lift the caliper off **(see illustration 2.2a)**. Leave the brake hose connected and support the caliper so the hose won't be strained.
4 Take the pads out of the caliper **(see illustrations)**.

Inspection

5 Inspect the pad spring and the steel shield that protects the caliper bracket. Replace the spring or shield it if it's rusted or damaged.
6 Refer to Chapter 1 and inspect the pads.
7 Look for signs of fluid leakage past the pistons. If this has occurred, overhaul the caliper (see Section 4).

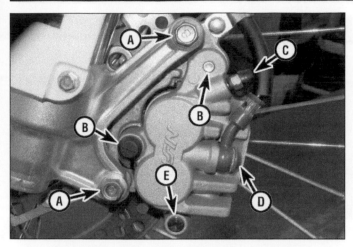

2.2a Front caliper details (2004 and later shown)

A   Caliper mounting bolts   D   Brake hose banjo bolt
B   Slide pins   E   Pad pin plug
C   Bleed valve

2.2b Loosen the pad pin (arrow) while the caliper is still bolted to the bracket . . .

8   Check the condition of the brake disc (see Section 5). If it's in need of replacement, follow the procedure in that Section to remove it. If it's okay, deglaze it with sandpaper or emery cloth, using a swirling motion.

Installation

9   Remove the cover from the master cylinder reservoir and drain out some fluid. Push the piston(s) into the caliper as far as possible, while checking the master cylinder reservoir to make sure it doesn't overflow. If you can't depress the pistons with thumb pressure, try using a C-clamp. If the pistons stick, remove the caliper and overhaul it as described in Section 4.

10   Install the spring, caliper shield and new pads. Coat the threads of the retaining pin(s) with non-permanent thread locking agent and install the retaining pins. Tighten the retaining pins to the torque listed in this Chapter's Specifications.

2.4a . . . and unscrew the pin with an Allen wrench after lifting the caliper off

11   Install the plug over the retaining pin and tighten it to the torque listed in this Chapter's Specifications.

12   Operate the brake lever several times to bring the pads into contact with the disc,

then check the brake fluid level in the master cylinder reservoir, adding the proper type of fluid as necessary (see Chapter 1).

13   Check the operation of the brake carefully before riding the motorcycle.

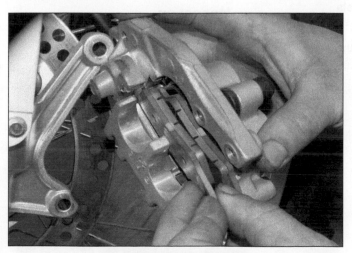

2.4b Slide out the outer pad . . .

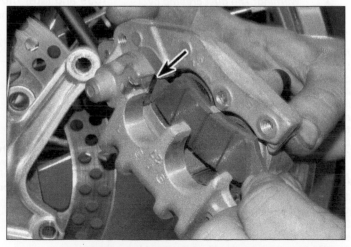

2.4c . . . then slide out the inner pad, noting how the pad ends engage the caliper (arrow)

3.5 Unbolt the caliper shield and lift it off

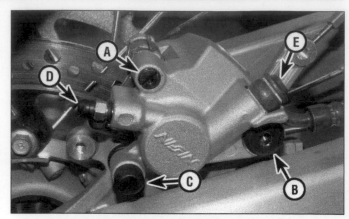

3.6a Rear caliper details (1999 and later shown)

A   Pad pin plug   D   Bleed valve
B   Caliper mounting bolt   E   Brake hose union bolt
C   Slide pin

3.6b Unscrew the pad pin plug . . .

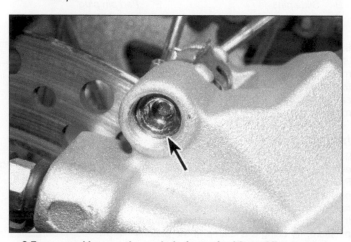

3.7a . . . and loosen the pad pin (arrow) with an Allen wrench

3   Rear brake pads - replacement

⚠ **Warning: The dust created by the brake system is harmful to your health. Yamaha hasn't used asbestos in brake parts for a number of years, but aftermarket parts may contain it. Even so, all brake dust should be considered harmful. Never blow it out with compressed air and don't inhale any of it. An approved filtering mask should be worn when working on the brakes. Use only clean brake fluid, brake system cleaner or denatured (rubbing) alcohol for cleaning brake parts.**

Removal

1998 models

1   Unbolt the caliper shield and lift it off.
2   Loosen the pad pins while the caliper is still bolted to the bracket.
3   Remove the rear wheel (see Section 11). Slide the caliper rearward off its rail on the swingarm. Leave the brake hose connected and support the caliper so the hose won't be strained.
4   Unscrew the pad pins and pull the pads out of the caliper.

1999 and later models

5   Remove the caliper shield (see illustration).
6   Unscrew the plug that covers the pad retaining pin (see illustrations).
7   Unscrew the pad retaining pin with an Allen wrench (see illustrations).
8   Pull the pads and shim out of the caliper (see illustrations).

Inspection

9   Inspection is the same as for front pads (see Section 2). Look inside the caliper to

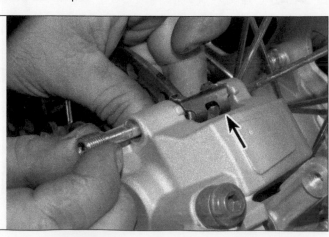

3.7b Push back the pad spring to relieve pressure on the pin and pull out the pin

3.8a Lift the inner pad up and out . . .

3.8b . . . then remove the outer pad together with its shim (arrow)

inspect the pad spring (see illustration). If it's worn, corroded or damaged, replace it (on some models, you'll need to remove the caliper for access).

Installation

10   Installation is the reverse of the removal steps. Lubricate the pad pin(s) with high temperature grease before installing them (see illustration).

11   Operate the brake pedal several times to bring the pads into contact with the disc, then check the brake fluid level in the master cylinder reservoir, adding the proper type of fluid as necessary (see Chapter 1).

12   Check the operation of the brakes carefully before riding the motorcycle.

4   Brake caliper - removal, overhaul and installation

⚠️ *Warning: If a caliper indicates the need for an overhaul (usually due to leaking fluid or sticky operation), all old brake fluid must be flushed from the system. Also, the dust created by the brake system is harmful to your health.*

Never blow it out with compressed air and don't inhale any of it. An approved filtering mask should be worn when working on the brakes. Do not, under any circumstances, use petroleum-based solvents to clean brake parts. Use clean brake fluid or denatured alcohol only!

Note: *If you are removing the caliper only to remove the front forks or rear swingarm, don't disconnect the hose from the caliper.*

Removal

1   Support the bike securely upright. Note: *If you're planning to disassemble the caliper, read through the overhaul procedure, paying particular attention to the steps involved in removing the pistons with compressed air. If you don't have access to an air compressor, you can use the bike's hydraulic system to force the pistons out instead. To do this, remove the pads and pump the brake lever or pedal. If one piston in a dual-piston caliper comes out before the other, push it back into its bore and hold it in with a C-clamp while pumping the brake lever to remove the remaining piston.*

Front caliper

2   Note: *Remember, if you're just removing the caliper to remove the forks, ignore*

3.9 Check the pad spring (arrow) and replace it if it's worn, damaged or corroded

this step. Disconnect the brake hose from the caliper. Remove the brake hose banjo fitting bolt and separate the hose from the caliper (see illustration). Discard the sealing washers. Plug the end of the hose or wrap a plastic bag tightly around it to prevent excessive fluid loss and contamination.

3   Unscrew the caliper mounting bolts and lift it off the fork leg, being careful not to

3.10 Lubricate the pad pin before you install it

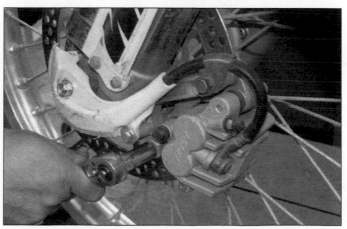

4.2 Unbolt the caliper from the bracket . . .

4.3 . . . and lift it off - check for signs of fluid leakage from the pistons (arrows)

4.6 Slide the caliper backward off its rail on the swingarm

strain or twist the brake hose if it's still connected **(see illustration)**.

Rear caliper

4   If you're planning to overhaul the caliper, remove its protective bracket and loosen the brake hose union bolt (it's easier to loosen the bolt while the caliper is mounted on the bike) **(see illustration 3.6a)**.

5   If you're working on a 1998 model, caliper removal is part of the pad removal procedure (see Section 3).

6   If you're working on a 1999 or later model, refer to Section 11 and remove the rear wheel. Slide the caliper bracket backward off its rail **(see illustration)**. Support the caliper so it doesn't hang by the brake hose.

7   Disconnect the brake hose from the caliper if you haven't already done so. Remove the brake hose banjo fitting bolt and separate the hose from the caliper. Discard the sealing washers. Plug the end of the hose or wrap a plastic bag tightly around it to prevent excessive fluid loss and contamination.

Overhaul

8   Remove the brake pads and anti-rattle spring from the caliper (see Section 2 or 3, if necessary). Clean the exterior of the caliper with denatured alcohol or brake system cleaner.

9   Slide the caliper off the bracket **(see illustration)**.

10   Pack a shop rag into the space that holds the brake pads. Use compressed air, directed into the caliper fluid inlet, to remove the piston(s). Use only enough air pressure to ease the piston(s) out of the bore. If a piston is blown out forcefully, even with the rag in place, it may be damaged.

⚠️ **Warning: Never place your fingers in front of the piston in an attempt to catch or protect it when applying compressed air, as serious injury could occur.**

11   Using a wood or plastic tool, remove the piston seals **(see illustration)**. Metal tools may cause bore damage.

12   Clean the pistons and the bores with denatured alcohol, clean brake fluid or brake system cleaner and blow dry them with filtered, unlubricated compressed air. Inspect the surfaces of the pistons for nicks and burrs and loss of plating. Check the caliper bores, too. If surface defects are present, the caliper must be replaced.

13   If the caliper is in bad shape, the master cylinder should also be checked.

14   Lubricate the piston seals with clean brake fluid and install them in their grooves in the caliper bore **(see illustration)**. Make sure they seat completely and aren't twisted.

15   Lubricate the dust seals with clean brake fluid and install them in their grooves, making sure they seat correctly.

16   Lubricate the piston (both pistons on front calipers) with clean brake fluid and install it into the caliper bore. Using your thumbs, push the piston all the way in, making sure it doesn't get cocked in the bore.

17   Pull the old pin boots out of the caliper and bracket. Coat new ones with silicone

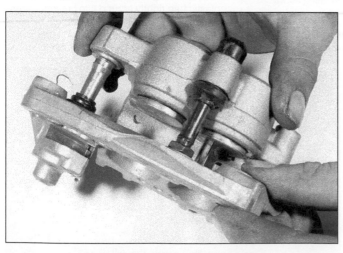

4.9 Slide the caliper off the bracket

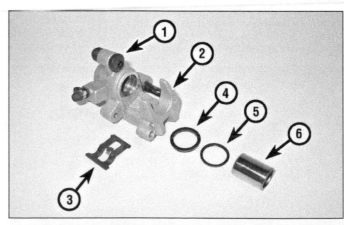

4.11 Brake caliper details (typical rear shown, front similar but with an additional piston)

1	Pin boot	4	Piston seal
2	Caliper body	5	Dust seal
3	Pad spring	6	Piston

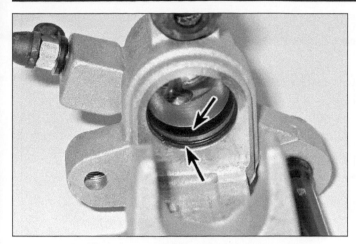

4.14 Fit the seals all the way into their grooves

4.17 Install the boot in the front caliper with its wide end facing the same direction as the piston(s)

grease and install them, making sure they seat completely **(see illustration)**.

18   Make sure the shields are in position on the caliper brackets.

Installation

19   Installation is the reverse of the removal steps, with the following additions:

a) *Apply silicone grease to the slider pins on the caliper bracket and caliper.*

b) *Use new sealing washers on the brake hose fitting. Position the brake hose fitting in the caliper notch or against the stop.*

c) *Tighten the caliper mounting bolts (if equipped), rear caliper shield bolts and brake line union bolt to the torque listed in this Chapter's Specifications.*

d) *If you're working on a 1998 model rear caliper, adjust the chain slack (see Chapter 1).*

20   Fill the master cylinder with the recommended brake fluid (see Chapter 1) and bleed the system (see Section 10). Check for leaks.

21   Check the operation of the brakes carefully before riding the motorcycle.

---

5   Brake disc(s) - inspection, removal and installation

Inspection

1   Support the bike securely upright. Place a jack beneath the bike and raise the wheel being checked off the ground. Be sure the bike is securely supported so it can't be knocked over.

2   Visually inspect the surface of the disc(s) for score marks and other damage. Light scratches are normal after use and won't affect brake operation, but deep grooves and heavy score marks will reduce braking efficiency and accelerate pad wear. If the discs are badly grooved they must be replaced (machining is not recommended).

3   To check disc runout, mount a dial indicator to a fork leg or the swingarm, with the plunger on the indicator touching the surface of the disc **(see illustration)**. Slowly turn the wheel and watch the indicator needle, comparing your reading with the limit listed in

this Chapter's Specifications. If the runout is greater than allowed, check the hub bearings for play (see Chapter 1). If the bearings are worn, replace them and repeat this check. If the disc runout is still excessive, the disc will have to be replaced.

4   The disc must not be allowed to wear down to a thickness less than the minimum allowable thickness, listed in this Chapter's Specifications. The thickness of the disc can be checked with a micrometer. If the thickness of the disc is less than the minimum allowable, it must be replaced. The minimum thickness is also stamped into the disc **(see illustration)**.

Removal

5   Remove the wheel (see Section 11).

Caution: Don't lay the wheel down and allow it to rest on the disc - the disc could become warped. Set the wheel on a bucket or wood blocks so the disc doesn't support the weight of the wheel.

6   Mark the relationship of the disc to the wheel, so it can be installed in the same posi-

5.3 Set up a dial indicator against the brake disc and turn the wheel to measure runout

5.4 Marks on the disc indicate the minimum thickness and direction of rotation (typical)

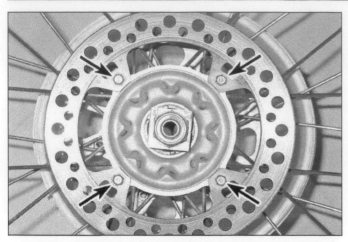

5.6 The rear brake disc is secured to the hub by four bolts (arrows); front discs have six bolts

5.7 When installing the disc, pay attention to the direction of rotation marking

tion. Remove the bolts that retain the disc to the wheel (see illustration). Loosen the bolts a little at a time, in a criss-cross pattern, to avoid distorting the disc.

Installation

7   Position the disc on the wheel, aligning the previously applied matchmarks (if you're reinstalling the original disc). On models so equipped, make sure the arrow (stamped on the disc) marking the direction of rotation is pointing in the proper direction (see illustration).

8   Install the bolts, tightening them a little at a time in a criss-cross pattern, until the torque listed in this Chapter's Specifications is reached. Clean off all grease from the brake disc using acetone or brake system cleaner.

9   Install the wheel.

10   Operate the brake lever or pedal several times to bring the pads into contact with the disc. Check the operation of the brakes carefully before riding the motorcycle.

6   Front brake master cylinder - removal, overhaul and installation

1   If the master cylinder is leaking fluid, or if the lever doesn't produce a firm feel when the brake is applied, and bleeding the brakes doesn't help, master cylinder overhaul is recommended. Before disassembling the master cylinder, read through the entire procedure and make sure that you have the correct rebuild kit. Also, you will need some new, clean brake fluid of the recommended type, some clean rags and internal snap-ring pliers. Note: *To prevent damage to the paint, plastic or graphics from spilled brake fluid, always cover surrounding areas when working on the master cylinder.*

2 Caution: Disassembly, overhaul and reassembly of the brake master cylinder must be done in a spotlessly clean work area to avoid contamination and possible failure of the brake hydraulic system components.

Removal

3   Place rags beneath the master cylinder to protect surrounding areas in case of brake fluid spills.

4   Brake fluid will run out of the upper brake hose during this step, so either have a container handy to place the end of the hose in, or have a plastic bag and rubber band handy to cover the end of the hose. The objective is to prevent excess loss of brake fluid, fluid spills and system contamination.

5   Remove the banjo fitting bolt and sealing washers from the master cylinder.

6   Remove the master cylinder mounting bolts (see illustration). Take the master cylinder off the handlebar.

Overhaul

7   Remove the master cylinder cover, retainer (if equipped) and diaphragm (see Chapter 1).

8   Remove the locknut from the underside of the lever pivot bolt, then unscrew the bolt (see illustrations). Remove the lever pivot cover (it's held on by the pivot bolt).

6.6 Make an alignment mark for the clamp if there isn't one, then remove the bolts - the UP mark and arrow face upward on installation

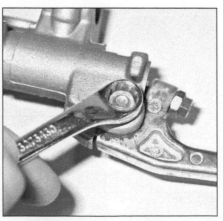

6.8a Remove the locknut . . .

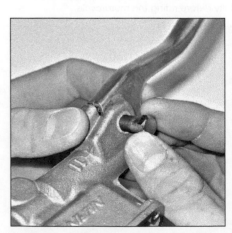

6.8b . . . and unscrew the pivot bolt to detach the lever

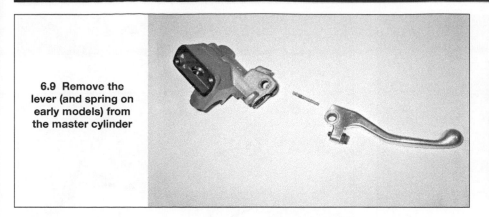

6.9 Remove the lever (and spring on early models) from the master cylinder

6.10 Remove the dust boot . . .

9   On early models, the lever is equipped with a spring (see illustration). This doesn't have to be removed to remove the hydraulic components, but make sure it doesn't get lost.

10   Remove the dust boot (see illustration).

11   Using snap-ring pliers, remove the snap-ring and slide out the piston assembly and the spring (see illustrations). Lay the parts out in the proper order to prevent confusion during reassembly.

12   Clean all of the parts with brake system cleaner (available at auto parts stores), isopropyl alcohol or clean brake fluid.

Caution: Do not, under any circumstances, use a petroleum-based solvent to clean brake parts. If compressed air is available, use it to dry the parts thoroughly (make sure it's filtered and unlubricated).

Check the master cylinder bore and piston for corrosion, scratches, nicks and score marks. If damage or wear can be seen, the master cylinder must be replaced with a new one. If the master cylinder is in poor condition, then the caliper should be checked as well.

13   If there's a baffle plate in the bottom of the reservoir, make sure it's securely held by its retainer (see illustration).

14   Yamaha supplies a new piston in its rebuild kits. If the cup seals are not installed on the new piston, install them, making sure the lips face away from the lever end of the piston (see illustration 6.11b). Use the new piston regardless of the condition of the old one.

15   Before reassembling the master cylinder, soak the piston and the rubber cup seals in clean brake fluid for ten or fifteen minutes. Lubricate the master cylinder bore with clean

brake fluid, then carefully insert the piston and related parts in the reverse order of disassembly. Make sure the lips on the cup seals do not turn inside out when they are slipped into the bore.

16   Depress the piston, then install the snap-ring (make sure the snap-ring is prop-

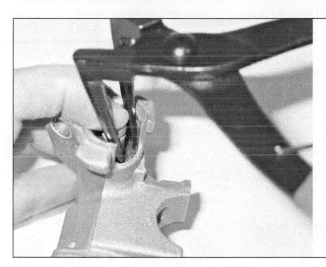

6.11a . . . then remove the snap-ring from the master cylinder bore

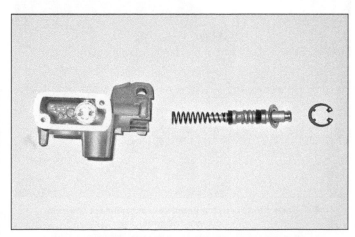

6.11b Remove the piston assembly from the bore

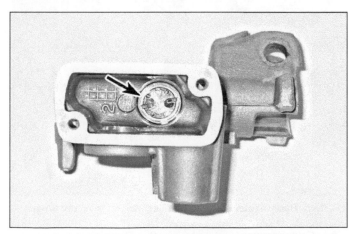

6.13 Make sure the baffle plate is securely retained in the bottom of the reservoir

6.18 The pivot cover fits like this, with the drain hole downward

7.4a Rear master cylinder details (separate reservoir type)

A   Clevis pin   C   Brake hose banjo bolt
B   Mounting bolts

erly seated in the groove with the sharp edge facing out). Install the rubber dust boot (make sure the lip is seated properly in the piston groove).

17   Install the brake lever, pivot cover and pivot bolt. Tighten the pivot bolt locknut.

Installation

18   Installation is the reverse of the removal steps, with the following additions:

  a) Attach the master cylinder to the han-dlebar. Align the upper gap between the master cylinder and clamp with the punch mark on the handlebar.
  b) Make sure the arrow and the word UP on the master cylinder clamp are pointing up (see illustration 6.6), then tighten the bolts to the torque listed in this Chapter's Specifications.

  c) Use new sealing washers at the brake hose banjo fitting. Tighten the union bolt to the torque listed in this Chapter's Specifications.
  d) Install the dust cover over the lever and pivot with its drain hole downward (see illustration).

19   Refer to Section 10 and bleed the air from the system.

---

7   Rear brake master cylinder - removal, overhaul and installation

1   If the master cylinder is leaking fluid, or if the pedal does not produce a firm feel when the brake is applied, and bleeding the brake does not help, master cylinder over-

haul is recommended. Before disassembling the master cylinder, read through the entire procedure and make sure that you have the correct rebuild kit. Also, you will need some new, clean brake fluid of the recommended type, some clean rags and internal snap-ring pliers.

2 Caution: Disassembly, overhaul and reassembly of the brake master cylinder must be done in a spotlessly clean work area to avoid contamination and possible failure of the brake hydraulic system components.

Removal

3   Support the bike securely upright.
4   Remove the cotter pin from the clevis pin on the master cylinder pushrod (see illustrations). Remove the clevis pin.
5   Have a container and some rags ready

7.4b Rear master cylinder details (integral reservoir type)

A   Cover screws   D   Clevis pin
B   Brake hose union bolt   E   Fluid level window
C   Mounting bolts

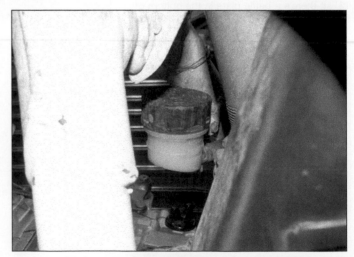

7.6 Early models use a separate reservoir for the rear master cylinder

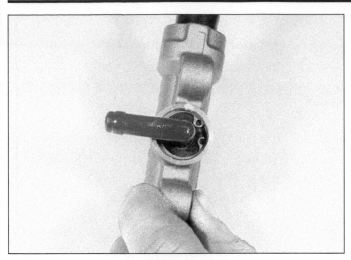

7.8a Remove the snap-ring . . .

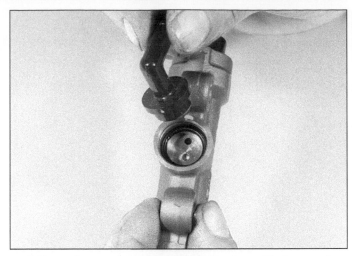

7.8b . . . then work the fluid feed fitting free of its bore and remove the O-ring

to catch spilling brake fluid. Using a six-point box wrench, unscrew the banjo fitting bolt from the top of the master cylinder. Discard the sealing washers on either side of the fitting.

6   On early models with a separate reservoir, squeeze the fluid feed hose clamp with pliers and slide the clamp up the hose (see illustration). Disconnect the hose from the fitting on the master cylinder. If necessary, unbolt the reservoir from the frame and take it out.

7   Remove the two master cylinder mounting bolts and detach the cylinder from the frame (see illustration 7.4a or 7.4b).

Overhaul

8   On early models, use a pair of snap-ring pliers to remove the snap-ring from the fluid inlet fitting and detach the fitting from the master cylinder. Remove the O-ring from the bore (see illustrations).

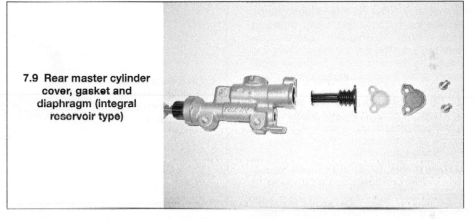

7.9 Rear master cylinder cover, gasket and diaphragm (integral reservoir type)

9   On later models, remove the cover, gasket and diaphragm from the master cylinder (see illustration).

10   Count the number of exposed threads on the end of the pushrod inside the clevis, or measure the distance between the cen-

terline of the lower mounting bolt hole and the clevis pin hole (see illustration). Write this number down for use on assembly. Hold the clevis with a pair of pliers and loosen the locknut, then unscrew the clevis and locknut from the pushrod (see illustration).

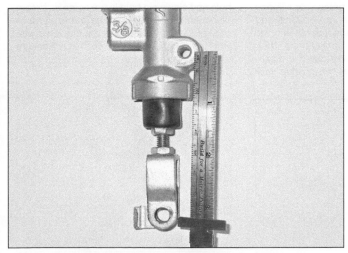

7.10a Measure the length of the pushrod

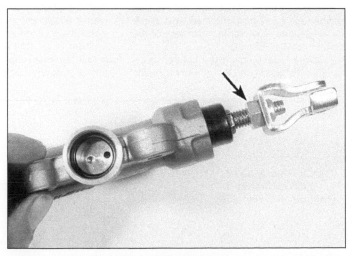

7.10b Loosen the locknut and unscrew the locknut and clevis from the pushrod

7.11a Take the dust boot off the pushrod

7.11b Remove the snap-ring from the master cylinder bore and withdraw the piston assembly and spring

11 Remove the rubber dust boot from the master cylinder bore (see illustration). Depress the pushrod and, using snap-ring pliers, remove the snap-ring (see illustration). Slide out the piston, the cup seal and spring. Lay the parts out in the proper order to prevent confusion during reassembly (see illustrations).

12 Clean all of the parts with brake system cleaner, isopropyl alcohol or clean brake fluid. If compressed air is available, use it to dry the parts thoroughly (make sure it's filtered and unlubricated).

Caution: Do not, under any circumstances, use a petroleum-based solvent to clean brake parts.

Check the master cylinder bore for corrosion, scratches, nicks and score marks. If damage is evident, the master cylinder must be replaced with a new one. If the master cylinder is in poor condition, then the caliper should be checked as well.

13 Yamaha supplies a new piston in its rebuild kits. If the cup seals are not installed on the new piston, install them, making sure the lips face away from the lever end of the piston. Use the new piston regardless of the condition of the old one.

14 Before reassembling the master cylinder, soak the piston and the rubber cup seals

7.11c Rear master cylinder details (separate reservoir type)

in clean brake fluid for ten or fifteen minutes. Lubricate the master cylinder bore with clean brake fluid, then carefully insert the parts in the reverse order of disassembly. Make sure the lips on the cup seals do not turn inside out when they are slipped into the bore.

15 Lubricate the end of the pushrod with PBC (poly butyl cuprysil) grease, or silicone grease designed for brake applications, and install the pushrod and stop washer into the cylinder bore. Depress the pushrod, then install the snap-ring (make sure the snap-ring is properly seated in the groove with

the sharp edge facing out) (see illustration). Install the rubber dust boot (make sure the lip is seated properly in the groove in the piston stop nut).

16 Install the locknut and clevis to the end of the pushrod, making sure the dimension between the master cylinder lower mounting hole and clevis pin hole is the same as before disassembly, or leaving the same number of exposed threads inside the clevis as was written down during removal. This will ensure

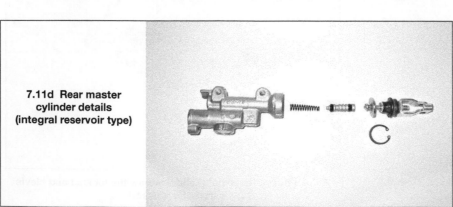

7.11d Rear master cylinder details (integral reservoir type)

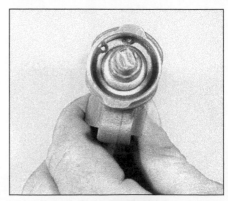

7.15 Make sure the snap-ring is securely seated in its groove

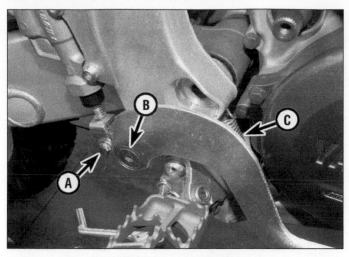
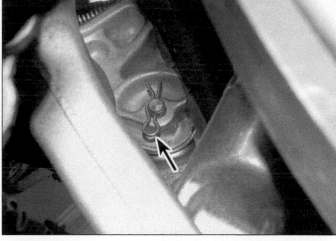

8.2 Brake pedal details

8.4a Remove the clip (if equipped) from the inner end of the pedal shaft

A   Clevis pin   C   Pedal spring
B   Pedal pivot bolt

the brake pedal will be positioned correctly.
17   Install the feed hose fitting, using a new O-ring. Install the snap-ring, making sure it seats properly in its groove.

Installation
18   On early models, install the fluid reservoir if it was removed. Connect the fluid feed hose to the fitting on the master cylinder and secure it with the clamp.
19   Position the master cylinder on the frame and install the bolts, tightening them to the torque listed in this Chapter's Specifications.
20   Connect the banjo fitting to the top of the master cylinder, using new sealing washers on each side of the fitting. Tighten the banjo fitting bolt to the torque listed in this Chapter's Specifications.
21   Connect the clevis to the brake pedal and secure the clevis pin with a new cotter pin.
22   Fill the fluid reservoir with the specified fluid (see Chapter 1) and bleed the system

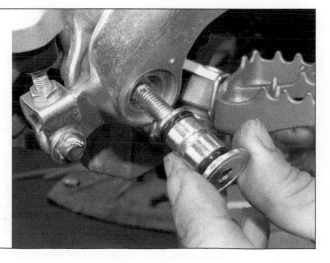

8.4b Unscrew the pedal shaft - use new O-rings on installation

following the procedure in Section 10.
23   Check the position of the brake pedal (see Chapter 1) and adjust it if necessary. Check the operation of the brakes carefully before riding the motorcycle.

8   Brake pedal - removal and installation

1   Support the bike securely upright so it can't be knocked over during this procedure.
2   Unhook the pedal return spring from the frame (see illustration).
3   Remove the cotter pin and clevis pin to detach the master cylinder pushrod from the pedal (see Section 7).
4   Remove the clip from the inner end of the pivot shaft (if equipped) (see illustration). Unscrew the brake pedal pivot shaft and remove the pedal (see illustration).

5   Inspect the pedal pivot shaft seals. If they're worn, damaged or appear to have been leaking, pry them out and press in new ones.
6   Installation is the reverse of the removal steps, with the following additions:
a)   Lubricate the pedal shaft or pivot bolt O-rings and the pivot hole with multi-purpose grease.
b)   Tighten the pedal pivot shaft to the torque listed in this Chapter's Specifications.
c)   Use a new cotter pin.
d)   Refer to Chapter 1 and adjust brake pedal height.

9   Brake hoses and lines - inspection and replacement

Inspection
1   Check the condition of the brake hoses at the intervals specified in Chapter 1.
2   Twist and flex the rubber hoses while looking for cracks, bulges and seeping fluid. Check extra carefully around the areas where the hoses connect with the metal fittings, as these are common areas for hose failure.

Replacement
3   The pressurized brake hoses have banjo fittings on each end of the hose. The fluid feed hose that connects the rear master cylinder reservoir to the master cylinder on early models is secured by spring clamps.
4   Cover the surrounding area with plenty of rags and unscrew the banjo bolt on either end of the hose. Detach the hose or line from any clips that may be present and remove

9.4a Make sure the front brake hose retainers (arrows) are securely attached

9.4b The rear brake hose runs along the swingarm - make sure its retainers are securely attached

the hose (see illustrations).

5   Position the new hose or line, making sure it isn't twisted or otherwise strained. On hoses equipped with banjo fittings, make sure the metal tube portion of the banjo fitting is located against the stop on the component it's connected to, if equipped. Install the banjo bolts, using new sealing washers on both sides of the fittings, and tighten them to the torque listed in this Chapter's Specifications.

6   Flush the old brake fluid from the system, refill the system with the recommended fluid (see Chapter 1) and bleed the air from the system (see Section 10). Check the operation of the brakes carefully before riding the motorcycle.

10 Brake system bleeding

1   Bleeding a brake is simply the process of removing all the air bubbles from the brake fluid reservoir, the lines and the brake caliper. Bleeding is necessary whenever a brake system hydraulic connection is loosened, when a component or hose is replaced, or

when the master cylinder or caliper is overhauled. Leaks in the system may also allow air to enter, but leaking brake fluid will reveal their presence and warn you of the need for repair.

2   To bleed a brake, you will need some new, clean brake fluid of the recommended type (see Chapter 1), a length of clear vinyl or plastic tubing, a small container partially filled with clean brake fluid, some rags and a wrench to fit the brake caliper bleeder valve.

3   Cover the fuel tank and other painted components to prevent damage in the event that brake fluid is spilled.

4   Remove the reservoir cover or cap and slowly pump the brake lever or pedal a few times, until no air bubbles can be seen floating up from the holes at the bottom of the reservoir. Doing this bleeds the air from the master cylinder end of the line. Reinstall the reservoir cover or cap.

5   Attach one end of the clear vinyl or plastic tubing to the brake caliper bleeder valve and submerge the other end in the brake fluid in the container (see illustrations 2.2a and 3.6a).

6   Check the fluid level in the reservoir. Do not allow the fluid level to drop below the lower mark during the bleeding process.

7   Carefully pump the brake lever or pedal three or four times and hold it while opening the caliper bleeder valve. When the valve is opened, brake fluid will flow out of the caliper into the clear tubing and the lever will move toward the handlebar or the pedal will move down.

8   Retighten the bleeder valve, then release the brake lever or pedal gradually. Repeat the process until no air bubbles are visible in the brake fluid leaving the caliper and the lever or pedal is firm when applied. Remember to add fluid to the reservoir as the level drops. Use only new, clean brake fluid of the recommended type. Never reuse the fluid lost during bleeding.

9   Be sure to check the fluid level in the master cylinder reservoir frequently.

10 Replace the reservoir cover or cap, wipe up any spilled brake fluid and check the entire system for leaks. Note: If bleeding is difficult, it may be necessary to let the brake fluid in the system stabilize for a few hours (it may be aerated). Repeat the bleeding procedure when the tiny bubbles in the system have settled out.

11 Wheels - inspection, removal and installation

Inspection

1   Clean the wheels thoroughly to remove mud and dirt that may interfere with the inspection procedure or mask defects. Make a general check of the wheels and tires as described in Chapter 1.

2   Support the motorcycle securely upright with the wheel to be checked in the air, then attach a dial indicator to the fork slider or the swingarm and position the stem against the side of the rim. Spin the wheel slowly and check the side-to-side (axial) runout of the rim, then compare your readings with the value listed in this Chapter's Specifications. In order to accurately check radial runout

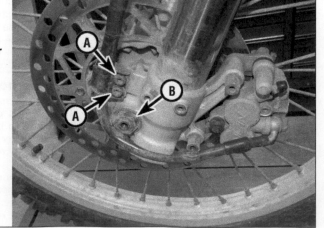

9.4c On 2004 and earlier models, the brake hose runs under the axle, through a groove in the bottom of the fork, to the caliper

A   Brake hose retainer bolts

B   Axle nut

11.8a The axle is secured by pinch bolts and has an external hex head . . .

11.8b . . . or an internal hex head . . .

with the dial indicator, the wheel would have to be removed from the machine and the tire removed from the wheel. With the axle clamped in a vise, the wheel can be rotated to check the runout.

3   An easier, though slightly less accurate, method is to attach a stiff wire pointer to the outer fork tube or the swingarm and position the end a fraction of an inch from the wheel (where the wheel and tire join). If the wheel is true, the distance from the pointer to the rim will be constant as the wheel is rotated. Repeat the procedure to check the runout of the rear wheel. **Note:** *If wheel runout is excessive, refer to Section 13, and check the wheel bearings very carefully before replacing the wheel.*

4   The wheels should also be visually inspected for cracks, flat spots on the rim and other damage. Individual spokes can be replaced. If other damage is evident, the wheel will have to be replaced with a new one. Never attempt to repair a damaged wheel.

5   Before installing the wheel, check the axle for straightness. If the axle is corroded,

first remove the corrosion with fine emery cloth. Set the axle on V-blocks and check it for runout with a dial indicator. If the axle exceeds the maximum allowable runout limit listed in this Chapter's Specifications, it must be replaced.

Removal

Front wheel

6   Support the bike from below with a jack beneath the engine. Securely prop the bike upright so it can't fall over when the wheel is removed.

7   If you're working on a 1008 model, remove the brake hose cover from the left fork leg, then loosen the brake hose retainer bolts.

8   Unscrew the axle pinch bolts **(see illustrations)**.

9   Unscrew the axle from the left fork leg. Hold the axle from turning by placing a socket on the hex (external hex) or a hex bit in the hex (internal hex), then remove the axle nut from the left side **(see illustration 11.8c)**.

11.8c . . . the left side of the axle is secured by pinch bolts and a nut

10   Remove the axle from the right fork leg. Support the wheel and pull the axle out. Lower the wheel away from the motorcycle, sliding the brake disc out from between the pads. Collect the wheel bearing spacers **(see illustrations)**.

11.10a If the wheel spacers are corroded like this, they should be cleaned up or replaced to prevent damage to the wheel bearing seals

11.10b This spacer is covered with sand, which should be removed to protect the seal

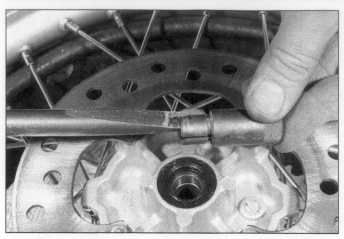

13.6a If you can't position a metal rod against the bearings, this tool can be used instead - place the split portion inside the bearing and pass the wedged rod through the hub into the split; tapping on the end of the rod will spread the split portion, locking it to the bearing, so the split portion and bearing can be driven out together

13.6b The split portion fits into the bearing like this - if it keeps slipping out when you tap on it, coat it with valve grinding compound

Rear wheel

11 Support the bike from below with a jack beneath the swingarm. Securely prop the bike upright so it can't fall over when the wheel is removed.

12 Remove the axle nut and chain adjuster block (see Chapter 1).

13 Support the wheel and pull the axle out, then disengage the drive chain from the rear sprocket. Slide the wheel back until the axle clears the swingarm, then lower the wheel away from the motorcycle.

Installation

14 Installation is the reverse of the removal steps, with the following additions:

a) If you're working on a front wheel, tighten the axle nut to the torque listed in this Chapter's Specifications, then tighten the pinch bolts to the torque listed in this Chapter's Specifications. **Note 1:** Move the bottom of the right fork leg to make sure it's floating in its "neutral" position on the axle (not binding towards or away from the wheel), then tighten the right-side pinch bolts to the torque listed in this Chapter's Specifications. **Note 2:** If you can't move the axle holder on the axle, wedge a small screwdriver into the gap in the holder to spread it, then try again.

c) If you're working on a rear wheel, adjust the drive chain slack (see Chapter 1), then tighten the axle nut to the torque listed in this Chapter's Specifications.

12 Wheels - alignment check

1 Misalignment of the wheels, which may be due to a cocked rear wheel or a bent frame or triple clamps, can cause strange and possibly serious handling problems. If the frame or triple clamps are at fault, repair by a frame specialist or replacement with new parts are the only alternatives.

2 To check the alignment you will need an assistant, a length of string or a perfectly straight piece of wood and a ruler graduated in 1/64 inch increments. A plumb bob or other suitable weight will also be required.

3 Support the motorcycle securely upright, then measure the width of both tires at their widest points. Subtract the smaller measurement from the larger measurement, then divide the difference by two. The result is the amount of offset that should exist between the front and rear tires on both sides.

4 If a string is used, have your assistant hold one end of it about half way between the floor and the rear axle, touching the rear sidewall of the tire.

5 Run the other end of the string forward and pull it tight so that it is roughly parallel to the floor. Slowly bring the string into contact with the front sidewall of the rear tire, then turn the front wheel until it is parallel with the string. Measure the distance from the front tire sidewall to the string.

6 Repeat the procedure on the other side of the motorcycle. The distance from the front tire sidewall to the string should be equal on both sides.

7 As was previously pointed out, a perfectly straight length of wood may be substituted for the string. The procedure is the same.

8 If the distance between the string and tire is greater on one side, or if the rear wheel appears to be cocked, refer to Chapter 6, and make sure the swingarm is tight.

9 If the front-to-back alignment is correct, the wheels still may be out of alignment vertically.

10 Using the plumb bob, or other suitable weight, and a length of string, check the rear wheel to make sure it is vertical. To do this, hold the string against the tire upper sidewall and allow the weight to settle just off the floor. When the string touches both the upper and lower tire sidewalls and is perfectly straight, the wheel is vertical. If it is not, place thin spacers under one leg of the centerstand.

11 Once the rear wheel is vertical, check the front wheel in the same manner. If both wheels are not perfectly vertical, the frame and/or major suspension components are bent.

13 Wheel bearings - inspection and maintenance

Front wheel bearings

1 Support the bike securely and remove the front wheel (see Section 11).

2 Set the wheel on blocks so as not to allow the weight of the wheel to rest on the brake disc.

3 Remove the spacers (if you haven't already done so) from the wheel (see Section 11).

4 Remove the seal from the right side of the wheel **(see illustration 11.10a or 11.10b)**.

5 Turn the wheel over. Pry the grease seal out of the left side.

 You can make the tool described in the next step by cutting a slot in the shaft of a bolt that just fits inside the bearings, and grinding a wedge shape on the end of a metal rod.

6 A common method of removing front wheel bearings is to insert a metal rod (preferably a brass drift punch) through the center

13.6c The tool can be used for front or rear wheel bearings - the wedged rod fits through the hub like this

13.16 Pull out the collar and pry the seal out of its bore

of one hub bearing and tap evenly around the inner race of the opposite bearing to drive it from the hub. The bearing spacer will also come out. On these motorcycles, it may not be possible to tilt the rod enough to catch the edge of the opposite bearing's inner race. In this case, use a bearing remover tool consisting of a shaft and remover head (see illustration). The head fits inside the bearing (see illustration), then the wedge end of the shaft is tapped into the groove in the head to expand the head and lock it inside the bearing. Tapping on the shaft from this point will force the bearing out of the hub (see illustration).

7 Lay the wheel on its other side and remove the remaining bearing using the same technique. **Note:** *The bearings must be replaced with new ones whenever they're removed, as they're almost certain to be damaged during removal.*

8 If you're installing bearings that aren't sealed on both sides, pack the new bearings with grease from the open side. Rotate the bearing to work the grease in between the bearing balls.

9 Thoroughly clean the hub area of the wheel. Install the bearing into the recess in the left side of the hub, with the sealed side facing out. Using a bearing driver or a socket large enough to contact the outer race of the bearing, drive it in until it seats.

10 Turn the wheel over and install the bearing spacer and the other bearing, driving the bearing into place as described in Step 9.

11 Coat the lip of a new grease seal with grease.

12 Install the grease seal; it should go in with thumb pressure but if not, use a seal driver, large socket or a flat piece of wood to drive it into place. Install the grease seal on the other side in the same way.

13 Clean off all grease from the brake disc using acetone or brake system cleaner. Install the wheel.

Rear wheel bearings

14 Refer to Section 11 and remove the rear wheel.

1998 models

15 1998 models use three rear wheel bearings. There are two bearings on the left side and one on the right.

16 Pull the collars out of the seals and pry the seals out (see illustration).

1999 and later models

17 Pull the collars out of the seals and pry the seals out (see illustration).

All models

18 Remove the bearings from the hub, using the tools and methods described in Step 7.

19 Thoroughly clean the hub area of the wheel.

20 Pack the single bearing with grease and install it into the recess in the hub, with the sealed side (if equipped) facing out. Using a bearing driver or a socket large enough to contact the outer race of the bearing, drive it in until it seats.

21 Turn the wheel over. Apply a coat of multi-purpose grease to the inside of the spacer and install it in the hub.

22 Pack the remaining bearings from the open side with grease, then install them in the hub (one at a time), driving the bearing in with a socket or bearing driver large enough to contact the outer race of the bearing. Drive the bearing in until it seats.

23 Install a new snap-ring on models so equipped.

24 Install a new grease seal in each side of the hub. It may go in with thumb pressure, but if not, use a seal driver, large socket or a flat piece of wood to drive it into place.

25 Install the collars in the grease seals.

26 Clean off all grease from the brake drum or disc using acetone or brake system cleaner. Install the wheel.

13.18 This seal removal tool is convenient, but a screwdriver will also work

14 Tires - removal and installation

1 To properly remove and install tires, you will need at least two motorcycle tire irons (three is better), some water and a tire pressure gauge.

2   Begin by removing the wheel from the motorcycle. If the tire is going to be re-used, mark it next to the valve stem, wheel balance weight or rim lock.

3   Deflate the tire by removing the valve stem core. When it is fully deflated, push the bead of the tire away from the rim on both sides. In some extreme cases, this can only be accomplished with a bead breaking tool, but most often it can be carried out with tire irons. Riding on a deflated tire to break the bead is not recommended, as damage to the rim and tire will occur.

4   Dismounting a tire is easier when the tire is warm, so an indoor tire change is recommended in cold climates. The rubber gets very stiff and is difficult to manipulate when cold.

5   Place the wheel on a thick pad or old blanket. This will help keep the wheel and tire from slipping around.

6   Once the bead is completely free of the rim, lubricate the inside edge of the rim and the tire bead. Some manufacturers recommend against the use of soap or other tire mounting lubricants, as the tire may shift on the rim; however, a soapy water solution will greatly ease the tire removal and installation process, and may even prevent damage to the tire and/or rim. Remove the locknut and push the tire valve through the rim.

7   Insert one of the tire irons under the bead of the tire at the valve stem and lift the bead up over the rim. This should be fairly easy. Take care not to pinch the tube as this is done. If it is difficult to pry the bead up, make sure that the rest of the bead opposite the valve stem is in the dropped center section of the rim.

8   Hold the tire iron down with the bead over the rim, then move about 1 or 2 inches to either side and insert the second tire iron. Be careful not to cut or slice the bead or the tire may split when inflated. Also, take care not to catch or pinch the inner tube as the second tire iron is levered over. For this reason, tire irons are recommended over screwdrivers or other implements.

9   With a small section of the bead up over the rim, one of the levers can be removed and reinserted 1 or 2 inches farther around the rim until about 1/4 of the tire bead is above the rim edge. Make sure that the rest of the bead is in the dropped center of the rim. At this point, the bead can usually be pulled up over the rim by hand.

10   Once all of the first bead is over the rim, the inner tube can be withdrawn from the tire and rim. Push in on the valve stem, lift up on the tire next to the stem, reach inside the tire and carefully pull out the tube. It is usually not necessary to completely remove the tire from the rim to repair the inner tube. It is sometimes recommended though, because checking for foreign objects in the tire is difficult while it is still mounted on the rim.

11   To remove the tire completely, make sure the bead is broken all the way around on the remaining edge, then stand the tire and wheel up on the tread and grab the wheel with one hand. Push the tire down over the same edge of the rim while pulling the rim away from the tire. If the bead is correctly positioned in the dropped center of the rim, the tire should roll off and separate from the rim very easily. If tire irons are used to work this last bead over the rim, the outer edge of the rim may be marred. If a tire iron is necessary, be sure to pad the rim as described earlier.

12   Refer to Section 15 for inner tube repair procedures.

13   Mounting a tire is basically the reverse of removal. Some tires have a balance mark and/or directional arrows molded into the tire sidewall. Look for these marks so that the tire can be installed properly. The dot should be aligned with the valve stem.

14   If the tire was not removed completely to repair or replace the inner tube, the tube should be inflated just enough to make it round. Sprinkle it with talcum powder, which acts as a dry lubricant, then carefully lift up the tire edge and install the tube with the valve stem next to the hole in the rim. Once the tube is in place, push the valve stem through the rim and start the locknut on the stem.

15   Lubricate the tire bead, then push it over the rim edge and into the dropped center section opposite the inner tube valve stem. Work around each side of the rim, carefully pushing the bead over the rim. The last section may have to be levered on with tire irons. If so, take care not to pinch the inner tube as this is done.

16   Once the bead is over the rim edge, check to see that the inner tube valve stem and the rim lock are pointing to the center of the hub. If they're angled slightly in either direction, rotate the tire on the rim to straighten it out. Run the locknut the rest of the way onto the stem and rim lock but don't tighten them completely.

17   Inflate the tube to approximately 1-1/2 times the pressure listed in the Chapter 1 Specifications and check to make sure the guidelines on the tire sidewalls are the same distance from the rim around the circumference of the tire.

⚠️ **Warning: Do not overinflate the tube or the tire may burst, causing serious injury.**

18   After the tire bead is correctly seated on the rim, allow the tire to deflate. Replace the valve core and inflate the tube to the recommended pressure, then tighten the valve stem locknut securely and tighten the cap. **Note:** *Some riders choose to leave off the valve stem locknut. They feel that installing the nut will make the tube more susceptible to failure at the valve stem if the tire happens to shift on the rim when riding.*
Tighten the locknut on the rim lock to the torque listed in the Chapter 1 Specifications.

15 Tubes - repair

1   Tire tube repair requires a patching kit that's usually available from motorcycle dealers, accessory stores or auto parts stores. Be sure to follow the directions supplied with the kit to ensure a safe repair. Patching should be done only when a new tube is unavailable. Replace the tube as soon as possible. Sudden deflation can cause loss of control and an accident.

2   To repair a tube, remove it from the tire, inflate and immerse it in a sink or tub full of water to pinpoint the leak. Mark the position of the leak, then deflate the tube. Dry it off and thoroughly clean the area around the puncture.

3   Most tire patching kits have a buffer to rough up the area around the hole for proper adhesion of the patch. Roughen an area slightly larger than the patch, then apply a thin coat of the patching cement to the roughened area. Allow the cement to dry until tacky, then apply the patch.

4   It may be necessary to remove a protective covering from the top surface of the patch after it has been attached to the tube. Keep in mind that tubes made from synthetic rubber may require a special patch and adhesive if a satisfactory bond is to be achieved.

5   Before replacing the tube, check the inside of the tire to make sure the object that caused the puncture is not still inside. Also check the outside of the tire, particularly the tread area, to make sure nothing is projecting through the tire that may cause another puncture. Check the rim for sharp edges or damage. Make sure the rubber trim band is in good condition and properly installed before inserting the tube.

TIRE CHANGING SEQUENCE - TUBED TIRES

 Deflate the tire and loosen the rim lock nut. After pushing the tire beads away from the rim flanges push the tire bead into the well of the rim at the point opposite the valve. Insert the tire lever adjacent to the valve and work the bead over the edge of the rim.

Use two levers to work the bead over the edge of the rim. Note the use of rim protectors

 Remove the inner tube from the tire

When the first bead is clear, remove the tire as shown. If the bead isn't too tight, rim protectors won't be necessary

To install, partially inflate the inner tube and insert it in the tire. Make sure there's a nut on the valve stem

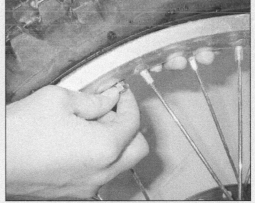

Work the first bead over the rim and feed the valve through the hole in the rim. Partially screw on the retaining nut to hold the valve in place. (The nut should be removed after the tire has been inflated)

Check that the inner tube is positioned correctly and work the second bead over the rim using the tire levers. Start at a point between the valve stem and the rim lock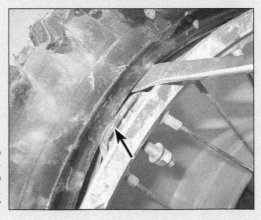

Work final area of the bead over the rim while pushing the valve inwards to ensure that the inner tube is not trapped. Also push the rim lock inwards and, if necessary, pry the tire bead over the rim lock. After inflating the tire, tighten the rim lock nut securely

Notes

Chapter 8
Frame and bodywork

Contents

Degrees of difficulty

Easy, suitable for novice with little experience	Fairly easy, suitable for beginner with some experience	Fairly difficult, suitable for competent DIY mechanic	Difficult, suitable for experienced DIY mechanic	Very difficult, suitable for expert DIY or professional

1 General information

This Chapter covers the procedures necessary to remove and install the fenders and other body parts. Since many service and repair operations on these motorcycles require removal of the fenders and/or other body parts, the procedures are grouped here and referred to from other Chapters.

In the case of damage to plastic body parts, it is usually necessary to remove the broken component and replace it with a new (or used) one. The material that the fenders and other plastic body parts are composed of doesn't lend itself to conventional repair techniques.

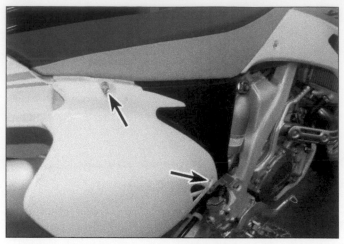

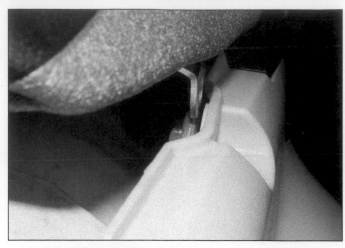

2.1a The upper bolt secures the seat and side cover - the lower bolt secures the side cover

2.1b Unscrew the seat/side cover bolt on each side

2 Seat - removal and installation

1 Remove one bolt from each side of the seat **(see illustrations)**.
2 Pull the seat back and down to detach its front hook and center slot **(see illustration)**. Lift the seat off.
3 Installation is the reverse of removal. Be sure to engage the front hook.

3 Side covers - removal and installation

1 Remove the seat (see Section 2).
2 Remove the side cover bolt and screw and take the side cover off **(see illustration 2.1a)**.
3 Installation is the reverse of the removal steps.

2.2 Detach the hook under the front of the seat from the retainer on the frame (left arrows) and slip the slot off the post (right arrows)

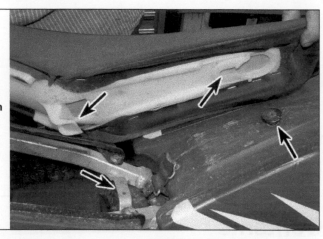

4 Footpegs - removal and installation

1 Support the bike securely so it can't be knocked over during this procedure.
2 To detach the footpeg from the pivot pin, note how the spring is installed, then

remove the cotter pin, washer and pivot pin **(see illustration)**. Separate the footpeg from the motorcycle. On WR models, the left footpeg includes a kickstand **(see illustration)**.
3 Installation is the reverse of removal, with the following addition: Use a new cotter pin and wrap its ends around the pivot pin.

4.2a Note how the spring is attached (upper arrow) - remove the cotter pin and washer (lower arrow) to detach the footpeg from the bracket

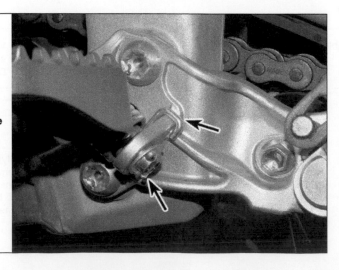

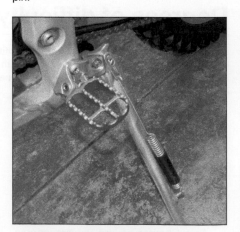

4.2b WR models are equipped with a kickstand

5.1a Remove the bolt . . .

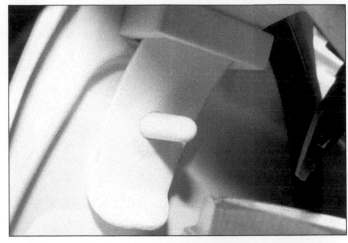

5.1b . . . and unhook the strap to detach the number plate

5 Number plate - removal and installation

1   Undo the number plate bolt and unhook its integral strap **(see illustrations)**.
2   On models so equipped, free the clutch cable and hot start cable from the clip inside the number plate **(see illustration)**.
3   Lift the number plate off its posts **(see illustration)**.
4   Installation is the reverse of the removal steps.

6 Radiator shrouds - removal and installation

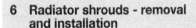

1   To remove the radiator shrouds, remove the seat (see Section 2). Undo the bolts and screws and remove the radiator shrouds from the motorcycle **(see illustration)**.
2   Installation is the reverse of removal.

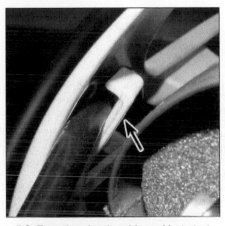

5.2 Free the clutch cable and hot start cable from the clip (if equipped)

7 Front fender - removal and installation

1   Remove the fender bolts **(see illustra-**

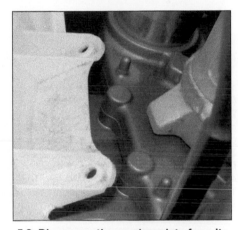

5.3 Disengage the number plate from its posts (if equipped)

tion). Lower the fender clear of the lower triple clamp and remove the washers.
2   Installation is the reverse of removal. Be sure to reinstall the grommets in their correct locations. Tighten the bolts securely, but don't overtighten them and strip the threads.

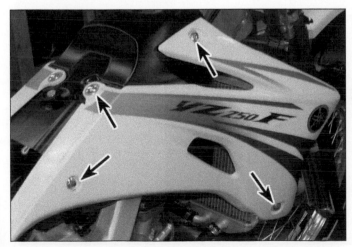

6.1 Remove the bolts or screws and their collars, then remove the radiator shroud

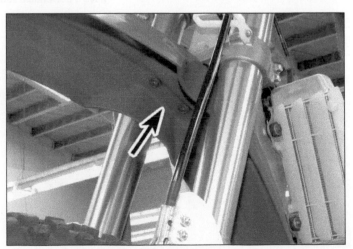

7.1 The front fender bolts and washers are accessible from below

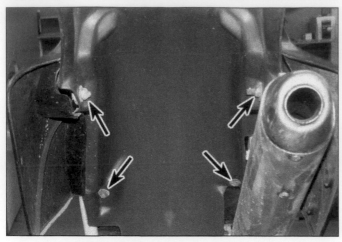

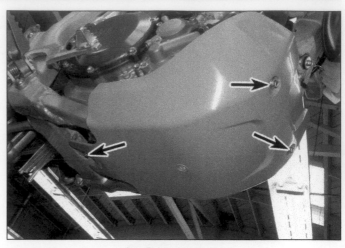

8.2 Typical rear fender mounting bolts

9.1 Skid plate bolts

8 Rear fender - removal and installation

1   Remove the seat and both side covers (see Sections 2 and 3).
2   Remove the fender mounting bolts (and grommets if equipped) and take the fender off **(see illustration)**.
3   If necessary, unbolt the inner fender from the air cleaner housing and lift it off.
4   Installation is the reverse of removal. Tighten the bolts securely, but don't over-tighten them and strip the threads.

9 Skid plate - removal and installation

1   Unbolt the skid plate from the motor-cycle **(see illustration)**.
2   Installation is the reverse of removal.

10 Sub-frame - removal and installation

1   Remove the seat (see Section 2).
2   Loosen the clamping band that secures the carburetor to the air cleaner housing (see Chapter 4).
3   Unbolt the sub-frame from the main frame and lift it off, together with the air cleaner housing and rear fender **(see illustrations)**.
4   Installation is the reverse of the removal steps.

11 Frame - general information, inspection and repair

1   All models use a semi-double cradle frame. On early models, the frame is made of round-section steel tubing. On later models, the frame uses aluminum spars. All models have a detachable sub-frame at the rear.

2   The frame shouldn't require attention unless accident damage has occurred. In most cases, frame replacement is the only satisfactory remedy for such damage. A few frame specialists have the jigs and other equipment necessary for straightening the frame to the required standard of accuracy, but even then there is no simple way of assessing to what extent the frame may have been overstressed.
3   After the motorcycle has accumulated a lot of running time, the frame should be examined closely for signs of cracking or splitting at the welded joints. Corrosion can also cause weakness at these joints. Loose engine mount bolts can cause ovaling or fracturing to the mounting bolt holes. Minor damage can often be repaired by welding, depending on the nature and extent of the damage.
4   Remember that a frame that is out of alignment will cause handling problems. If misalignment is suspected as the result of an accident, it will be necessary to strip the machine completely so the frame can be thoroughly checked.

10.3a Remove the lower sub-frame bolt from the right side . . .

10.3b . . . then remove the lower left bolt and upper bolt and pull the sub-frame off

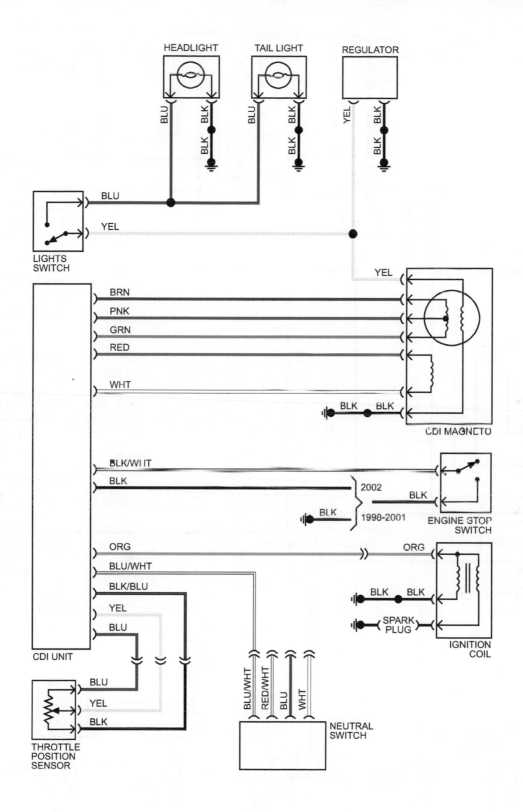

1998-2002 Yamaha WR models

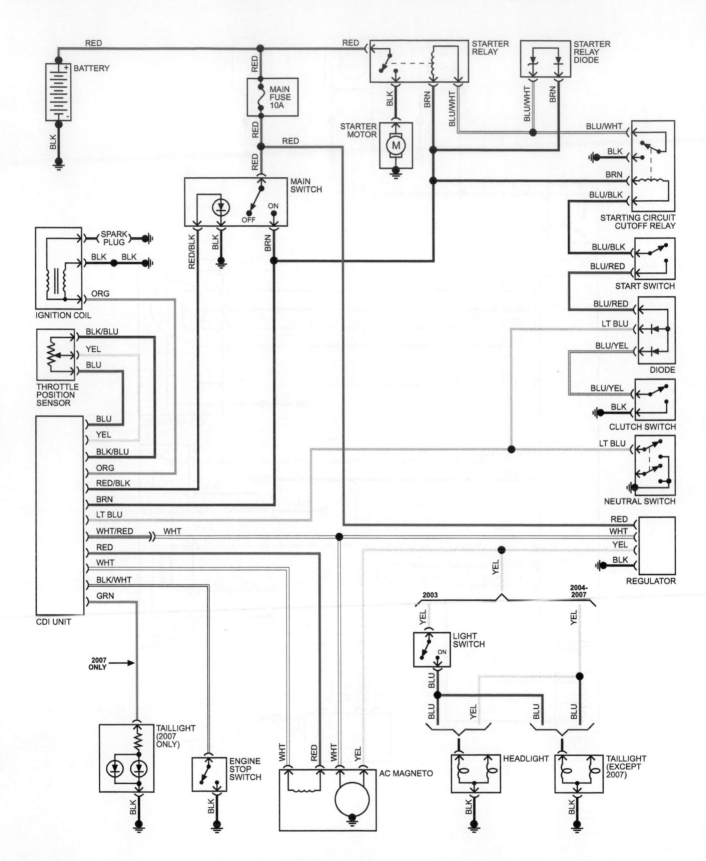

2003-2007 Yamaha WR models

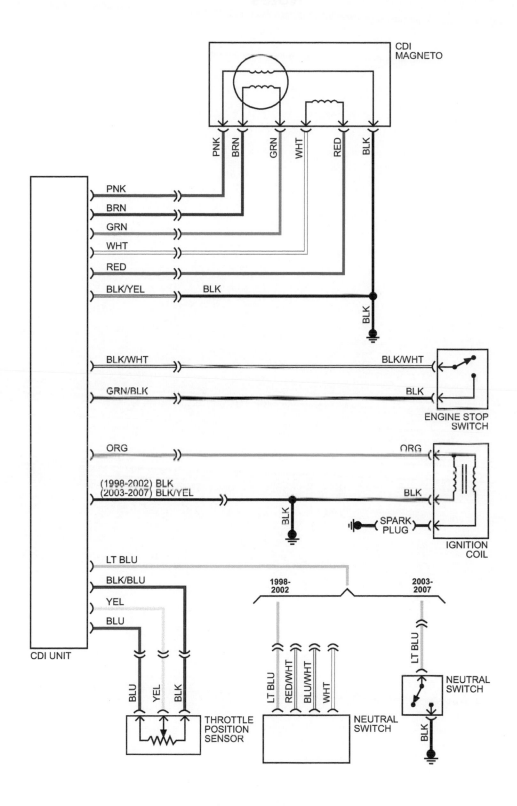

Yamaha YZ models

Notes

Dimensions and weights

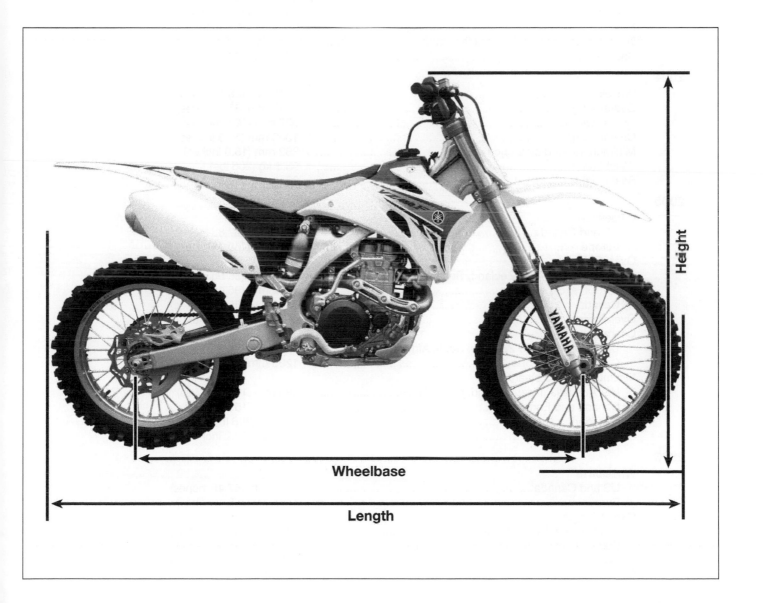

Height

Wheelbase

Length

YZ250

2001 and 2002

Wheelbase	1475 mm (58.1 inches)
Overall length	2165 mm (85.2 inches)
Overall width	827 mm (32.6 inches)
Overall height	1303 mm (51.3 inches)
Minimum ground clearance	382 mm (15.0 inches)
Weight (with oil and full fuel tank)	104.5 kg (230 lbs)
Seat height	998 mm (39.3 inches)

2003

Wheelbase	1475 mm (58.1 inches)
Overall length	2165 mm (85.2 inches)
Overall width	827 mm (32.6 inches)
Overall height	1303 mm (51.3 inches)
Minimum ground clearance	382 mm (15.0 inches)
Weight (with oil and full fuel tank)	101.5 kg (224 lbs)
Seat height	995 mm (39.2 inches)

2004

Wheelbase	1475 mm (58.1 inches)
Overall length	2165 mm (85.2 inches)
Overall width	827 mm (32.6 inches)
Overall height	1303 mm (51.3 inches)
Minimum ground clearance	382 mm (15.0 inches)
Weight (dry)	93.5 kg (206 lbs)
Seat height	995 mm (39.2 inches)

2005

Wheelbase	
US and Canada	1469 mm (57.83 inches)
Europe	1475 mm (58.1 inches)
Overall length	
US, Australia, New Zealand, South Africa, France	2161 mm (85.08 inches)
Europe except France	2162 mm (85.12 inches)
Canada	2165 mm (85.24 inches)
Overall width	827 mm (32.6 inches)
Overall height	
US, Australia, New Zealand, South Africa	1293 mm (50.91 inches)
Canada and Europe	1295 mm (50.98 inches)
Minimum ground clearance	
US, Australia, New Zealand, South Africa	380 mm (14.96 inches)
Canada and Europe	382 mm (15.0 inches)
Weight (dry)	93.5 kg (206 lbs)
Seat height	995 mm (39.2 inches)

2006

Wheelbase	
US and Canada	1485 mm (57.46 inches)
Europe	1475 mm (58.1 inches)
Overall length	
US, Canada, South Africa	2176 mm (85.67 inches)
Europe except France	2170 mm (85.43 inches)
Canada	2169 mm (85.39 inches)
Overall width	815 mm (32.09 inches)

Overall height
   US, Canada, South Africa .... 1292 mm (50.87 inches)
   Europe .... 1294 mm (50.94 Inches)
Minimum ground clearance
   US, Canada, South Africa .... 369 mm (14.53 inches)
   Canada and Europe .... 371 mm (14.61 inches)
Weight (dry) .... 93.5 kg (206 lbs)
Seat height
   US, Canada, South Africa .... 983 mm (38.7 inches)
   Europe except France .... 994 mm (39.13 inches)
   France .... 993 mm (39.03 inches).

2007

Wheelbase .... 1469 mm (57.83 inches)
Overall length
   US, Canada, Australia, New Zealand, South Africa .... 2160 mm (85.04 inches)
   Europe except France .... 2166 mm (85.28 inches)
   France .... 2165 mm (85.24 inches)
Overall width .... 825 mm (32.48 inches)
Overall height
   US, Canada, Australia, New Zealand, South Africa .... 1299 mm (51.14 inches)
   Europe .... 1302 mm (51.26 inches)
Minimum ground clearance
   US, Canada, Australia, New Zealand, South Africa .... 370 mm (14.57 inches)
   Europe .... 372 mm (14.65 inches)
Weight (dry) .... 92.7 kg (204 lbs)
Seat height
   US, Canada, Australia, New Zealand, South Africa .... 984 mm (38.74 inches)
   Europe except France .... 997 mm (39.3 Inches)
   France .... 996 mm (39.21 inches)

2008

Wheelbase .... 1469 mm (57.83 inches)
Overall length
   All except Europe .... 2162 mm (85.12 inches)
   Europe .... 2165 mm (85.24 inches)
Overall width .... 825 mm (32.48 inches)
Overall height .... 1300 mm (51.18 inches)
Minimum ground clearance
   All except Europe .... 370 mm (14.57 inches)
   Europe .... 372 mm (14.65 inches)
Weight (dry) .... 92.7 kg (204 lbs)
Seat height
   All except Europe .... 985 mm (38.78 inches)
   Europe .... 997 mm (39.3 inches)

YZ400

1998

Wheelbase .... 1495 mm (58.9 inches)
Overall length .... 2176 mm (85.7 inches)
Overall width .... 827 mm (32.6 inches)
Overall height .... 1303 mm (51.3 inches)
Minimum ground clearance .... 373 mm (14.7 inches)
Weight (with oil and full fuel tank) .... 115 kg (254 lbs)
Seat height .... 991 mm (39.0 inches)

YZ400 (continued)

1999

Wheelbase	1495 mm (58.9 inches)
Overall length	2176 mm (85.7 inches)
Overall width	827 mm (32.6 inches)
Overall height	1303 mm (51.3 inches)
Minimum ground clearance	373 mm (14.7 inches)
Weight (with oil and full fuel tank)	114.5 kg (254 lbs)
Seat height	998 mm (39.3 inches)

YZ426

2000

Wheelbase	1490 mm (58.7 inches)
Overall length	2171 mm (85.5 inches)
Overall width	827 mm (32.6 inches)
Overall height	1303 mm (51.3 inches)
Minimum ground clearance	373 mm (14.7 inches)
Weight (with oil and full fuel tank)	113.5 kg (254 lbs)
Seat height	998 mm (39.3 inches)

2001 and 2002

Wheelbase	1490 mm (58.7 inches)
Overall length	2171 mm (85.5 inches)
Overall width	827 mm (32.6 inches)
Overall height	1303 mm (51.3 inches)
Minimum ground clearance	373 mm (14.7 inches)
Weight (with oil and full fuel tank)	113.0 kg (249 lbs)
Seat height	998 mm (39.3 inches)

YZ450

2003

Wheelbase	1485 mm (58.5 inches)
Overall length	2171 mm (85.5 inches)
Overall width	827 mm (32.6 inches)
Overall height	1303 mm (51.3 inches)
Minimum ground clearance	371 mm (14.6 inches)
Weight (with oil and full fuel tank)	113.0 kg (249 lbs)
Seat height	995 mm (39.2 inches)

2004

Wheelbase	1485 mm (58.5 inches)
Overall length	2171 mm (85.5 inches)
Overall width	827 mm (32.6 inches)
Overall height	1303 mm (51.3 inches)
Minimum ground clearance	371 mm (14.6 inches)
Weight (dry)	100.0 kg (220 lbs)
Seat height	995 mm (39.2 inches)

2005

Wheelbase	1493 mm (58.8 inches)

Overall length
    US, Australia, New Zealand, South Africa......................... 2190 mm (86.2 inches)
    Canada and Europe .... 2195 mm (86.4 inches)
Overall width .... 827 mm (32.6 inches)
Overall height
    US, Australia, New Zealand, South Africa......................... 1300mm (51.2 inches)
    Canada and France... 1303 mm (51.3 inches)
    Europe .... 1302 mm (51.3 inches)
Minimum ground clearance
    US, Australia, New Zealand, South Africa .... 379 mm (14.9 inches)
    Canada and Europe .... 381 mm (15.0 inches)
Weight (dry)... 100.0 kg (220 lbs)
Seat height
    US, Australia, New Zealand, South Africa......................... 999 mm (39.3 inches)
    Canada and Europe .... 1001 mm (39.4 inches)

2006

Wheelbase
    All except France .... 1495 mm (58.86 inches)
    France .... 1494 mm (58.82 inches)
Overall length
    All except Europe.. 2192 mm (86.3 inches)
    Europe .... 2197 mm (86.5 inches)
Overall width .... 815 mm (32.09 inches)
Overall height
    US, Canada, Australia, New Zealand, South Africa.......... 1298 mm (51.1 Inches)
    Europe except France... 1300 mm (51.18 inches)
    France .... 1301 mm (51.22 inches)
Minimum ground clearance
    US, Canada, Australia, Now Zealand, South Africa .... 370 mm (14.57 inches)
    Europe except France... 373 mm (14.09 inches)
    France .... 374 mm (14.72 inches)
Weight (dry)... 99.8 kg (220 lbs)
Seat height
    All except Europe.. 986 mm (38.82 inches)
    Europe .... 998 mm (39.29 inches)

2007

Wheelbase .... 1494 mm (58.82 inches)
Overall length
    US, Canada, Australia, New Zealand, South Africa.......... 2191 mm (86.26 inches)
    Europe except France... 2196 mm (86.49 inches)
    France .... 2197 mm (86.5 inches)
Overall width .... 825 mm (32.48 inches)
Overall height
    US, Canada, Australia, New Zealand, South Africa.......... 1306 mm (51.42 inches)
    Europe except France... 1308 mm (51.5 inches)
    France .... 1309 mm (51.54 inches)
Minimum ground clearance
    All except Europe.. 373 mm (14.69 inches)
    Europe .... 376 mm (14.8 inches)
Weight (dry)... 99.8 kg (220 lbs)
Seat height
    All except Europe.. 989 mm (38.94 inches)
    Europe .... 1002 mm (39.45 inches)

YZ450 (continued)

2008

Wheelbase	1495 mm (58.86 inches)
Overall length	
All except Europe	2194 mm (86.38 inches)
Europe	2195 mm (86.42 inches)
Overall width	825 mm (32.48 inches)
Overall height	1306 mm (51.42 inches)
Minimum ground clearance	
All except Europe	374 mm (14.72 inches)
Europe	375 mm (14.76 inches)
Weight (dry)	99.5 kg (219 lbs)
Seat height	
All except Europe	990 mm (38.98 inches)
Europe	1001 mm (39.41 inches)

WR250

2001 and 2002

Wheelbase	1475 mm (58.1 inches)
Overall length	2165 mm (85.2 inches)
Overall width	827 mm (32.6 inches)
Overall height	1303 mm (51.3 inches)
Minimum ground clearance	382 mm (15.0 inches)
Weight (with oil and full fuel tank)	110.0 kg (242.5 lbs)
Seat height	998 mm (39.3 inches)

2003

Wheelbase	1475 mm (58.1 inches)
Overall length	2165 mm (85.2 inches)
Overall width	827 mm (32.6 inches)
Overall height	1303 mm (51.3 inches)
Minimum ground clearance	382 mm (15.0 inches)
Weight (with oil and full fuel tank)	115.1 kg (253.7 lbs)
Seat height	998 mm (39.3 inches)

2004

Wheelbase	1475 mm (58.1 inches)
Overall length	2165 mm (85.2 inches)
Overall width	827 mm (32.6 inches)
Overall height	1303 mm (51.3 inches)
Minimum ground clearance	382 mm (15.0 inches)
Weight (dry)	105.5 kg (23.6 lbs)
Seat height	998 mm (39.3 inches)

2005

Wheelbase	1470 mm (57.87 inches)
Overall length	
US, Canada, South Africa	2190 mm (86.22 inches)
Europe, Australia, New Zealand	2180 mm (85.83 inches)
Overall width	830 mm (32.68 inches)

Overall height
    US, Canada, South Africa .... 1290 mm (50.79 inches)
    Europe, Australia, New Zealand.................................... 1295 mm (50.98 inches)
Minimum ground clearance
    US, Canada, South Africa .... 370 mm (14.57 inches)
    Europe, Australia, New Zealand.................................... 375 mm (14.76 inches)
Weight (dry) .... 105.5 kg (232.6 lbs)
Seat height
    US, Canada, South Africa .... 985 mm (38.78 inches)
    Europe, Australia, New Zealand.................................... 990 mm (38.98 inches)

2006

Wheelbase .... 1470 mm (57.87 inches)
Overall length
    US, Canada, South Africa .... 2190 mm (86.22 inches)
    Europe, Australia, New Zealand.................................... 2180 mm (85.83 inches)
Overall width .... 830 mm (32.68 inches)
Overall height
    US, Canada, South Africa .... 1290 mm (50.79 inches)
    Europe, Australia, New Zealand.................................... 1295 mm (50.98 inches)
Minimum ground clearance
    US, Canada, South Africa .... 370 mm (14.57 inches)
    Europe, Australia, New Zealand.................................... 375 mm (14.76 inches)
Weight (dry) .... 106.5 kg (234.8 lbs)
Seat height
    US, Canada, South Africa .... 985 mm (38.78 inches)
    Europe, Australia, New Zealand.................................... 990 mm (38.98 inches)

2007

Wheelbase .... 1480 mm (58.27 inches)
Overall length
    US, Canada, South Africa .... 2165 mm (85.04 inches)
    Europe, Australia, New Zealand.................................... 2180 mm (85.83 inches)
Overall width .... 825 mm (32.48 inches)
Overall height
    US, Canada, South Africa .... 1300 mm (51.18 inches)
    Europe, Australia, New Zealand.................................... 1305 mm (51.38 inches)
Minimum ground clearance
    US, Canada, South Africa .... 365 mm (14.37 inches)
    Europe, Australia, New Zealand.................................... 370 mm (14.57 inches)
Weight (dry) .... 106.0 kg (233.7 lbs)
Seat height
    US, Canada, South Africa .... 980 mm (38.58 inches)
    Europe, Australia, New Zealand.................................... 99 mm (38.98 inches)

2008

Wheelbase .... 1480 mm (58.27 inches)
Overall length .... 2165 mm (85.24 inches)
Overall width .... 825 mm (32.48 inches)
Overall height .... 1300 mm (51.18 inches)
Minimum ground clearance .... 365 mm (14.37 inches)
Weight (dry) .... 106.0 kg (233.7 lbs)
Seat height .... 980 mm (38.58 inches)

WR400

1998

Wheelbase	1495 mm (58.9 inches)
Overall length	2191 mm (86.3 inches)
Overall width	827 mm (32.6 inches)
Overall height	1303 mm (51.3 inches)
Minimum ground clearance	373 mm (14.7 inches)
Weight (with oil and full fuel tank)	123 kg (271 lbs)
Seat height	991 mm (39.0 inches)

1999

Wheelbase	1495 mm (58.9 inches)
Overall length	2191 mm (86.3 inches)
Overall width	827 mm (32.6 inches)
Overall height	1303 mm (51.3 inches)
Minimum ground clearance	373 mm (14.7 inches)
Weight (with oil and full fuel tank)	122.5 kg (270 lbs)
Seat height	998 mm (39.3 inches)

2000

Wheelbase	1490 mm (58.9 inches)
Overall length	2171 mm (86.3 inches)
Overall width	827 mm (32.6 inches)
Overall height	1303 mm (51.3 inches)
Minimum ground clearance	373 mm (14.7 inches)
Weight (with oil and full fuel tank)	122 kg (270 lbs)
Seat height	998 mm (39.3 inches)

WR426

2001 and 2002

Wheelbase	1490 mm (58.9 inches)
Overall length	2171 mm (86.3 inches)
Overall width	827 mm (32.6 inches)
Overall height	1303 mm (51.3 inches)
Minimum ground clearance	373 mm (14.7 inches)
Weight (with oil and full fuel tank)	121.5 kg (268 lbs)
Seat height	998 mm (39.3 inches)

WR450

2003

Wheelbase	1485 mm (58.5 inches)
Overall length	2171 mm (86.3 inches)
Overall width	827 mm (32.6 inches)
Overall height	1303 mm (51.3 inches)
Minimum ground clearance	371 mm (14.6 inches)
Weight (with oil and full fuel tank)	122 kg (269 lbs)
Seat height	998 mm (39.3 inches)

2004

Wheelbase	1485 mm (58.5 inches)
Overall length	2171 mm (86.3 inches)
Overall width	827 mm (32.6 inches)
Overall height	1303 mm (51.3 inches)

Minimum ground clearance	....	371 mm (14.6 inches)
Weight (dry)	....	112.5 kg (248 lbs)
Seat height	....	998 mm (39.3 inches)

2005

Wheelbase	....	1490 mm (58.66 inches)
Overall length		
US, Canada, South Africa	....	2200 mm (86.61 inches)
Europe, Australia, New Zealand	....	2195 mm (86.42 inches)
Overall width	....	830 mm (32.68 inches)
Overall height	....	1295 mm (50.98 inches)
Minimum ground clearance	....	370 mm (14.57 inches)
Weight (dry)	....	112.5 kg (248 lbs)
Seat height		
US, Canada, South Africa	....	985 mm (38.78 inches)
Europe, Australia, New Zealand	....	990 mm (38.98 inches)

2006

Wheelbase	....	1490 mm (58.66 inches)
Overall length		
US, Canada, South Africa	....	2200 mm (86.61 inches)
Europe, Australia, New Zealand	....	2195 mm (86.42 inches)
Overall width	....	830 mm (32.68 inches)
Overall height	....	1295 mm (50.98 inches)
Minimum ground clearance	....	370 mm (14.57 inches)
Weight (dry)	....	113 kg (249.1 lbs)
Seat height		
US, Canada, South Africa	....	985 mm (38.78 inches)
Europe, Australia, New Zealand	....	990 mm (38.98 inches)

2007

Wheelbase	....	1485 mm (58.46 inches)
Overall length		
US, Canada, South Africa	....	2175 mm (85.63 inches)
Europe, Australia, New Zealand	....	2190 mm (86.22 inches)
Overall width	....	825 mm (32.48 inches)
Overall height		
US, Canada, South Africa	....	1295 mm (50.98 inches)
Europe, Australia, New Zealand	....	1300 mm (51.18 inches)
Minimum ground clearance		
US, Canada, South Africa	....	365 mm (14.37 inches)
Europe, Australia, New Zealand	....	370 mm (14.57 inches)
Weight (dry)	....	112.5 kg (248 lbs)
Seat height		
US, Canada, South Africa	....	980 mm (38.58 inches)
Europe, Australia, New Zealand	....	990 mm (38.98 inches)

2008

Wheelbase	....	1485 mm (58.46 inches)
Overall length	....	2175 mm (85.63 inches)
Overall width	....	825 mm (32.48 inches)
Overall height	....	1295 mm (50.98 inches)
Minimum ground clearance	....	365 mm (14.37 inches)
Weight (dry)	....	112.5 kg (248 lbs)
Seat height	....	980 mm (38.58 inches)

Buying tools

A good set of tools is a fundamental requirement for servicing and repairing a motorcycle. Although there will be an initial expense in building up enough tools for servicing, this will soon be offset by the savings made by doing the job yourself. As experience and confidence grow, additional tools can be added to enable the repair and overhaul of the motorcycle. Many of the special tools are expensive and not often used so it may be preferable to rent them, or for a group of friends or motorcycle club to join in the purchase.

As a rule, it is better to buy more expensive, good quality tools. Cheaper tools are likely to wear out faster and need to be replaced more often, nullifying the original savings.

> ⚠ **Warning: To avoid the risk of a poor quality tool breaking in use, causing injury or damage to the component being worked on, always aim to purchase tools which meet the relevant national safety standards.**

The following lists of tools do not represent the manufacturer's service tools, but serve as a guide to help the owner decide which tools are needed for this level of work. In addition, items such as an electric drill, hacksaw, files, soldering iron and a workbench equipped with a vise, may be needed. Although not classed as tools, a selection of bolts, screws, nuts, washers and pieces of tubing always come in useful.

For more information about tools, refer to the Haynes *Motorcycle Workshop Practice Techbook* (Bk. No. 3470).

Manufacturer's service tools

Inevitably certain tasks require the use of a service tool. Where possible an alternative tool or method of approach is recommended, but sometimes there is no option if personal injury or damage to the component is to be avoided. Where required, service tools are referred to in the relevant procedure.

Service tools can usually only be purchased from a motorcycle dealer and are identified by a part number. Some of the commonly-used tools, such as rotor pullers, are available in aftermarket form from mail-order motorcycle tool and accessory suppliers.

Maintenance and minor repair tools

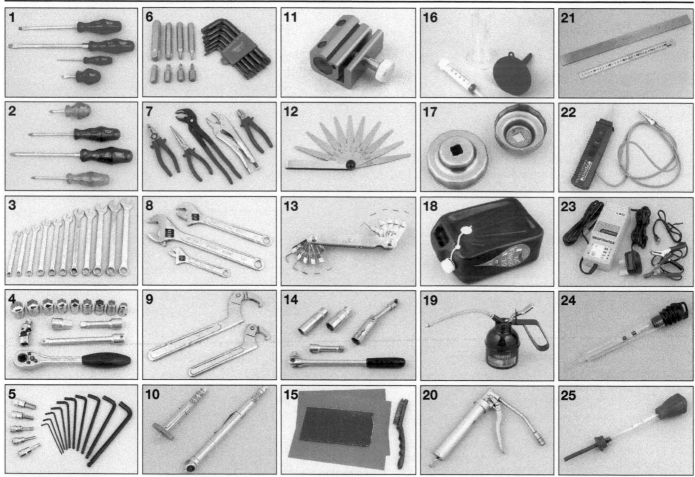

1 Set of flat-bladed screwdrivers
2 Set of Phillips head screwdrivers
3 Combination open-end and box wrenches
4 Socket set (3/8 inch or 1/2 inch drive)
5 Set of Allen keys or bits

6 Set of Torx keys or bits
7 Pliers, cutters and self-locking grips (vise grips)
8 Adjustable wrenches
9 C-spanners
10 Tread depth gauge and tire pressure gauge

11 Cable oiler clamp
12 Feeler gauges
13 Spark plug gap measuring tool
14 Spark plug wrench or deep plug sockets
15 Wire brush and emery paper

16 Calibrated syringe, measuring cup and funnel
17 Oil filter adapters
18 Oil drainer can or tray
19 Pump type oil can
20 Grease gun

21 Straight-edge and steel rule
22 Continuity tester
23 Battery charger
24 Hydrometer (for battery specific gravity check)
25 Anti-freeze tester (for liquid-cooled engines)

Repair and overhaul tools

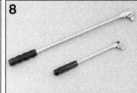

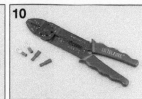

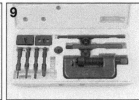

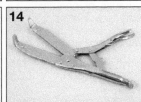

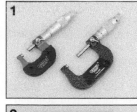

1 Torque wrench
(small and mid-ranges)
2 Conventional, plastic or
soft-faced hammers
3 Impact driver set

4 Vernier caliper
5 Snap-ring pliers
(internal and external, or
combination)
6 Set of cold chisels
and punches

7 Selection of pullers
8 Breaker bars
9 Chain breaking/
riveting tool set
10 Wire stripper and
crimper tool

11 Multimeter (measures
amps, volts and ohms)
12 Stroboscope (for
dynamic timing checks)
13 Hose clamp
(wingnut type shown)

14 Clutch holding tool
15 One-man brake/clutch
bleeder kit

Special tools

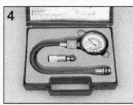

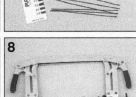

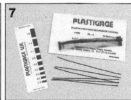

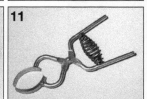

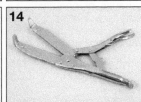

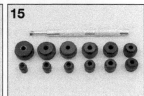

1 Micrometers
(external type)
2 Telescoping gauges
3 Dial gauge

4 Cylinder
compression gauge
5 Vacuum gauges (left) or
manometer (right)
6 Oil pressure gauge

7 Plastigage kit
8 Valve spring compressor
(4-stroke engines)
9 Piston pin drawbolt tool

10 Piston ring removal and
installation tool
11 Piston ring clamp
12 Cylinder bore hone
(stone type shown)

13 Stud extractor
14 Screw extractor set
15 Bearing driver set

1 Workshop equipment and facilities

The workbench

● Work is made much easier by raising the bike up on a ramp - components are much more accessible if raised to waist level. The hydraulic or pneumatic types seen in the dealer's workshop are a sound investment if you undertake a lot of repairs or overhauls **(see illustration 1.1)**.

1.1 Hydraulic motorcycle ramp

● If raised off ground level, the bike must be supported on the ramp to avoid it falling. Most ramps incorporate a front wheel locating clamp which can be adjusted to suit different diameter wheels. When tightening the clamp, take care not to mark the wheel rim or damage the tire - use wood blocks on each side to prevent this.

● Secure the bike to the ramp using tie-downs **(see illustration 1.2)**. If the bike has only a sidestand, and hence leans at a dangerous angle when raised, support the bike on an auxiliary stand.

1.2 Tie-downs are used around the passenger footrests to secure the bike

● Auxiliary (paddock) stands are widely available from mail order companies or motorcycle dealers and attach either to the wheel axle or swingarm pivot **(see illustration 1.3)**. If the motorcycle has a centerstand, you can support it under the crankcase to prevent it toppling while either wheel is removed **(see illustration 1.4)**.

1.3 This auxiliary stand attaches to the swingarm pivot

1.4 Always use a block of wood between the engine and jack head when supporting the engine in this way

Fumes and fire

● Refer to the Safety first! page at the beginning of the manual for full details. Make sure your workshop is equipped with a fire extinguisher suitable for fuel-related fires (Class B fire - flammable liquids) - it is not sufficient to have a water-filled extinguisher.

● Always ensure adequate ventilation is available. Unless an exhaust gas extraction system is available for use, ensure that the engine is run outside of the workshop.

● If working on the fuel system, make sure the workshop is ventilated to avoid a build-up of fumes. This applies equally to fume build-up when charging a battery. Do not smoke or allow anyone else to smoke in the workshop.

Fluids

● If you need to drain fuel from the tank, store it in an approved container marked as suitable for the storage of gasoline **(see illustration 1.5)**. Do not store fuel in glass jars

1.5 Use an approved can only for storing gasoline

or bottles.

● Use proprietary engine degreasers or solvents which have a high flash-point, such as kerosene, for cleaning off oil, grease and dirt - never use gasoline for cleaning. Wear rubber gloves when handling solvent and engine degreaser. The fumes from certain solvents can be dangerous - always work in a well-ventilated area.

Dust, eye and hand protection

● Protect your lungs from inhalation of dust particles by wearing a filtering mask over the nose and mouth. Many frictional materials still contain asbestos which is dangerous to your health. Protect your eyes from spouts of liquid and sprung components by wearing a pair of protective

1.6 A fire extinguisher, goggles, mask and protective gloves should be at hand in the workshop

goggles **(see illustration 1.6)**.

● Protect your hands from contact with solvents, fuel and oils by wearing rubber gloves. Alternatively apply a barrier cream to your hands before starting work. If handling hot components or fluids, wear suitable gloves to protect your hands from scalding and burns.

What to do with old fluids

● Old cleaning solvent, fuel, coolant and oils should not be poured down domestic drains or onto the ground. Package the fluid up in old oil containers, label it accordingly, and take it to a garage or disposal facility. Contact your local disposal company for location of such sites.

Note: It is illegal to dump oil down the drain. Check with your local auto parts store, disposal facility or environmental agency to see if they accept the oil for recycling.

2 Fasteners -
screws, bolts and nuts

Fastener types and applications

Bolts and screws

● Fastener head types are either of hexagonal, Torx or splined design, with internal and external versions of each type **(see illustrations 2.1 and 2.2)**; splined head fasteners are not in common use on motorcycles. The conventional slotted or Phillips head design is used for certain screws. Bolt or screw length is always measured from the underside of the head to the end of the item **(see illustration 2.11)**.

2.1 Internal hexagon/Allen (A), Torx (B) and splined (C) fasteners, with corresponding bits

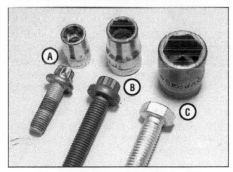

2.2 External Torx (A), splined (B) and hexagon (C) fasteners, with corresponding sockets

● Certain fasteners on the motorcycle have a tensile marking on their heads, the higher the marking the stronger the fastener. High tensile fasteners generally carry a 10 or higher marking. Never replace a high tensile fastener with one of a lower tensile strength.

Washers (see illustration 2.3)

● Plain washers are used between a fastener head and a component to prevent damage to the component or to spread the load when torque is applied. Plain washers can also be used as spacers or shims in certain assemblies. Copper or aluminum plain washers are often used as sealing washers on drain plugs.

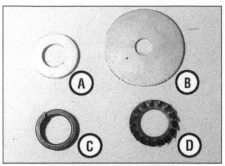

2.3 Plain washer (A), penny washer (B), spring washer (C) and serrated washer (D)

● The split-ring spring washer works by applying axial tension between the fastener head and component. If flattened, it is fatigued and must be replaced. If a plain (flat) washer is used on the fastener, position the spring washer between the fastener and the plain washer.

● Serrated star type washers dig into the fastener and component faces, preventing loosening. They are often used on electrical ground connections to the frame.

● Cone type washers (sometimes called Belleville) are conical and when tightened apply axial tension between the fastener head and component. They must be installed with the diched side against the component and often carry an OUTSIDE marking on their outer face. If flattened, they are fatigued and must be replaced.

● Tab washers are used to lock plain nuts or bolts on a shaft. A portion of the tab washer is bent up hard against one flat of the nut or bolt to prevent it loosening. Due to the tab washer being deformed in use, a new tab washer should be used every time it is removed.

● Wave washers are used to take up endfloat on a shaft. They provide light springing and prevent excessive side-to-side play of a component. Can be found on rocker arm shafts.

Nuts and cotter pins

● Conventional plain nuts are usually six-sided **(see illustration 2.4)**. They are sized by thread diameter and pitch. High tensile nuts carry a number on one end to denote their tensile strength.

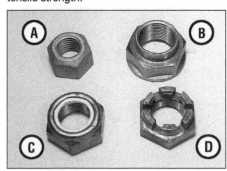

2.4 Plain nut (A), shouldered locknut (B), nylon insert nut (C) and castellated nut (D)

● Self-locking nuts either have a nylon insert, or two spring metal tabs, or a shoulder which is staked into a groove in the shaft - their advantage over conventional plain nuts is a resistance to loosening due to vibration. The nylon insert type can be used a number of times, but must be replaced when the friction of the nylon insert is reduced, i.e. when the nut spins freely on the shaft. The spring tab type can be reused unless the tabs are damaged. The shouldered type must be replaced every time it is removed.

● Cotter pins are used to lock a castellated nut to a shaft or to prevent loosening of a plain nut. Common applications are wheel axles and brake torque arms. Because the cotter pin arms are deformed to lock around the nut a new cotter pin must always be used on installation - always use the correct size cotter pin which will fit snugly in the shaft hole. Make sure the cotter pin arms are correctly located around the nut **(see illustrations 2.5 and 2.6)**.

2.5 Bend cotter pin arms as shown (arrows) to secure a castellated nut

2.6 Bend cotter pin arms as shown to secure a plain nut

Caution: If the castellated nut slots do not align with the shaft hole after tightening to the torque setting, tighten the nut until the next slot aligns with the hole - never loosen the nut to align its slot.

● R-pins (shaped like the letter R), or slip pins as they are sometimes called, are sprung and can be reused if they are otherwise in good condition. Always install R-pins with their closed end facing forwards **(see illustration 2.7)**.

2.7 Correct fitting of R-pin. Arrow indicates forward direction

Snap-rings (see illustration 2.8)

● Snap-rings (sometimes called circlips) are used to retain components on a shaft or in a housing and have corresponding external or internal ears to permit removal. Parallel-sided (machined) snap-rings can be installed either way round in their groove, whereas stamped snap-rings (which have a chamfered edge on one face) must be installed with the chamfer facing away from the direction of thrust load (see illustration 2.9).

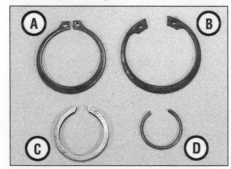

2.8 External stamped snap-ring (A), internal stamped snap-ring (B), machined snap-ring (C) and wire snap-ring (D)

● Always use snap-ring pliers to remove and install snap-rings; expand or compress them just enough to remove them. After installation, rotate the snap-ring in its groove to ensure it is securely seated. If installing a snap-ring on a splined shaft, always align its opening with a shaft channel to ensure the snap-ring ends are well supported and unlikely to catch (see illustration 2.10).

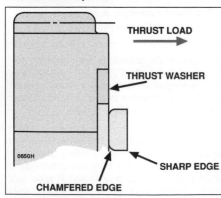

2.9 Correct fitting of a stamped snap-ring

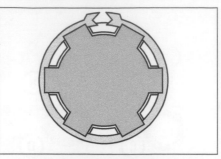

2.10 Align snap-ring opening with shaft channel

● Snap-rings can wear due to the thrust of components and become loose in their grooves, with the subsequent danger of becoming dislodged in operation. For this reason, replacement is advised every time a snap-ring is disturbed.

● Wire snap-rings are commonly used as piston pin retaining clips. If a removal tang is provided, long-nosed pliers can be used to dislodge them, otherwise careful use of a small flat-bladed screwdriver is necessary. Wire snap-rings should be replaced every time they are disturbed.

Thread diameter and pitch

● Diameter of a male thread (screw, bolt or stud) is the outside diameter of the threaded portion (see illustration 2.11). Most motorcycle manufacturers use the ISO (International Standards Organization) metric system expressed in millimeters. For example, M6 refers to a 6 mm diameter thread. Sizing is the same for nuts, except that the thread diameter is measured across the valleys of the nut.

● Pitch is the distance between the peaks of the thread (see illustration 2.11). It is expressed in millimeters, thus a common bolt size may be expressed as 6.0 x 1.0 mm (6 mm thread diameter and 1 mm pitch). Generally pitch increases in proportion to thread diameter, although there are always exceptions.

● Thread diameter and pitch are related for conventional fastener applications and the accompanying table can be used as a guide. Additionally, the AF (Across Flats), wrench or socket size dimension of the bolt or nut (see illustration 2.11) is linked to thread and pitch specification. Thread pitch can be measured with a thread gauge (see illustration 2.12).

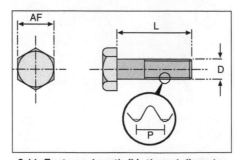

2.11 Fastener length (L), thread diameter (D), thread pitch (P) and head size (AF)

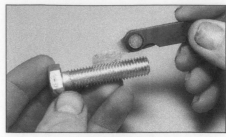

2.12 Using a thread gauge to measure pitch

AF size	Thread diameter x pitch (mm)
8 mm	M5 x 0.8
8 mm	M6 x 1.0
10 mm	M6 x 1.0
12 mm	M8 x 1.25
14 mm	M10 x 1.25
17 mm	M12 x 1.25

● The threads of most fasteners are of the right-hand type, ie they are turned clockwise to tighten and counterclockwise to loosen. The reverse situation applies to left-hand thread fasteners, which are turned counterclockwise to tighten and clockwise to loosen. Left-hand threads are used where rotation of a component might loosen a conventional right-hand thread fastener.

Seized fasteners

● Corrosion of external fasteners due to water or reaction between two dissimilar metals can occur over a period of time. It will build up sooner in wet conditions or in countries where salt is used on the roads during the winter. If a fastener is severely corroded it is likely that normal methods of removal will fail and result in its head being ruined. When you attempt removal, the fastener thread should be heard to crack free and unscrew easily - if it doesn't, stop there before damaging something.

● A smart tap on the head of the fastener will often succeed in breaking free corrosion which has occurred in the threads (see illustration 2.13).

● An aerosol penetrating fluid (such as WD-40) applied the night beforehand may work its way down into the thread and ease removal. Depending on the location, you may be able to make up a modeling-clay well around the fastener head and fill it with penetrating fluid.

2.13 A sharp tap on the head of a fastener will often break free a corroded thread

● If you are working on an engine internal component, corrosion will most likely not be a problem due to the well lubricated environment. However, components can be very tight and an impact driver is a useful tool in freeing them (see illustration 2.14).

2.14 Using an impact driver to free a fastener

● Where corrosion has occurred between dissimilar metals (e.g. steel and aluminum alloy), the application of heat to the fastener head will create a disproportionate expansion rate between the two metals and break the seizure caused by the corrosion. Whether heat can be applied depends on the location of the fastener - any surrounding components likely to be damaged must first be removed (see illustration 2.15). Heat can be applied using a paint stripper heat gun or clothes iron, or by immersing the component in boiling water - wear protective gloves to prevent scalding or burns to the hands.

2.15 Using heat to free a seized fastener

● As a last resort, it is possible to use a hammer and cold chisel to work the fastener head unscrewed (see illustration 2.16). This will damage the fastener, but more importantly extreme care must be taken not to damage the surrounding component.

Caution: Remember that the component being secured is generally of more value than the bolt, nut or screw - when the fastener is freed, do not unscrew it with force, instead work the fastener back and forth when resistance is felt to prevent thread damage.

2.16 Using a hammer and chisel to free a seized fastener

Broken fasteners and damaged heads

● If the shank of a broken bolt or screw is accessible you can grip it with self-locking grips. The knurled wheel type stud extractor tool or self-gripping stud puller tool is particularly useful for removing the long studs which screw into the cylinder mouth surface of the crankcase or bolts and screws from which the head has broken off (see illustration 2.17). Studs can also be removed by locking two nuts together on the threaded end of the stud and using a wrench on the lower nut (see illustration 2.18).

2.17 Using a stud extractor tool to remove a broken crankcase stud

2.18 Two nuts can be locked together to unscrew a stud from a component

● A bolt or screw which has broken off below or level with the casing must be extracted using a screw extractor set. Centerpunch the fastener to centralize the drill bit, then drill a hole in the fastener (see illustration 2.19). Select a drill bit which is approximately half

2.19 When using a screw extractor, first drill a hole in the fastener . . .

to three-quarters the diameter of the fastener and drill to a depth which will accommodate the extractor. Use the largest size extractor possible, but avoid leaving too small a wall thickness otherwise the extractor will merely force the fastener walls outwards wedging it in the casing thread.

● If a spiral type extractor is used, thread it counterclockwise into the fastener. As it is screwed in, it will grip the fastener and unscrew it from the casing (see illustration 2.20).

2.20 . . . then thread the extractor counterclockwise into the fastener

● If a taper type extractor is used, tap it into the fastener so that it is firmly wedged in place. Unscrew the extractor (counter-clockwise) to draw the fastener out.

> ⚠ *Warning: Stud extractors are very hard and may break off in the fastener if care is not taken - ask a machine shop about spark erosion if this happens.*

● Alternatively, the broken bolt/screw can be drilled out and the hole retapped for an oversize bolt/screw or a diamond-section thread insert. It is essential that the drilling is carried out squarely and to the correct depth, otherwise the casing may be ruined - if in doubt, entrust the work to a machine shop.

● Bolts and nuts with rounded corners cause the correct size wrench or socket to slip when force is applied. Of the types of wrench/socket available always use a six-point type rather than an eight or twelve-point type - better grip

2.21 Comparison of surface drive box wrench (left) with 12-point type (right)

is obtained. Surface drive wrenches grip the middle of the hex flats, rather than the corners, and are thus good in cases of damaged heads **(see illustration 2.21)**.

● Slotted-head or Phillips-head screws are often damaged by the use of the wrong size screwdriver. Allen-head and Torx-head screws are much less likely to sustain damage. If enough of the screw head is exposed you can use a hacksaw to cut a slot in its head and then use a conventional flat-bladed screwdriver to remove it. Alternatively use a hammer and cold chisel to tap the head of the fastener around to loosen it. Always replace damaged fasteners with new ones, preferably Torx or Allen-head type.

A dab of valve grinding compound between the screw head and screw-driver tip will often give a good grip.

Thread repair

● Threads (particularly those in aluminum alloy components) can be damaged by overtightening, being assembled with dirt in the threads, or from a component working loose and vibrating. Eventually the thread will fail completely, and it will be impossible to tighten the fastener.

● If a thread is damaged or clogged with old locking compound it can be renovated with a thread repair tool (thread chaser) **(see illustrations 2.22 and 2.23)**; special thread

2.22 A thread repair tool being used to correct an internal thread

2.23 A thread repair tool being used to correct an external thread

chasers are available for spark plug hole threads. The tool will not cut a new thread, but clean and true the original thread. Make sure that you use the correct diameter and pitch tool. Similarly, external threads can be cleaned up with a die or a thread restorer file **(see illustration 2.24)**.

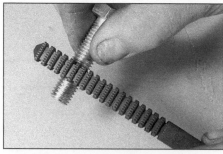

2.24 Using a thread restorer file

● It is possible to drill out the old thread and retap the component to the next thread size. This will work where there is enough surrounding material and a new bolt or screw can be obtained. Sometimes, however, this is not possible - such as where the bolt/screw passes through another component which must also be suitably modified, also in cases where a spark plug or oil drain plug cannot be obtained in a larger diameter thread size.

● The diamond-section thread insert (often known by its popular trade name of Heli-Coil) is a simple and effective method of replacing the thread and retaining the original size. A kit can be purchased which contains the tap, insert and installing tool **(see illustration 2.25)**. Drill out the damaged thread with the size drill specified **(see illustration 2.26)**. Carefully retap the thread **(see illustration 2.27)**. Install the

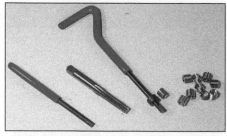

2.25 Obtain a thread insert kit to suit the thread diameter and pitch required

2.26 To install a thread insert, first drill out the original thread . . .

2.27 . . . tap a new thread . . .

2.28 . . . fit insert on the installing tool . . .

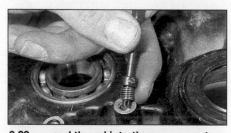

2.29 . . . and thread into the component . . .

2.30 . . . break off the tang when complete

insert on the installing tool and thread it slowly into place using a light downward pressure **(see illustrations 2.28 and 2.29)**. When positioned between a 1/4 and 1/2 turn below the surface withdraw the installing tool and use the break-off tool to press down on the tang, breaking it off **(see illustration 2.30)**.

● There are epoxy thread repair kits on the market which can rebuild stripped internal threads, although this repair should not be used on high load-bearing components.

Thread locking and sealing compounds

● Locking compounds are used in locations where the fastener is prone to loosening due to vibration or on important safety-related items which might cause loss of control of the motorcycle if they fail. It is also used where important fasteners cannot be secured by other means such as lockwashers or cotter pins.

● Before applying locking compound, make sure that the threads (internal and external) are clean and dry with all old compound removed. Select a compound to suit the component being secured - a non-permanent general locking and sealing type is suitable for most applications, but a high strength type is needed for permanent fixing of studs in castings. Apply a drop or two of the compound to the first few threads of the fastener, then thread it into place and tighten to the specified torque. Do not apply excessive thread locking compound otherwise the thread may be damaged on subsequent removal.

● Certain fasteners are impregnated with a dry film type coating of locking compound on their threads. Always replace this type of fastener if disturbed.

● Anti-seize compounds, such as copper-based greases, can be applied to protect threads from seizure due to extreme heat and corrosion. A common instance is spark plug threads and exhaust system fasteners.

3 Measuring tools and gauges

Feeler gauges

● Feeler gauges (or blades) are used for measuring small gaps and clearances (see illustration 3.1). They can also be used to measure endfloat (sideplay) of a component on a shaft where access is not possible with a dial gauge.

● Feeler gauge sets should be treated with care and not bent or damaged. They are etched with their size on one face. Keep them clean and very lightly oiled to prevent corrosion build-up.

3.1 Feeler gauges are used for measuring small gaps and clearances - thickness is marked on one face of gauge

● When measuring a clearance, select a gauge which is a light sliding fit between the two components. You may need to use two gauges together to measure the clearance accurately.

Micrometers

● A micrometer is a precision tool capable of measuring to 0.01 or 0.001 of a millimeter. It should always be stored in its case and not in the general toolbox. It must be kept clean and never dropped, otherwise its frame or measuring anvils could be distorted resulting in inaccurate readings.

● External micrometers are used for measuring outside diameters of components and have many more applications than internal micrometers. Micrometers are available in different size ranges, typically 0 to 25 mm, 25 to 50 mm, and upwards in 25 mm steps; some large micrometers have interchangeable anvils to allow a range of measurements to be taken. Generally the largest precision measurement you are likely to take on a motorcycle is the piston diameter.

● Internal micrometers (or bore micrometers) are used for measuring inside diameters, such as valve guides and cylinder bores. Telescoping gauges and small hole gauges are used in conjunction with an external micrometer, whereas the more expensive internal micrometers have their own measuring device.

External micrometer

Note: *The conventional analogue type instrument is described. Although much easier to read, digital micrometers are considerably more expensive.*

● Always check the calibration of the micrometer before use. With the anvils closed (0 to 25 mm type) or set over a test gauge

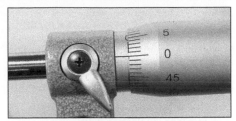

3.2 Check micrometer calibration before use

(for the larger types) the scale should read zero (see illustration 3.2); make sure that the anvils (and test piece) are clean first. Any discrepancy can be adjusted by referring to the instructions supplied with the tool. Remember that the micrometer is a precision measuring tool - don't force the anvils closed, use the ratchet (4) on the end of the micrometer to close it. In this way, a measured force is always applied.

● To use, first make sure that the item being measured is clean. Place the anvil of the micrometer (1) against the item and use the thimble (2) to bring the spindle (3) lightly into contact with the other side of the item (see illustration 3.3). Don't tighten the thimble down because this will damage the micrometer - instead use the ratchet (4) on the end of the micrometer. The ratchet mechanism applies a measured force preventing damage to the instrument.

● The micrometer is read by referring to the linear scale on the sleeve and the annular scale on the thimble. Read off the sleeve first to obtain the base measurement, then add the fine measurement from the thimble to obtain the overall reading. The linear scale on the sleeve represents the measuring range of the micrometer (eg 0 to 25 mm). The annular scale

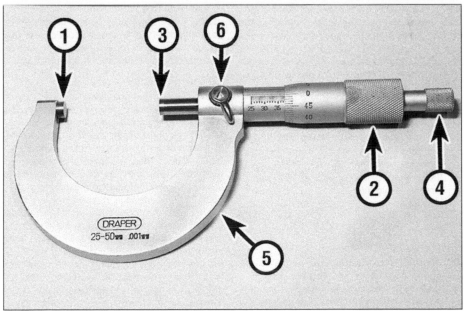

3.3 Micrometer component parts

1 Anvil	3 Spindle	5 Frame
2 Thimble	4 Ratchet	6 Locking lever

on the thimble will be in graduations of 0.01 mm (or as marked on the frame) - one full revolution of the thimble will move 0.5 mm on the linear scale. Take the reading where the datum line on the sleeve intersects the thimble's scale. Always position the eye directly above the scale otherwise an inaccurate reading will result.

In the example shown the item measures 2.95 mm (see illustration 3.4):

Linear scale	2.00 mm
Linear scale	0.50 mm
Annular scale	0.45 mm
Total figure	**2.95 mm**

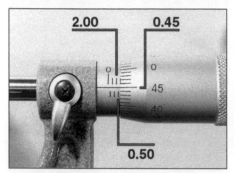

3.4 Micrometer reading of 2.95 mm

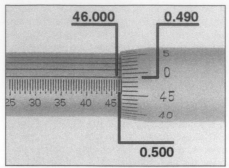

3.5 Micrometer reading of 46.99 mm on linear and annular scales . . .

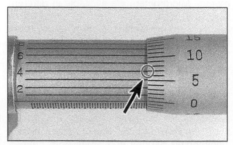

3.6 . . . and 0.004 mm on vernier scale

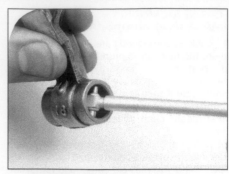

3.7 Expand the telescoping gauge in the bore, lock its position . . .

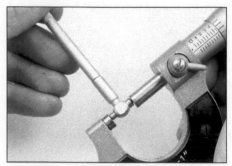

3.8 . . . then measure the gauge with a micrometer

3.9 Expand the small hole gauge in the bore, lock its position . . .

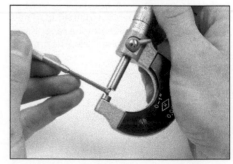

3.10 . . . then measure the gauge with a micrometer

Most micrometers have a locking lever (6) on the frame to hold the setting in place, allowing the item to be removed from the micrometer.
● Some micrometers have a vernier scale on their sleeve, providing an even finer measurement to be taken, in 0.001 increments of a millimeter. Take the sleeve and thimble measurement as described above, then check which graduation on the vernier scale aligns with that of the annular scale on the thimble **Note:** *The eye must be perpendicular to the scale when taking the vernier reading - if necessary rotate the body of the micrometer to ensure this.* Multiply the vernier scale figure by 0.001 and add it to the base and fine measurement figures.

In the example shown the item measures 46.994 mm (see illustrations 3.5 and 3.6):

Linear scale (base)	46.000 mm
Linear scale (base)	00.500 mm
Annular scale (fine)	00.490 mm
Vernier scale	00.004 mm
Total figure	**46.994 mm**

Internal micrometer

● Internal micrometers are available for measuring bore diameters, but are expensive and unlikely to be available for home use. It is suggested that a set of telescoping gauges and small hole gauges, both of which must be used with an external micrometer, will suffice for taking internal measurements on a motorcycle.
● Telescoping gauges can be used to

measure internal diameters of components. Select a gauge with the correct size range, make sure its ends are clean and insert it into the bore. Expand the gauge, then lock its position and withdraw it from the bore (see illustration 3.7). Measure across the gauge ends with a micrometer (see illustration 3.8).
● Very small diameter bores (such as valve guides) are measured with a small hole gauge. Once adjusted to a slip-fit inside the component, its position is locked and the gauge withdrawn for measurement with a micrometer (see illustrations 3.9 and 3.10).

Vernier caliper

Note: *The conventional linear and dial gauge type instruments are described. Digital types are easier to read, but are far more expensive.*
● The vernier caliper does not provide the precision of a micrometer, but is versatile in being able to measure internal and external diameters. Some types also incorporate a depth gauge. It is ideal for measuring clutch plate friction material and spring free lengths.
● To use the conventional linear scale vernier, loosen off the vernier clamp screws (1) and set its jaws over (2), or inside (3), the item to be measured (see illustration 3.11). Slide the jaw into contact, using the thumb-wheel (4) for fine movement of the sliding scale (5) then tighten the clamp screws (1). Read off the main scale (6) where the zero on the sliding scale (5) intersects it, taking the whole number to the left of the zero; this provides the base measurement. View along the sliding scale and select the division which lines up exactly

with any of the divisions on the main scale, noting that the divisions usually represents 0.02 of a millimeter. Add this fine measurement to the base measurement to obtain the total reading.

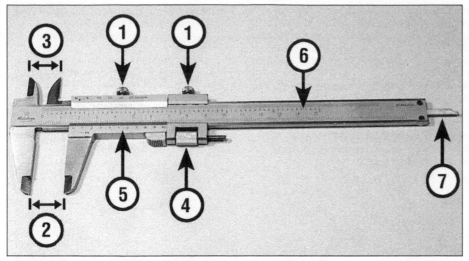

3.11 Vernier component parts (linear gauge)

1 Clamp screws
2 External jaws
3 Internal jaws
4 Thumbwheel
5 Sliding scale
6 Main scale
7 Depth gauge

In the example shown the item measures 55.92 mm **(see illustration 3.12)**:

Base measurement	55.00 mm
Fine measurement	00.92 mm
Total figure	**55.92 mm**

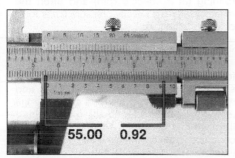

3.12 Vernier gauge reading of 55.92 mm

● Some vernier calipers are equipped with a dial gauge for fine measurement. Before use, check that the jaws are clean, then close them fully and check that the dial gauge reads zero. If necessary adjust the gauge ring accordingly. Slacken the vernier clamp screw (1) and set its jaws over (2), or inside (3), the item to be measured **(see illustration 3.13)**. Slide the jaws into contact, using the thumbwheel (4) for fine movement. Read off the main scale (5) where the edge of the sliding scale (6) intersects it, taking the whole number to the left of the zero; this provides the base measurement. Read off the needle position on the dial gauge (7) scale to provide the fine measurement; each division represents 0.05 of a millimeter. Add this fine measurement to the base measurement to obtain the total reading.

In the example shown the item measures 55.95 mm **(see illustration 3.14)**:

Base measurement	55.00 mm
Fine measurement	00.95 mm
Total figure	**55.95 mm**

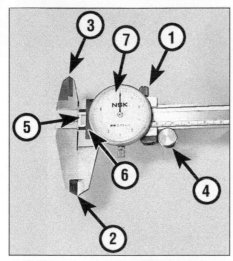

3.13 Vernier component parts (dial gauge)

1 Clamp screw
2 External jaws
3 Internal jaws
4 Thumbwheel
5 Main scale
6 Sliding scale
7 Dial gauge

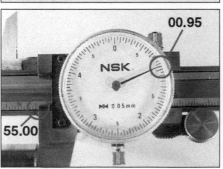

3.14 Vernier gauge reading of 55.95 mm

Plastigage

● Plastigage is a plastic material which can be compressed between two surfaces to measure the oil clearance between them. The width of the compressed Plastigage is measured against a calibrated scale to determine the clearance.

● Common uses of Plastigage are for measuring the clearance between crankshaft journal and main bearing inserts, between crankshaft journal and big-end bearing inserts, and between camshaft and bearing surfaces. The following example describes big-end oil clearance measurement.

● Handle the Plastigage material carefully to prevent distortion. Using a sharp knife, cut a length which corresponds with the width of the bearing being measured and place it carefully across the journal so that it is parallel with the shaft **(see illustration 3.15)**. Carefully install both bearing shells and the connecting rod. Without rotating the rod on the journal tighten its bolts or nuts (as applicable) to the specified torque. The connecting rod and bearings are then disassembled and the crushed Plastigage examined.

3.15 Plastigage placed across shaft journal

● Using the scale provided in the Plastigage kit, measure the width of the material to determine the oil clearance **(see illustration 3.16)**. Always remove all traces of Plastigage after use using your fingernails.

Caution: Arriving at the correct clearance demands that the assembly is torqued correctly, according to the settings and sequence (where applicable) provided by the motorcycle manufacturer.

3.16 Measuring the width of the crushed Plastigage

Dial gauge or DTI (Dial Test Indicator)

● A dial gauge can be used to accurately measure small amounts of movement. Typical uses are measuring shaft runout or shaft endfloat (sideplay) and setting piston position for ignition timing on two-strokes. A dial gauge set usually comes with a range of different probes and adapters and mounting equipment.

● The gauge needle must point to zero when at rest. Rotate the ring around its periphery to zero the gauge.

● Check that the gauge is capable of reading the extent of movement in the work. Most gauges have a small dial set in the face which records whole millimeters of movement as well as the fine scale around the face periphery which is calibrated in 0.01 mm divisions. Read off the small dial first to obtain the base measurement, then add the measurement from the fine scale to obtain the total reading.

Base measurement	1.00 mm
Fine measurement	0.48 mm
Total figure	**1.48 mm**

3.17 Dial gauge reading of 1.48 mm

In the example shown the gauge reads 1.48 mm (see illustration 3.17):

● If measuring shaft runout, the shaft must be supported in vee-blocks and the gauge mounted on a stand perpendicular to the shaft. Rest the tip of the gauge against the center of the shaft and rotate the shaft slowly while watching the gauge reading (see illustration 3.18). Take several measurements along the length of the shaft and record the

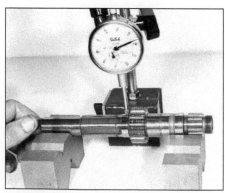

3.18 Using a dial gauge to measure shaft runout

maximum gauge reading as the amount of runout in the shaft. **Note:** The reading obtained will be total runout at that point - some manufacturers specify that the runout figure is halved to compare with their specified runout limit.

● Endfloat (sideplay) measurement requires that the gauge is mounted securely to the surrounding component with its probe touching the end of the shaft. Using hand pressure, push and pull on the shaft noting the maximum endfloat recorded on the gauge (see illustration 3.19).

3.19 Using a dial gauge to measure shaft endfloat

● A dial gauge with suitable adapters can be used to determine piston position BTDC on two-stroke engines for the purposes of ignition timing. The gauge, adapter and suitable length probe are installed in the place of the spark plug and the gauge zeroed at TDC. If the piston position is specified as 1.14 mm BTDC, rotate the engine back to 2.00 mm BTDC, then slowly forwards to 1.14 mm BTDC.

Cylinder compression gauges

● A compression gauge is used for measuring cylinder compression. Either the rubber-cone type or the threaded adapter type can be used. The latter is preferred to ensure a perfect seal against the cylinder head. A 0 to 300 psi (0 to 20 Bar) type gauge (for gasoline engines) will be suitable for motorcycles.

● The spark plug is removed and the gauge either held hard against the cylinder head (cone type) or the gauge adapter screwed into the cylinder head (threaded type) (see illustration 3.20). Cylinder compression is measured with the engine turning over, but not running - carry out the compression test as described in

3.20 Using a rubber-cone type cylinder compression gauge

Troubleshooting Equipment. The gauge will hold the reading until manually released.

Oil pressure gauge

● An oil pressure gauge is used for measuring engine oil pressure. Most gauges come with a set of adapters to fit the thread of the take-off point (see illustration 3.21). If the take-off point specified by the motorcycle manufacturer is an external oil pipe union, make sure that the specified replacement union is used to prevent oil starvation.

3.21 Oil pressure gauge and take-off point adapter (arrow)

● Oil pressure is measured with the engine running (at a specific rpm) and often the manufacturer will specify pressure limits for a cold and hot engine.

Straight-edge and surface plate

● If checking the gasket face of a component for warpage, place a steel rule or precision straight-edge across the gasket face and measure any gap between the straight-edge and component with feeler gauges (see illustration 3.22). Check diagonally across the component and between mounting holes (see illustration 3.23).

3.22 Use a straight-edge and feeler gauges to check for warpage

3.23 Check for warpage in these directions

● Checking individual components for warpage, such as clutch plain (metal) plates, requires a perfectly flat plate or piece of plate glass and feeler gauges.

4 Torque and leverage

What is torque?

● Torque describes the twisting force around a shaft. The amount of torque applied is determined by the distance from the center of the shaft to the end of the lever and the amount of force being applied to the end of the lever; distance multiplied by force equals torque.

● The manufacturer applies a measured torque to a bolt or nut to ensure that it will not loosen in use and to hold two components securely together without movement in the joint. The actual torque setting depends on the thread size, bolt or nut material and the composition of the components being held.

● Too little torque may cause the fastener to loosen due to vibration, whereas too much torque will distort the joint faces of the component or cause the fastener to shear off. Always stick to the specified torque setting.

Using a torque wrench

● Check the calibration of the torque wrench and make sure it has a suitable range for the job. Torque wrenches are available in Nm (Newton-meters), kgf m (kilograms force meter), lbf ft (pounds-feet), lbf in (inch-pounds). Do not confuse lbf ft with lbf in.

● Adjust the tool to the desired torque on the scale (see illustration 4.1). If your torque wrench is not calibrated in the units specified, carefully convert the figure (see Conversion Factors). A manufacturer sometimes gives a torque setting as a range (8 to 10 Nm) rather than a single figure - in this case set the tool midway between the two settings. The same torque may be expressed as 9 Nm ± 1 Nm. Some torque wrenches have a method of locking the setting so that it isn't inadvertently altered during use.

4.1 Set the torque wrench index mark to the setting required, in this case 12 Nm

● Install the bolts/nuts in their correct location and secure them lightly. Their threads must be clean and free of any old locking compound. Unless specified the threads and flange should be dry - oiled threads are necessary in certain circumstances and the manufacturer will take this into account in the specified torque figure. Similarly, the manufacturer may also specify the application of thread-locking compound.

● Tighten the fasteners in the specified sequence until the torque wrench clicks, indicating that the torque setting has been reached. Apply the torque again to double-check the setting. Where different thread diameter fasteners secure the component, as a rule tighten the larger diameter ones first.

● When the torque wrench has been finished with, release the lock (where applicable) and fully back off its setting to zero - do not leave the torque wrench tensioned. Also, do not use a torque wrench for loosening a fastener.

Angle-tightening

● Manufacturers often specify a figure in degrees for final tightening of a fastener. This usually follows tightening to a specific torque setting.

● A degree disc can be set and attached to the socket (see illustration 4.2) or a protractor can be used to mark the angle of movement on the bolt/nut head and the surrounding casting (see illustration 4.3).

4.2 Angle tightening can be accomplished with a torque-angle gauge . . .

4.3 . . . or by marking the angle on the surrounding component

Loosening sequences

● Where more than one bolt/nut secures a component, loosen each fastener evenly a little at a time. In this way, not all the stress of the joint is held by one fastener and the components are not likely to distort.

● If a tightening sequence is provided, work in the REVERSE of this, but if not, work from the outside in, in a criss-cross sequence (see illustration 4.4).

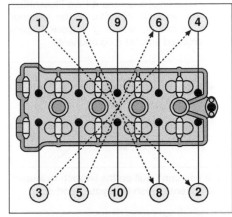

4.4 When loosening, work from the outside inwards

Tightening sequences

● If a component is held by more than one fastener it is important that the retaining bolts/nuts are tightened evenly to prevent uneven stress build-up and distortion of sealing faces. This is especially important on high-compression joints such as the cylinder head.

● A sequence is usually provided by the manufacturer, either in a diagram or actually marked in the casting. If not, always start in the center and work outwards in a criss-cross pattern (see illustration 4.5). Start off by securing all bolts/nuts finger-tight, then set the torque wrench and tighten each fastener by a small amount in sequence until the final torque is reached. By following this practice,

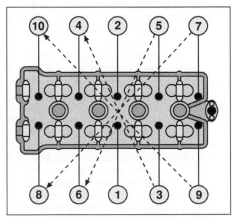

4.5 When tightening, work from the inside outwards

the joint will be held evenly and will not be distorted. Important joints, such as the cylinder head and big-end fasteners often have two- or three-stage torque settings.

Applying leverage

● Use tools at the correct angle. Position a socket or wrench on the bolt/nut so that you pull it towards you when loosening. If this can't be done, push the wrench without curling your fingers around it **(see illustration 4.6)** - the wrench may slip or the fastener loosen suddenly, resulting in your fingers being crushed against a component.

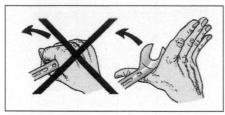

4.6 If you can't pull on the wrench to loosen a fastener, push with your hand open

● Additional leverage is gained by extending the length of the lever. The best way to do this is to use a breaker bar instead of the regular length tool, or to slip a length of tubing over the end of the wrench or socket.
● If additional leverage will not work, the fastener head is either damaged or firmly corroded in place (see *Fasteners*).

5 Bearings

Bearing removal and installation

Drivers and sockets

● Before removing a bearing, always inspect the casing to see which way it must be driven out - some casings will have retaining plates or a cast step. Also check for any identifying markings on the bearing and, if installed to a certain depth, measure this at this stage. Some roller bearings are sealed on one side - take note of the original installed position.
● Bearings can be driven out of a casing using a bearing driver tool (with the correct size head) or a socket of the correct diameter. Select the driver head or socket so that it contacts the outer race of the bearing, not the balls/rollers or inner race. Always support the casing around the bearing housing with wood blocks, otherwise there is a risk of fracture. The bearing is driven out with a few blows on the driver or socket from a heavy mallet. Unless access is severely restricted (as with wheel bearings), a pin-punch is not recommended unless it is moved around the bearing to keep it square in its housing.

● The same equipment can be used to install bearings. Make sure the bearing housing is supported on wood blocks and line up the bearing in its housing. Install the bearing as noted on removal - generally they are installed with their marked side facing outwards. Tap the bearing squarely into its housing using a driver or socket which bears only on the bearing's outer race - contact with the bearing balls/rollers or inner race will destroy it **(see illustrations 5.1 and 5.2)**.
● Check that the bearing inner race and balls/rollers rotate freely.

5.1 Using a bearing driver against the bearing's outer race

5.2 Using a large socket against the bearing's outer race

Pullers and slide-hammers

● Where a bearing is pressed on a shaft a puller will be required to extract it **(see illustration 5.3)**. Make sure that the puller clamp or legs fit securely behind the bearing and are unlikely to slip out. If pulling a bearing

5.3 This bearing puller clamps behind the bearing and pressure is applied to the shaft end to draw the bearing off

off a gear shaft for example, you may have to locate the puller behind a gear pinion if there is no access to the race and draw the gear pinion off the shaft as well **(see illustration 5.4)**.

Caution: Ensure that the puller's center bolt locates securely against the end of the shaft and will not slip when pressure is applied. Also ensure that puller does not damage the shaft end.

5.4 Where no access is available to the rear of the bearing, it is sometimes possible to draw off the adjacent component

● Operate the puller so that its center bolt exerts pressure on the shaft end and draws the bearing off the shaft.
● When installing the bearing on the shaft, tap only on the bearing's inner race - contact with the balls/rollers or outer race will destroy the bearing. Use a socket or length of tubing as a drift which fits over the shaft end **(see illustration 5.5)**.

5.5 When installing a bearing on a shaft use a piece of tubing which bears only on the bearing's inner race

● Where a bearing locates in a blind hole in a casing, it cannot be driven or pulled out as described above. A slide-hammer with knife-edged bearing puller attachment will be required. The puller attachment passes through the bearing and when tightened expands to fit firmly behind the bearing **(see illustration 5.6)**. By operating the slide-hammer part of the tool the bearing is jarred out of its housing **(see illustration 5.7)**.
● It is possible, if the bearing is of reasonable weight, for it to drop out of its housing if the casing is heated as described opposite. If

5.6 Expand the bearing puller so that it locks behind the bearing . . .

5.7 . . . attach the slide hammer to the bearing puller

this method is attempted, first prepare a work surface which will enable the casing to be tapped face down to help dislodge the bearing - a wood surface is ideal since it will not damage the casing's gasket surface. Wearing protective gloves, tap the heated casing several times against the work surface to dislodge the bearing under its own weight (see illustration 5.8).

5.8 Tapping a casing face down on wood blocks can often dislodge a bearing

● Bearings can be installed in blind holes using the driver or socket method described above.

Drawbolts

● Where a bearing or bushing is set in the eye of a component, such as a suspension linkage arm or connecting rod small-end, removal by drift may damage the component. Furthermore, a rubber bushing in a shock absorber eye cannot successfully be driven out of position. If access is available to a hydraulic press, the task is straightforward. If not, a drawbolt can be fabricated to extract the bearing or bushing.

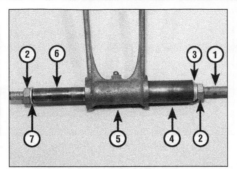

5.9 Drawbolt component parts assembled on a suspension arm

1 Bolt or length of threaded bar
2 Nuts
3 Washer (external diameter greater than tubing internal diameter)
4 Tubing (internal diameter sufficient to accommodate bearing)
5 Suspension arm with bearing
6 Tubing (external diameter slightly smaller than bearing)
7 Washer (external diameter slightly smaller than bearing)

5.10 Drawing the bearing out of the suspension arm

● To extract the bearing/bushing you will need a long bolt with nut (or piece of threaded bar with two nuts), a piece of tubing which has an internal diameter larger than the bearing/ bushing, another piece of tubing which has an external diameter slightly smaller than the bearing/bushing, and a selection of washers (see illustrations 5.9 and 5.10). Note that the pieces of tubing must be of the same length, or longer, than the bearing/bushing.
● The same kit (without the pieces of tubing) can be used to draw the new bearing/bushing back into place (see illustration 5.11).

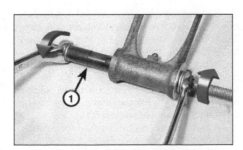

5.11 Installing a new bearing (1) in the suspension arm

Temperature change

● If the bearing's outer race is a tight fit in the casing, the aluminum casing can be heated to release its grip on the bearing. Aluminum will expand at a greater rate than the steel bearing outer race. There are several ways to do this, but avoid any localized extreme heat (such as a blow torch) - aluminum alloy has a low melting point.
● Approved methods of heating a casing are using a domestic oven (heated to 100°C/200°F) or immersing the casing in boiling water (see illustration 5.12). Low temperature range localized heat sources such as a paint stripper heat gun or clothes iron can also be used (see illustration 5.13). Alternatively, soak a rag in boiling water, wring it out and wrap it around the bearing housing.

> ⚠ **Warning: All of these methods require care in use to prevent scalding and burns to the hands. Wear protective gloves when handling hot components.**

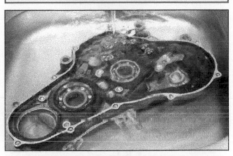

5.12 A casing can be immersed in a sink of boiling water to aid bearing removal

5.13 Using a localized heat source to aid bearing removal

● If heating the whole casing note that plastic components, such as the neutral switch, may suffer - remove them beforehand.
● After heating, remove the bearing as described above. You may find that the expansion is sufficient for the bearing to fall out of the casing under its own weight or with a light tap on the driver or socket.
● If necessary, the casing can be heated to aid bearing installation, and this is sometimes the recommended procedure if the motorcycle manufacturer has designed the housing and bearing fit with this intention.

● Installation of bearings can be eased by placing them in a freezer the night before installation. The steel bearing will contract slightly, allowing easy insertion in its housing. This is often useful when installing steering head outer races in the frame.

Bearing types and markings

● Plain shell bearings, ball bearings, needle roller bearings and tapered roller bearings will all be found on motorcycles (see illustrations 5.14 and 5.15). The ball and roller types are usually caged between an inner and outer race, but uncaged variations may be found.

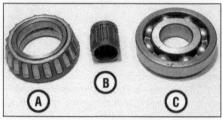

5.14 Shell bearings are either plain or grooved. They are usually identified by color code (arrow)

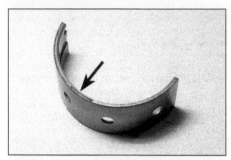

5.15 Tapered roller bearing (A), needle roller bearing (B) and ball journal bearing (C)

● Shell bearings (often called inserts) are usually found at the crankshaft main and connecting rod big-end where they are good at coping with high loads. They are made of a phosphor-bronze material and are impregnated with self-lubricating properties.

● Ball bearings and needle roller bearings consist of a steel inner and outer race with the balls or rollers between the races. They require constant lubrication by oil or grease and are good at coping with axial loads. Taper roller bearings consist of rollers set in a tapered cage set on the inner race; the outer race is separate. They are good at coping with axial loads and prevent movement along the shaft - a typical application is in the steering head.

● Bearing manufacturers produce bearings to ISO size standards and stamp one face of the bearing to indicate its internal and external diameter, load capacity and type (see illustration 5.16).

● Metal bushings are usually of phosphor-bronze material. Rubber bushings are used in suspension mounting eyes. Fiber bushings have also been used in suspension pivots.

5.16 Typical bearing marking

Bearing troubleshooting

● If a bearing outer race has spun in its housing, the housing material will be damaged. You can use a bearing locking compound to bond the outer race in place if damage is not too severe.

● Shell bearings will fail due to damage of their working surface, as a result of lack of lubrication, corrosion or abrasive particles in the oil (see illustration 5.17). Small particles of dirt in the oil may embed in the bearing material whereas larger particles will score the bearing and shaft journal. If a number of short journeys are made, insufficient heat will be generated to drive off condensation which has built up on the bearings.

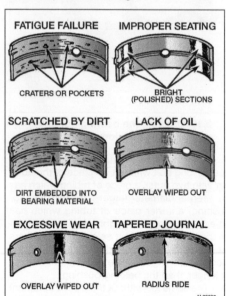

5.17 Typical bearing failures

● Ball and roller bearings will fail due to lack of lubrication or damage to the balls or rollers. Tapered-roller bearings can be damaged by overloading them. Unless the bearing is sealed on both sides, wash it in kerosene to remove all old grease then allow it to dry. Make a visual inspection looking to dented balls or rollers, damaged cages and worn or pitted races (see illustration 5.18).

● A ball bearing can be checked for wear by listening to it when spun. Apply a film of light oil to the bearing and hold it close to the ear - hold the outer race with one hand and spin the

5.18 Example of ball journal bearing with damaged balls and cages

5.19 Hold outer race and listen to inner race when spun

inner race with the other hand (see illustration 5.19). The bearing should be almost silent when spun; if it grates or rattles it is worn.

6 Oil seals

Oil seal removal and installation

● Oil seals should be replaced every time a component is dismantled. This is because the seal lips will become set to the sealing surface and will not necessarily reseal.

● Oil seals can be pried out of position using a large flat-bladed screwdriver (see illustration 6.1). In the case of crankcase seals, check first that the seal is not lipped on the inside, preventing its removal with the crankcases joined.

6.1 Pry out oil seals with a large flat-bladed screwdriver

● New seals are usually installed with their marked face (containing the seal reference code) outwards and the spring side towards the fluid being retained. In certain cases, such as a two-stroke engine crankshaft seal, a double lipped seal may be used due to there being fluid or gas on each side of the joint.

● Use a bearing driver or socket which bears only on the outer hard edge of the seal to install it in the casing - tapping on the inner edge will damage the sealing lip.

Oil seal types and markings

● Oil seals are usually of the single-lipped type. Double-lipped seals are found where a liquid or gas is on both sides of the joint.

● Oil seals can harden and lose their sealing ability if the motorcycle has been in storage for a long period - replacement is the only solution.

● Oil seal manufacturers also conform to the ISO markings for seal size - these are molded into the outer face of the seal (see illustration 6.2).

6.2 These oil seal markings indicate inside diameter, outside diameter and seal thickness

7 Gaskets and sealants

Types of gasket and sealant

● Gaskets are used to seal the mating surfaces between components and keep lubricants, fluids, vacuum or pressure contained within the assembly. Aluminum gaskets are sometimes found at the cylinder joints, but most gaskets are paper-based. If the mating surfaces of the components being joined are undamaged the gasket can be installed dry, although a dab of sealant or grease will be useful to hold it in place during assembly.

● RTV (Room Temperature Vulcanizing) silicone rubber sealants cure when exposed to moisture in the atmosphere. These sealants are good at filling pits or irregular gasket faces, but will tend to be forced out of the joint under very high torque. They can be used to replace a paper gasket, but first make sure that the width of the paper gasket is not essential to the shimming of internal components. RTV sealants should not be used on components containing gasoline.

● Non-hardening, semi-hardening and hard setting liquid gasket compounds can be used with a gasket or between a metal-to-metal joint. Select the sealant to suit the application: universal non-hardening sealant can be used on virtually all joints; semi-hardening on joint faces which are rough or damaged; hard setting sealant on joints which require a permanent bond and are subjected to high temperature and pressure. **Note:** *Check first if the paper gasket has a bead of sealant*

impregnated in its surface before applying additional sealant.

● When choosing a sealant, make sure it is suitable for the application, particularly if being applied in a high-temperature area or in the vicinity of fuel. Certain manufacturers produce sealants in either clear, silver or black colors to match the finish of the engine. This has a particular application on motorcycles where much of the engine is exposed.

● Do not over-apply sealant. That which is squeezed out on the outside of the joint can be wiped off, whereas an excess of sealant on the inside can break off and clog oilways.

Breaking a sealed joint

● Age, heat, pressure and the use of hard setting sealant can cause two components to stick together so tightly that they are difficult to separate using finger pressure alone. Do not resort to using levers unless there is a pry point provided for this purpose (see illustration 7.1) or else the gasket surfaces will be damaged.

● Use a soft-faced hammer (see illustration 7.2) or a wood block and conventional hammer to strike the component near the mating surface. Avoid hammering against cast extremities since they may break off. If this method fails, try using a wood wedge between the two components.

> **Caution: If the joint will not separate, double-check that you have removed all the fasteners.**

7.1 If a pry point is provided, apply gentle pressure with a flat-bladed screwdriver

7.2 Tap around the joint with a soft-faced mallet if necessary - don't strike cooling fins

Removal of old gasket and sealant

● Paper gaskets will most likely come away complete, leaving only a few traces stuck

Most components have one or two hollow locating dowels between the two gasket faces. If a dowel cannot be removed, do not resort to gripping it with pliers - it will almost certainly be distorted. Install a close-fitting socket or Phillips screwdriver into the dowel and then grip the outer edge of the dowel to free it.

on the sealing faces of the components. It is imperative that all traces are removed to ensure correct sealing of the new gasket.

● Very carefully scrape all traces of gasket away making sure that the sealing surfaces are not gouged or scored by the scraper (see illustrations 7.3, 7.4 and 7.5). Stubborn deposits can be removed by spraying with an aerosol gasket remover. Final preparation of

7.3 Paper gaskets can be scraped off with a gasket scraper tool . . .

7.4 . . . a knife blade . . .

7.5 . . . or a household scraper

7.6 Fine abrasive paper is wrapped around a flat file to clean up the gasket face

7.7 A kitchen scourer can be used on stubborn deposits

the gasket surface can be made with very fine abrasive paper or a plastic kitchen scourer **(see illustrations 7.6 and 7.7)**.

● Old sealant can be scraped or peeled off components, depending on the type originally used. Note that gasket removal compounds are available to avoid scraping the components clean; make sure the gasket remover suits the type of sealant used.

8 Chains

Breaking and joining "endless" final drive chains

● Drive chains for many larger bikes are continuous and do not have a clip-type connecting link. The chain must be broken using a chain breaker tool and the new chain securely riveted together using a new soft rivet-type link. Never use a clip-type connecting link instead of a rivet-type link, except in an emergency. Various chain breaking and riveting tools are available, either as separate tools or combined as illustrated in the accompanying photographs - read the instructions supplied with the tool carefully.

⚠ **Warning: The need to rivet the new link pins correctly cannot be overstressed - loss of control of the motorcycle is very likely to result if the chain breaks in use.**

● Rotate the chain and look for the soft link. The soft link pins look like they have been

8.1 Tighten the chain breaker to push the pin out of the link . . .

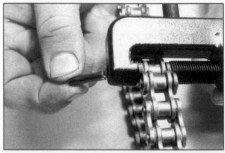

8.2 . . . withdraw the pin, remove the tool . . .

8.3 . . . and separate the chain link

deeply center-punched instead of peened over like all the other pins **(see illustration 8.9)** and its sideplate may be a different color. Position the soft link midway between the sprockets and assemble the chain breaker tool over one of the soft link pins **(see illustration 8.1)**. Operate the tool to push the pin out through the chain **(see illustration 8.2)**. On an O-ring chain, remove the O-rings **(see illustration 8.3)**. Carry out the same procedure on the other soft link pin.

Caution: Certain soft link pins (particularly on the larger chains) may require their ends to be filed or ground off before they can be pressed out using the tool.

● Check that you have the correct size and strength (standard or heavy duty) new soft link - do not reuse the old link. Look for the size marking on the chain sideplates **(see illustration 8.10)**.

● Position the chain ends so that they are

8.4 Insert the new soft link, with O-rings, through the chain ends . . .

8.5 . . . install the O-rings over the pin ends . . .

8.6 . . . followed by the sideplate

engaged over the rear sprocket. On an O-ring chain, install a new O-ring over each pin of the link and insert the link through the two chain ends **(see illustration 8.4)**. Install a new O-ring over the end of each pin, followed by the sideplate (with the chain manufacturer's marking facing outwards) **(see illustrations 8.5 and 8.6)**. On an unsealed chain, insert the link through the two chain ends, then install the sideplate with the chain manufacturer's marking facing outwards.

● Note that it may not be possible to install the sideplate using finger pressure alone. If using a joining tool, assemble it so that the plates of the tool clamp the link and press the sideplate over the pins **(see illustration 8.7)**. Otherwise, use two small sockets placed over

8.7 Push the sideplate into position using a clamp

8.8 Assemble the chain riveting tool over one pin at a time and tighten it fully

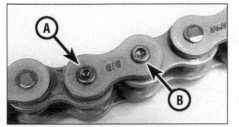

8.9 Pin end correctly riveted (A), pin end unriveted (B)

the rivet ends and two pieces of the wood between a C-clamp. Operate the clamp to press the sideplate over the pins.

● Assemble the joining tool over one pin (following the manufacturer's instructions) and tighten the tool down to spread the pin end securely (see illustrations 8.8 and 8.9). Do the same on the other pin.

> ⚠️ **Warning: Check that the pin ends are secure and that there is no danger of the sideplate coming loose. If the pin ends are cracked the soft link must be replaced.**

Final drive chain sizing

● Chains are sized using a three digit number, followed by a suffix to denote the chain type (see illustration 8.10). Chain type is either standard or heavy duty (thicker sideplates), and also unsealed or O-ring/X-ring type.

● The first digit of the number relates to the pitch of the chain, ie the distance from the center of one pin to the center of the next pin (see illustration 8.11). Pitch is expressed in eighths of an inch, as follows:

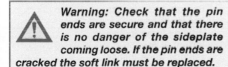

8.10 Typical chain size and type marking

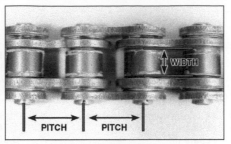

8.11 Chain dimensions

Sizes commencing with a 4 (for example 428) have a pitch of 1/2 inch (12.7 mm)

Sizes commencing with a 5 (for example 520) have a pitch of 5/8 inch (15.9 mm)

Sizes commencing with a 6 (for example 630) have a pitch of 3/4 inch (19.1 mm)

● The second and third digits of the chain size relate to the width of the rollers, for example the 525 shown has 5/16 inch (7.94 mm) rollers (see illustration 8.11).

9 Hoses

Clamping to prevent flow

● Small-bore flexible hoses can be clamped to prevent fluid flow while a component is worked on. Whichever method is used, ensure that the hose material is not permanently distorted or damaged by the clamp.

a) A brake hose clamp available from auto parts stores (see illustration 9.1).
b) A wingnut type hose clamp (see illustration 9.2).

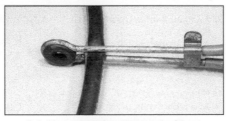

9.1 Hoses can be clamped with an automotive brake hose clamp . . .

9.2 . . . a wingnut type hose clamp . . .

c) Two sockets placed on each side of the hose and held with straight-jawed self-locking pliers (see illustration 9.3).
d) Thick card stock on each side of the hose held between straight-jawed self-locking pliers (see illustration 9.4).

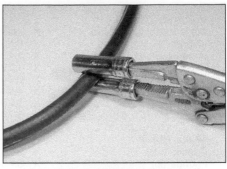

9.3 . . . two sockets and a pair of self-locking grips . . .

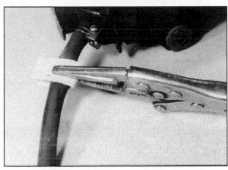

9.4 . . . or thick card and self-locking grips

Freeing and fitting hoses

● Always make sure the hose clamp is moved well clear of the hose end. Grip the hose with your hand and rotate it while pulling it off the union. If the hose has hardened due to age and will not move, slit it with a sharp knife and peel its ends off the union (see illustration 9.5).

● Resist the temptation to use grease or soap on the unions to aid installation; although it helps the hose slip over the union it will equally aid the escape of fluid from the joint. It is preferable to soften the hose ends in hot water and wet the inside surface of the hose with water or a fluid which will evaporate.

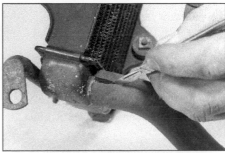

9.5 Cutting a coolant hose free with a sharp knife

Conversion Factors

Length (distance)

Inches (in)	X	25.4	= Millimeters (mm)	X 0.0394	= Inches (in)
Feet (ft)	X	0.305	= Meters (m)	X 3.281	= Feet (ft)
Miles	X	1.609	= Kilometers (km)	X 0.621	= Miles

Volume (capacity)

Cubic inches (cu in; in³)	X	16.387	= Cubic centimeters (cc; cm³)	X 0.061	= Cubic inches (cu in; in³)
Imperial pints (Imp pt)	X	0.568	= Liters (l)	X 1.76	= Imperial pints (Imp pt)
Imperial quarts (Imp qt)	X	1.137	= Liters (l)	X 0.88	= Imperial quarts (Imp qt)
Imperial quarts (Imp qt)	X	1.201	= US quarts (US qt)	X 0.833	= Imperial quarts (Imp qt)
US quarts (US qt)	X	0.946	= Liters (l)	X 1.057	= US quarts (US qt)
Imperial gallons (Imp gal)	X	4.546	= Liters (l)	X 0.22	= Imperial gallons (Imp gal)
Imperial gallons (Imp gal)	X	1.201	= US gallons (US gal)	X 0.833	= Imperial gallons (Imp gal)
US gallons (US gal)	X	3.785	= Liters (l)	X 0.264	= US gallons (US gal)

Mass (weight)

Ounces (oz)	X	28.35	= Grams (g)	X 0.035	= Ounces (oz)
Pounds (lb)	X	0.454	= Kilograms (kg)	X 2.205	= Pounds (lb)

Force

Ounces-force (ozf; oz)	X	0.278	= Newtons (N)	X 3.6	= Ounces-force (ozf; oz)
Pounds-force (lbf; lb)	X	4.448	= Newtons (N)	X 0.225	= Pounds-force (lbf; lb)
Newtons (N)	X	0.1	= Kilograms-force (kgf; kg)	X 9.81	= Newtons (N)

Pressure

Pounds-force per square inch (psi; lbf/in²; lb/in²)	X	0.070	= Kilograms-force per square centimeter (kgf/cm²; kg/cm²)	X 14.223	= Pounds-force per square inch (psi; lbf/in²; lb/in²)
Pounds-force per square inch (psi; lbf/in²; lb/in²)	X	0.068	= Atmospheres (atm)	X 14.696	= Pounds-force per square inch (psi; lbf/in²; lb/in²)
Pounds-force per square inch (psi; lbf/in²; lb/in²)	X	0.069	= Bars	X 14.5	= Pounds-force per square inch (psi; lbf/in²; lb/in²)
Pounds-force per square inch (psi; lbf/in²; lb/in²)	X	6.895	= Kilopascals (kPa)	X 0.145	= Pounds-force per square inch (psi; lbf/in²; lb/in²)
Kilopascals (kPa)	X	0.01	= Kilograms-force per square centimeter (kgf/cm²; kg/cm²)	X 98.1	= Kilopascals (kPa)

Torque (moment of force)

Pounds-force inches (lbf in; lb in)	X	1.152	= Kilograms-force centimeter (kgf cm; kg cm)	X 0.868	= Pounds-force inches (lbf in; lb in)
Pounds-force inches (lbf in; lb in)	X	0.113	= Newton meters (Nm)	X 8.85	= Pounds-force inches (lbf in; lb in)
Pounds-force inches (lbf in; lb in)	X	0.083	= Pounds-force feet (lbf ft; lb ft)	X 12	= Pounds-force inches (lbf in; lb in)
Pounds-force feet (lbf ft; lb ft)	X	0.138	= Kilograms-force meters (kgf m; kg m)	X 7.233	= Pounds-force feet (lbf ft; lb ft)
Pounds-force feet (lbf ft; lb ft)	X	1.356	= Newton meters (Nm)	X 0.738	= Pounds-force feet (lbf ft; lb ft)
Newton meters (Nm)	X	0.102	= Kilograms-force meters (kgf m; kg m)	X 9.804	= Newton meters (Nm)

Vacuum

Inches mercury (in. Hg)	X	3.377	= Kilopascals (kPa)	X 0.2961	= Inches mercury
Inches mercury (in. Hg)	X	25.4	= Millimeters mercury (mm Hg)	X 0.0394	= Inches mercury

Power

Horsepower (hp)	X	745.7	= Watts (W)	X 0.0013	= Horsepower (hp)

Velocity (speed)

Miles per hour (miles/hr; mph)	X	1.609	= Kilometers per hour (km/hr; kph)	X 0.621	= Miles per hour (miles/hr; mph)

Fuel consumption*

Miles per gallon, Imperial (mpg)	X	0.354	= Kilometers per liter (km/l)	X 2.825	= Miles per gallon, Imperial (mpg)
Miles per gallon, US (mpg)	X	0.425	= Kilometers per liter (km/l)	X 2.352	= Miles per gallon, US (mpg)

Temperature

Degrees Fahrenheit = (°C x 1.8) + 32   Degrees Celsius (Degrees Centigrade; °C) = (°F - 32) x 0.56

*It is common practice to convert from miles per gallon (mpg) to liters/100 kilometers (l/100km), where mpg (Imperial) x l/100 km = 282 and mpg (US) x l/100 km = 235

A number of chemicals and lubricants are available for use in motorcycle maintenance and repair. They include a wide variety of products ranging from cleaning solvents and degreasers to lubricants and protective sprays for rubber, plastic and vinyl.

• **Contact point/spark plug cleaner** is a solvent used to clean oily film and dirt from points, grim from electrical connectors and oil deposits from spark plugs. It is oil free and leaves no residue. It can also be used to remove gum and varnish from carburetor jets and other orifices.

• **Carburetor cleaner** is similar to contact point/spark plug cleaner but it usually has a stronger solvent and may leave a slight oily residue. It is not recommended for cleaning electrical components or connections.

• **Brake system cleaner** is used to remove brake dust, grease and brake fluid from the brake system, where clean surfaces are absolutely necessary. It leaves no residue and often eliminates brake squeal caused by contaminants.

• **Silicone-based lubricants** are used to protect rubber parts such as hoses and grommets, and are used as lubricants for hinges and locks.

• **Multi-purpose grease** is an all purpose lubricant used wherever grease is more practical than a liquid lubricant such as oil. Some multi-purpose grease is colored white and specially formulated to be more resistant to water than ordinary grease.

• **Gear oil** (sometimes called gear lube) is a specially designed oil used in transmissions and final drive units, as well as other areas where high friction, high temperature lubrication is required. It is available in a number of viscosities (weights) for various applications.

• **Motor oil** is the lubricant formulated for use in engines. It normally contains a wide variety of additives to prevent corrosion and reduce foaming and wear. Motor oil comes in various weights (viscosity ratings) from 0 to 50. The recommended weight of the oil depends on the season, temperature and the demands on the engine. Light oil is used in cold climates and under light load conditions. Heavy oil is used in hot climates and where high loads are encountered. Multi-viscosity oils are designed to have characteristics of both light and heavy oils and are available in a number of weights from 0W-20 to 20W-50.

• **Gasoline additives** perform several functions, depending on their chemical makeup. They usually contain solvents that help dissolve gum and varnish that build up on carburetor and inlet parts. They also serve to break down carbon deposits that form on the inside surfaces of the combustion chambers. Some additives contain upper cylinder lubricants for valves and piston rings.

• **Brake and clutch fluid** is a specially formulated hydraulic fluid that can withstand the heat and pressure encountered in break/clutch systems. Care must be taken that this fluid does not come in contact with painted surfaces or plastics. An opened container should always be resealed to prevent contamination by water or dirt.

• **Chain lubricants** are formulated especially for use on motorcycle final drive chains. A good chain lube should adhere well and have good penetrating qualities to be effective as a lubricant inside the chain and on the side plates, pins and rollers. Most chain lubes are either the foaming type or quick drying type and are usually marketed as sprays. Take care to use a lubricant marked as being suitable for O-ring chains.

• **Degreasers** are heavy duty solvents used to remove grease and grime that may accumulate on the engine and frame components. They can be sprayed or brushed on and, depending on the type, are rinsed with either water or solvent.

• **Solvents** are used alone or in combination with degreasers to clean parts and assemblies during repair and overhaul. The home mechanic should use only solvents that are non-flammable and that do not produce irritating fumes.

• **Gasket sealing compounds** may be used in conjunction with gaskets, to improve their sealing capabilities, or alone, to seal metal-to-metal joints. Many gasket sealers can withstand extreme heat, some are impervious to gasoline and lubricants, while others are capable of filling and sealing large cavities. Depending on the intended use, gasket sealers either dry hard or stay relatively soft and pliable. They are usually applied by hand, with a brush or are sprayed on the gasket sealing surfaces.

• **Thread locking compound** is an adhesive locking compound that prevents threaded fasteners from loosening because of vibration. It is available in a variety of types for different applications.

• **Moisture dispersants** are usually sprays that can be used to dry out electrical components such as the fuse block and wiring connectors. Some types an also be used as treatment for rubber and as a lubricant for hinges, cables and locks.

• **Waxes and polishes** are used to help protect painted and plated surfaces from the weather. Different types of pain may require the use of different types of wax polish. Some polishes utilize a chemical or abrasive cleaner to help remove the top layer of oxidized (dull) paint on older vehicles. In recent years, many non-wax polishes (that contain a wide variety of chemicals such as polymers and silicones) have been introduced. These non-wax polishes are usually easier to apply and last longer than conventional waxes and polishes.

Preparing for storage

Before you start

If repairs or an overhaul is needed, see that this is carried out now rather than left until you want to ride the bike again.

Give the bike a good wash and scrub all dirt from its underside. Make sure the bike dries completely before preparing for storage.

Engine

● Remove the spark plug(s) and lubricate the cylinder bores with approximately a teaspoon of motor oil using a spout-type oil can (see illustration 1). Reinstall the spark plug(s). Crank the engine over a couple of times to coat the piston rings and bores with oil. If the bike has a kickstart, use this to turn the engine over. If not, flick the kill switch to the OFF position and crank the engine over on the starter (see illustration 2). If the nature of the ignition system prevents the starter operating with the kill switch in the OFF position, remove the spark plugs and fit them back in

their caps; ensure that the plugs are grounded against the cylinder head when the starter is operated (see illustration 3).

Warning: It is important that the plugs are grounded away from the spark plug holes otherwise there is a risk of atomized fuel from the cylinders igniting.

HAYNES HINT *On a single cylinder four-stroke engine, you can seal the combustion chamber completely by positioning the piston at TDC on the compression stroke.*

● Drain the carburetor(s) otherwise there is a risk of jets becoming blocked by gum deposits from the fuel (see illustration 4).

● If the bike is going into long-term storage, consider adding a fuel stabilizer to the fuel in the tank. If the tank is drained completely, corrosion of its internal surfaces may occur if left unprotected for a long period. The tank can be treated with a rust preventative especially for this purpose. Alternatively, remove the tank and pour half a liter of motor oil into it, install the filler cap and shake the tank to coat its internals with oil before draining off the excess. The same effect can also be achieved by spraying WD40 or a similar water-dispersant around the inside of the tank via its flexible nozzle.

● Make sure the cooling system contains the correct mix of antifreeze. Antifreeze also contains important corrosion inhibitors.

● The air intakes and exhaust can be sealed off by covering or plugging the openings. Ensure that you do not seal in any condensation; run the engine until it is hot, then switch off and allow to cool. Tape a

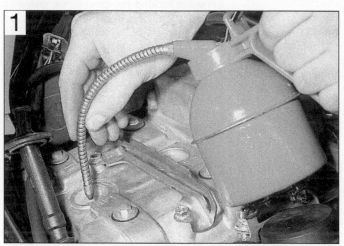

Squirt a drop of motor oil into each cylinder

Flick the kill switch to OFF . . .

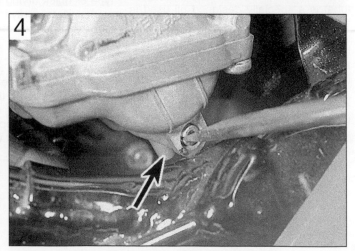

. . . and ensure that the metal bodies of the plugs (arrows) are grounded against the cylinder head

Connect a hose to the carburetor float chamber drain stub (arrow) and unscrew the drain screw

5

Exhausts can be sealed off with a plastic bag

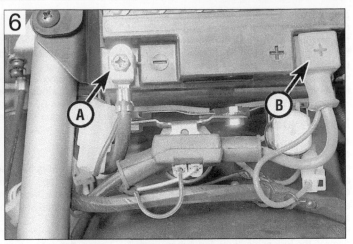

6

Disconnect the negative lead (A) first, followed by the positive lead (B)

piece of thick plastic over the silencer end(s) **(see illustration 5)**. Note that some advocate pouring a tablespoon of motor oil into the silencer(s) before sealing them off.

Battery

● Remove it from the bike - in extreme cases of cold the battery may freeze and crack its case **(see illustration 6)**.
● Check the electrolyte level and top up if necessary (conventional refillable batteries). Clean the terminals.
● Store the battery off the motorcycle and away from any sources of fire. Position a wooden block under the battery if it is to sit on the ground.
● Give the battery a trickle charge for a few hours every month **(see illustration 7)**.

Tires

● Place the bike on its centerstand or an auxiliary stand which will support the motorcycle in an upright position. Position wood blocks under the tires to keep them off the ground and to provide insulation from damp. If the bike is being put into long-term

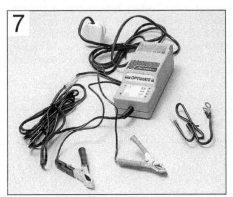

7

Use a suitable battery charger - this kit also assesses battery condition

storage, ideally both tires should be off the ground; not only will this protect the tires, but will also ensure that no load is placed on the steering head or wheel bearings.
● Deflate each tire by 5 to 10 psi, no more or the beads may unseat from the rim, making subsequent inflation difficult on tubeless tires.

Pivots and controls

● Lubricate all lever, pedal, stand and footrest pivot points. If grease nipples are fitted to the rear suspension components, apply lubricant to the pivots.
● Lubricate all control cables.

Cycle components

● Apply a wax protectant to all painted and plastic components. Wipe off any excess, but don't polish to a shine. Where fitted, clean the screen with soap and water.
● Coat metal parts with Vaseline (petroleum jelly). When applying this to the fork tubes, do not compress the forks otherwise the seals will rot from contact with the Vaseline.
● Apply a vinyl cleaner to the seat.

Storage conditions

● Aim to store the bike in a shed or garage which does not leak and is free from damp.
● Drape an old blanket or bedspread over the bike to protect it from dust and direct contact with sunlight (which will fade paint). Beware of tight-fitting plastic covers which may allow condensation to form and settle on the bike.

Getting back on the road

Engine and transmission

● Change the oil and replace the oil filter. If this was done prior to storage, check that the oil hasn't emulsified - a thick whitish substance which occurs through condensation.
● Remove the spark plugs. Using a spout-type oil can, squirt a few drops of oil into the cylinder(s). This will provide initial lubrication as the piston rings and bores comes back into contact. Service the spark plugs, or buy new ones, and install them in the engine.

● Check that the clutch isn't stuck on. The plates can stick together if left standing for some time, preventing clutch operation. Engage a gear and try rocking the bike back and forth with the clutch lever held against the handlebar. If this doesn't work on cable-operated clutches, hold the clutch lever back against the handlebar with a strong rubber band or cable tie for a couple of hours **(see illustration 8)**.
● If the air intakes or silencer end(s) were blocked off, remove the plug or cover used.
● If the fuel tank was coated with a rust

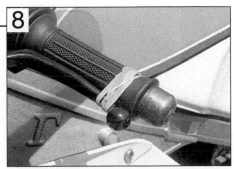

8

Hold the clutch lever back against the handlebar with rubber bands or a cable tie

preventative, oil or a stabilizer added to the fuel, drain and flush the tank and dispose of the fuel sensibly. If no action was taken with the fuel tank prior to storage, it is advised that the old fuel is disposed of since it will go bad over a period of time. Refill the fuel tank with fresh fuel.

Frame and running gear

● Oil all pivot points and cables.
● Check the tire pressures. They will definitely need inflating if pressures were reduced for storage.
● Lubricate the final drive chain (where applicable).
● Remove any protective coating applied to the fork tubes (stanchions) since this may well destroy the fork seals. If the fork tubes weren't protected and have picked up rust spots, remove them with very fine abrasive paper and refinish with metal polish.
● Check that both brakes operate correctly. Apply each brake hard and check that it's not possible to move the motorcycle forwards, then check that the brake frees off again once released. Brake caliper pistons can stick due to corrosion around the piston head, or on the sliding caliper types, due to corrosion of the slider pins. If the brake doesn't free after repeated operation, take the caliper off for examination. Similarly drum brakes can stick due to a seized operating cam, cable or rod linkage.
● If the motorcycle has been in long-term storage, replace the brake fluid and clutch fluid (where applicable).
● Depending on where the bike has been stored, the wiring, cables and hoses may have been nibbled by rodents. Make a visual check and investigate disturbed wiring loom tape.

Battery

● If the battery has been previously removed and given top up charges it can simply be reconnected. Remember to connect the positive cable first and the negative cable last.
● On conventional refillable batteries, if the battery has not received any attention, remove it from the motorcycle and check its electrolyte level. Top up if necessary then charge the battery. If the battery fails to hold a charge and a visual check show heavy white sulfation of the plates, the battery is probably defective and must be replaced. This is particularly likely if the battery is old. Confirm battery condition with a specific gravity check.
● On sealed (MF) batteries, if the battery has not received any attention, remove it from the motorcycle and charge it according to the information on the battery case - if the battery fails to hold a charge it must be replaced.

Starting procedure

● If a kickstart is fitted, turn the engine over a couple of times with the ignition OFF to distribute oil around the engine. If no kickstart is fitted, flick the engine kill switch OFF and the ignition ON and crank the engine over a couple of times to work oil around the upper cylinder components. If the nature of the ignition system is such that the starter won't work with the kill switch OFF, remove the spark plugs, fit them back into their caps and ground their bodies on the cylinder head. Reinstall the spark plugs afterwards.
● Switch the kill switch to RUN, operate the choke and start the engine. If the engine won't start don't continue cranking the engine - not only will this flatten the battery, but the starter motor will overheat. Switch the ignition off and try again later. If the engine refuses to start, go through the troubleshooting procedures in this manual. **Note:** *If the bike has been in storage for a long time, old fuel or a carburetor blockage may be the problem. Gum deposits in carburetors can block jets - if a carburetor cleaner doesn't prove successful the carburetors must be dismantled for cleaning.*
● Once the engine has started, check that the lights, turn signals and horn work properly.
● Treat the bike gently for the first ride and check all fluid levels on completion. Settle the bike back into the maintenance schedule.

This Section provides an easy reference-guide to the more common faults that are likely to afflict your machine. Obviously, the opportunities are almost limitless for faults to occur as a result of obscure failures, and to try and cover all eventualities would require a book. Indeed, a number have been written on the subject.

Successful troubleshooting is not a mysterious 'black art' but the application of a bit of knowledge combined with a systematic and logical approach to the problem. Approach any troubleshooting by first accurately identifying the symptom and then checking through the list of possible causes, starting with the simplest or most obvious and progressing in stages to the most complex. Take nothing for granted, but above all apply liberal quantities of common sense.

The main symptom of a fault is given in the text as a major heading below which are listed the various systems or areas which may contain the fault. Details of each possible cause for a fault and the remedial action to be taken are given. Further information should be sought in the relevant Chapter.

1 Engine doesn't start or is difficult to start
- [ ] Starter motor doesn't rotate
- [ ] Starter motor rotates but engine does not turn over
- [ ] Starter works but engine won't turn over (seized)
- [ ] No fuel flow
- [ ] Engine flooded
- [ ] No spark or weak spark
- [ ] Compression low
- [ ] Stalls after starting
- [ ] Rough idle

2 Poor running at low speed
- [ ] Spark weak
- [ ] Fuel/air mixture incorrect
- [ ] Compression low
- [ ] Poor acceleration

3 Poor running or no power at high speed
- [ ] Firing incorrect
- [ ] Fuel/air mixture incorrect
- [ ] Compression low
- [ ] Knocking or pinging
- [ ] Miscellaneous causes

4 Overheating
- [ ] Engine overheats
- [ ] Firing incorrect
- [ ] Fuel/air mixture incorrect
- [ ] Compression too high
- [ ] Engine load excessive
- [ ] Lubrication inadequate
- [ ] Miscellaneous causes

5 Clutch problems
- [ ] Clutch slipping
- [ ] Clutch not disengaging completely

6 Gearchanging problems
- [ ] Doesn't go into gear, or lever doesn't return
- [ ] Jumps out of gear
- [ ] Overselects

7 Abnormal engine noise
- [ ] Knocking or pinging
- [ ] Piston slap or rattling
- [ ] Valve noise
- [ ] Other noise

8 Abnormal driveline noise
- [ ] Clutch noise
- [ ] Transmission noise
- [ ] Final drive noise

9 Abnormal frame and suspension noise
- [ ] Front end noise
- [ ] Shock absorber noise
- [ ] Brake noise

10 Oil pressure warning light comes on
- [ ] Engine lubrication system
- [ ] Electrical system

11 Excessive exhaust smoke
- [ ] White smoke
- [ ] Black smoke

12 Poor handling or stability
- [ ] Handlebar hard to turn
- [ ] Handlebar shakes or vibrates excessively
- [ ] Handlebar pulls to one side
- [ ] Poor shock absorbing qualities

13 Braking problems
- [ ] Brakes are spongy, don't hold
- [ ] Brake lever or pedal pulsates
- [ ] Brakes drag

14 Electrical problems
- [ ] Battery dead or weak
- [ ] Battery overcharged

1 Engine doesn't start or is difficult to start

Starter motor doesn't rotate

☐ Engine kill switch OFF.
☐ Fuse blown. Check main fuse and starter circuit fuse (Chapter 9).
☐ Battery voltage low. Check and recharge battery (Chapter 9).
☐ Starter motor defective. Make sure the wiring to the starter is secure. Make sure the starter relay clicks when the start button is pushed. If the relay clicks, then the fault is in the wiring or motor.
☐ Starter relay faulty. Check it according to the procedure in Chap-ter 9.
☐ Starter switch not contacting. The contacts could be wet, corroded or dirty. Disassemble and clean the switch (Chapter 9).
☐ Wiring open or shorted. Check all wiring connections and harnesses to make sure that they are dry, tight and not corroded. Also check for broken or frayed wires that can cause a short to ground (see wiring diagram, Chapter 9).
☐ Ignition (main) switch defective. Check the switch according to the procedure in Chapter 9. Replace the switch with a new one if it is defective.
☐ Engine kill switch defective. Check for wet, dirty or corroded contacts. Clean or replace the switch as necessary (Chapter 9).
☐ Faulty neutral, side stand or clutch switch. Check the wiring to each switch and the switch itself according to the procedures in Chapter 9.

Starter motor rotates but engine does not turn over

☐ Starter motor clutch defective. Inspect and repair or replace (Chapter 2).
☐ Damaged idler or starter gears. Inspect and replace the damaged parts (Chapter 2).

Starter works but engine won't turn over (seized)

☐ Seized engine caused by one or more internally damaged components. Failure due to wear, abuse or lack of lubrication. Damage can include seized valves, followers/rocker arms, camshafts, pistons, crankshaft, connecting rod bearings, or transmission gears or bearings. Refer to Chapter 2 for engine disassembly.

No fuel flow

☐ No fuel in tank.
☐ Fuel tank breather hose obstructed.
☐ Fuel filter is blocked (see Chapter 1).

Engine flooded

☐ Starting technique incorrect. Under normal circumstances the machine should start with little or no throttle. When the engine is cold, the choke should be operated and the engine started without opening the throttle. When the engine is at operating temperature, only a very slight amount of throttle should be necessary.

No spark or weak spark

☐ Ignition switch OFF.
☐ Engine kill switch turned to the OFF position.
☐ Battery voltage low. Check and recharge the battery as necessary (Chapter 9).
☐ Spark plugs dirty, defective or worn out. Locate reason for fouled plugs using spark plug condition chart and follow the plug maintenance procedures (Chapter 1).
☐ Spark plug caps or secondary (HT) wiring faulty. Check condition. Replace either or both components if cracks or deterioration are evident (Chapter 5).

☐ Spark plug caps not making good contact. Make sure that the plug caps fit snugly over the plug ends.
☐ Ignition HT coils defective. Check the coils, referring to Chapter 5.
☐ IC igniter unit defective. Refer to Chapter 5 for details.
☐ Pick-up coil defective. Check the unit, referring to Chapter 5 for details.
☐ Ignition or kill switch shorted. This is usually caused by water, corrosion, damage or excessive wear. The switches can be disassembled and cleaned with electrical contact cleaner. If cleaning does not help, replace the switches (Chapter 9).
☐ Wiring shorted or broken between:

a) Ignition (main) switch and engine kill switch (or blown fuse)
b) IC igniter unit and engine kill switch
c) IC igniter unit and ignition HT coils
d) Ignition HT coils and spark plugs
e) IC igniter unit and ignition pick-up coil

☐ Make sure that all wiring connections are clean, dry and tight. Look for chafed and broken wires (Chapters 5 and 9).

Compression low

☐ Spark plugs loose. Remove the plugs and inspect their threads. Reinstall and tighten to the specified torque (Chapter 1).
☐ Cylinder head not sufficiently tightened down. If the cylinder head is suspected of being loose, then there's a chance that the gasket or head is damaged if the problem has persisted for any length of time. The head bolts should be tightened to the proper torque in the correct sequence (Chapter 2).
☐ Improper valve clearance. This means that the valve is not closing completely and compression pressure is leaking past the valve. Check and adjust the valve clearances (Chapter 1).
☐ Cylinder and/or piston worn. Excessive wear will cause compression pressure to leak past the rings. This is usually accompanied by worn rings as well. A top-end overhaul is necessary (Chapter 2).
☐ Piston rings worn, weak, broken, or sticking. Broken or sticking piston rings usually indicate a lubrication or fuelling problem that causes excess carbon deposits or seizures to form on the pistons and rings. Top-end overhaul is necessary (Chapter 2).
☐ Piston ring-to-groove clearance excessive. This is caused by excessive wear of the piston ring lands. Piston replacement is necessary (Chapter 2).
☐ Cylinder head gasket damaged. If a head is allowed to become loose, or if excessive carbon build-up on the piston crown and combustion chamber causes extremely high compression, the head gasket may leak. Retorquing the head is not always sufficient to restore the seal, so gasket replacement is necessary (Chapter 2).
☐ Cylinder head warped. This is caused by overheating or improperly tightened head bolts. Machine shop resurfacing or head replacement is necessary (Chapter 2).
☐ Valve spring broken or weak. Caused by component failure or wear; the springs must be replaced (Chapter 2).
☐ Valve not seating properly. This is caused by a bent valve (from over-revving or improper valve adjustment), burned valve or seat (improper fuelling) or an accumulation of carbon deposits on the seat (from fuelling or lubrication problems). The valves must be cleaned and/or replaced and the seats serviced if possible (Chapter 2).

1 Engine doesn't start or is difficult to start (continued)

Stalls after starting

☐ Improper choke action (carbureted models). Make sure the choke linkage shaft is getting a full stroke and staying in the out position (Chapter 4).
☐ Ignition malfunction. See Chapter 5.
☐ Carburetor or fuel injection malfunction. See Chapter 4.
☐ Fuel contaminated. The fuel can be contaminated with either dirt or water, or can change chemically if the machine is allowed to sit for several months or more. Drain the tank (Chapter 4).
☐ Intake air leak. Check for loose intake manifold retaining clips and damaged/disconnected vacuum hoses (Chapter 4).
☐ Engine idle speed incorrect. Turn idle adjusting screw until the engine idles at the specified rpm (Chapter 1).

Rough idle

☐ Ignition malfunction. See Chapter 5.
☐ Idle speed incorrect. See Chapter 1.
☐ Carburetors not synchronized (twin carburetor models). Adjust them with vacuum gauge or manometer set as described in Chapter 4.
☐ Carburetor or fuel injection malfunction. See Chapter 4.
☐ Fuel contaminated. The fuel can be contaminated with either dirt or water, or can change chemically if the machine is allowed to sit for several months or more. Drain the tank (Chapter 4).
☐ Intake air leak. Check for loose intake manifold retaining clips and damaged/disconnected vacuum hoses. Replace the intake ducts if they are split or deteriorated (Chapter 4).
☐ Air filter clogged. Replace the air filter element (Chapter 1).

2 Poor running at low speeds

Spark weak

☐ Battery voltage low. Check and recharge battery (Chapter 9).
☐ Spark plugs fouled, defective or worn out. Refer to Chapter 1 for spark plug maintenance.
☐ Spark plug cap or HT wiring defective. Refer to Chapters 1 and 5 for details on the ignition system.
☐ Spark plug caps not making contact.
☐ Incorrect spark plugs. Wrong type, heat range or cap configuration. Check and install correct plugs listed in Chapter 1.
☐ IC igniter unit or ECM faulty. See Chapter 5.
☐ Pick-up coil defective. See Chapter 5.
☐ Ignition coils defective. See Chapter 5.

Fuel/air mixture incorrect

☐ Pilot screws incorrectly set (Chapter 4)
☐ Pilot jet or air passage blocked. Remove and overhaul the carburetors (Chapter 4).
☐ Air filter clogged, poorly sealed or missing (Chapter 1).
☐ Air filter housing poorly sealed. Look for cracks, holes or loose clamps and replace or repair defective parts.
☐ Fuel tank breather hose obstructed.
☐ Intake air leak. Check for loose intake manifold retaining clips and damaged/disconnected vacuum hoses. Replace the intake ducts if they are split or deteriorated (Chapter 4).

Compression low

☐ Spark plugs loose. Remove the plugs and inspect their threads. Reinstall and tighten to the specified torque (Chapter 1).
☐ Cylinder head not sufficiently tightened down. If the cylinder head is suspected of being loose, then there's a chance that the gasket and head are damaged if the problem has persisted for any length of time. The head bolts should be tightened to the proper torque in the correct sequence (Chapter 2).
☐ Improper valve clearance. This means that the valve is not closing completely and compression pressure is leaking past the valve. Check and adjust the valve clearances (Chapter 1).

☐ Cylinder and/or piston worn. Excessive wear will cause compression pressure to leak past the rings. This is usually accompanied by worn rings as well. A top-end overhaul is necessary (Chapter 2).
☐ Piston rings worn, weak, broken, or sticking. Broken or sticking piston rings usually indicate a lubrication or fuelling problem that causes excess carbon deposits or seizures to form on the pistons and rings. Top-end overhaul is necessary (Chapter 2).
☐ Piston ring-to-groove clearance excessive. This is caused by excessive wear of the piston ring lands. Piston replacement is necessary (Chapter 2).
☐ Cylinder head gasket damaged. If a head is allowed to become loose, or if excessive carbon build-up on the piston crown and combustion chamber causes extremely high compression, the head gasket may leak. Retorquing the head is not always sufficient to restore the seal, so gasket replacement is necessary (Chapter 2).
☐ Cylinder head warped. This is caused by overheating or improperly tightened head bolts. Machine shop resurfacing or head replacement is necessary (Chapter 2).
☐ Valve spring broken or weak. Caused by component failure or wear; the springs must be replaced (Chapter 2).
☐ Valve not seating properly. This is caused by a bent valve (from over-revving or improper valve adjustment), burned valve or seat (improper fuelling) or an accumulation of carbon deposits on the seat (from fuelling, lubrication problems). The valves must be cleaned and/or replaced and the seats serviced if possible (Chapter 2).

Poor acceleration

☐ Fuel system fault. Remove and overhaul the carburetors or check the fuel injection system (Chapter 4).
☐ Engine oil viscosity too high. Using a heavier oil than that recommended in Chapter 1 can damage the oil pump or lubrication system and cause drag on the engine.
☐ Brakes dragging. Usually caused by debris which has entered the brake piston seals, or from a warped disc or drum or bent axle. Repair as necessary (Chapter 7).

3 Poor running or no power at high speed

Firing incorrect

☐ Air filter restricted. Clean or replace filter (Chapter 1).

☐ Spark plugs fouled, defective or worn out. See Chapter 1 for spark plug maintenance.

☐ Spark plug caps or HT wiring defective. See Chapters 1 and 5 for details of the ignition system.

☐ Spark plug caps not in good contact. See Chapter 5.

☐ Incorrect spark plugs. Wrong type, heat range or cap configuration. Check and install correct plugs listed in Chapter 1.

☐ IC igniter unit or ECM defective. See Chapter 5.

☐ Pick-up coil defective. See Chapter 5.

☐ Ignition coils defective. See Chapter 5.

Fuel/air mixture incorrect

☐ Fuel system fault. Remove and overhaul the carburetors or check the fuel injection system (Chapter 4).

☐ Air filter clogged, poorly sealed, or missing (Chapter 1).

☐ Air filter housing poorly sealed. Look for cracks, holes or loose clamps, and replace or repair defective parts.

☐ Fuel tank breather hose obstructed.

☐ Intake air leak. Check for loose intake manifold retaining clips and damaged/disconnected vacuum hoses. Replace the intake ducts if they are split or deteriorated (Chapter 4).

Compression low

☐ Spark plugs loose. Remove the plugs and inspect their threads. Reinstall and tighten to the specified torque (Chapter 1).

☐ Cylinder head not sufficiently tightened down. If the cylinder head is suspected of being loose, then there's a chance that the gasket and head are damaged if the problem has persisted for any length of time. The head bolts should be tightened to the proper torque in the correct sequence (Chapter 2).

☐ Improper valve clearance. This means that the valve is not closing completely and compression pressure is leaking past the valve. Check and adjust the valve clearances (Chapter 1).

☐ Cylinder and/or piston worn. Excessive wear will cause compression pressure to leak past the rings. This is usually accompanied by worn rings as well. A top-end overhaul is necessary (Chapter 2).

☐ Piston rings worn, weak, broken, or sticking. Broken or sticking piston rings usually indicate a lubrication or fueling problem that causes excess carbon deposits or seizures to form on the pistons and rings. Top-end overhaul is necessary (Chapter 2).

☐ Piston ring-to-groove clearance excessive. This is caused by excessive wear of the piston ring lands. Piston replacement is necessary (Chapter 2).

☐ Cylinder head gasket damaged. If a head is allowed to become loose, or if excessive carbon build-up on the piston crown and combustion chamber causes extremely high compression, the head gasket may leak. Retorquing the head is not always sufficient to restore the seal, so gasket replacement is necessary (Chapter 2).

☐ Cylinder head warped. This is caused by overheating or improperly tightened head bolts. Machine shop resurfacing or head replacement is necessary (Chapter 2).

☐ Valve spring broken or weak. Caused by component failure or wear; the springs must be replaced (Chapter 2).

☐ Valve not seating properly. This is caused by a bent valve (from over-revving or improper valve adjustment), burned valve or seat (improper fuelling) or an accumulation of carbon deposits on the seat (from fuelling or lubrication problems). The valves must be cleaned and/or replaced and the seats serviced if possible (Chapter 2).

Knocking or pinging

☐ Carbon build-up in combustion chamber. Use of a fuel additive that will dissolve the adhesive bonding the carbon particles to the crown and chamber is the easiest way to remove the build-up. Otherwise, the cylinder head will have to be removed and decarbonized (Chapter 2).

☐ Incorrect or poor quality fuel. Old or improper grades of fuel can cause detonation. This causes the piston to rattle, thus the knocking or pinging sound. Drain old fuel and always use the recommended fuel grade.

☐ Spark plug heat range incorrect. Uncontrolled detonation indicates the plug heat range is too hot. The plug in effect becomes a glow plug, raising cylinder temperatures. Install the proper heat range plug (Chapter 1).

☐ Improper air/fuel mixture. This will cause the cylinders to run hot, which leads to detonation. An intake air leak can cause this imbalance. See Chapter 4.

Miscellaneous causes

☐ Throttle valve doesn't open fully. Adjust the throttle grip freeplay (Chapter 1).

☐ Clutch slipping. May be caused by loose or worn clutch components. Refer to Chapter 2 for clutch overhaul procedures.

☐ Engine oil viscosity too high. Using a heavier oil than the one recommended in Chapter 1 can damage the oil pump or lubrication system and cause drag on the engine.

☐ Brakes dragging. Usually caused by debris which has entered the brake piston seals, or from a warped disc, out-of-round drum or bent axle. Repair as necessary.

4 Overheating

Engine overheats

☐ Coolant level low. Check and add coolant (Chapter 1).
☐ Leak in cooling system. Check cooling system hoses and radiator for leaks and other damage. Repair or replace parts as necessary (Chapter 3).
☐ Thermostat sticking open or closed. Check and replace as described in Chapter 3.
☐ Faulty pressure cap. Remove the cap and have it pressure tested (Chapter 3).
☐ Coolant passages clogged. Have the entire system drained and flushed, then refill with fresh coolant.
☐ Water pump defective. Remove the pump and check the components (Chapter 3).
☐ Clogged radiator fins. Clean them by blowing compressed air through the fins from the backside.
☐ Cooling fan or fan switch fault (Chapter 3).

Firing incorrect

☐ Spark plugs fouled, defective or worn out. See Chapter 1 for spark plug maintenance.
☐ Incorrect spark plugs.
☐ IC igniter unit or ECM defective. See Chapter 5.
☐ Pick-up coil faulty. See Chapter 5.
☐ Faulty ignition coils. See Chapter 5.

Fuel/air mixture incorrect

☐ Fuel system fault. Remove and overhaul the carburetors or check the fuel injection system (Chapter 4).
☐ Air filter clogged, poorly sealed, or missing (Chapter 1).
☐ Air filter housing poorly sealed. Look for cracks, holes or loose clamps, and replace or repair defective parts.
☐ Fuel tank breather hose obstructed.
☐ Intake air leak. Check for loose intake manifold retaining clips and damaged/disconnected vacuum hoses. Replace the intake manifold(s) if they are split or deteriorated (Chapter 4).

Compression too high

☐ Carbon build-up in combustion chamber. Use of a fuel additive that will dissolve the adhesive bonding the carbon particles to the piston crown and chamber is the easiest way to remove the build-up. Otherwise, the cylinder head will have to be removed and decarbonized (Chapter 2).
☐ Improperly machined head surface or installation of incorrect gasket during engine assembly.

Engine load excessive

☐ Clutch slipping. Can be caused by damaged, loose or worn clutch components. Refer to Chapter 2 for overhaul procedures.
☐ Engine oil level too high. The addition of too much oil will cause pressurization of the crankcase and inefficient engine operation. Check Specifications and drain to proper level (Chapter 1).
☐ Engine oil viscosity too high. Using a heavier oil than the one recommended in Chapter 1 can damage the oil pump or lubrication system as well as cause drag on the engine.
☐ Brakes dragging. Usually caused by debris which has entered the brake piston seals, or from a warped disc or drum or bent axle. Repair as necessary.

Lubrication inadequate

☐ Engine oil level too low. Friction caused by intermittent lack of lubrication or from oil that is overworked can cause overheating. The oil provides a definite cooling function in the engine. Check the oil level (Chapter 1).
☐ Poor quality engine oil or incorrect viscosity or type. Oil is rated not only according to viscosity but also according to type. Some oils are not rated high enough for use in this engine. Check the Specifications section and change to the correct oil (Chapter 1).

Miscellaneous causes

☐ Modification to exhaust system. Most aftermarket exhaust systems cause the engine to run leaner, which makes it run hotter.

5 Clutch problems

Clutch slipping

☐ Clutch cable freeplay incorrectly adjusted (cable clutch models) (Chapter 1).
☐ Friction plates worn or warped. Overhaul the clutch assembly (Chapter 2).
☐ Plain plates warped (Chapter 2).
☐ Clutch springs broken or weak. Old or heat-damaged (from slipping clutch) springs should be replaced with new ones (Chapter 2).
☐ Clutch pushrod bent. Check and, if necessary, replace (Chapter 2).
☐ Clutch center or housing unevenly worn. This causes improper engagement of the plates. Replace the damaged or worn parts (Chapter 2).
☐ Wrong type of oil used. Make sure oil with anti-friction additives (such as molybdenum disulfide) is NOT used in an engine/transmission with a wet clutch.

Clutch not disengaging completely

☐ Clutch cable freeplay incorrectly adjusted (Chapter 1).
☐ Air in clutch hydraulic system (hydraulic clutch models). Bleed the system (Chapter 2).

☐ Worn master or slave cylinder (hydraulic clutch models). Inspect and repair or replace as necessary (Chapter 2).
☐ Clutch plates warped or damaged. This will cause clutch drag, which in turn will cause the machine to creep. Overhaul the clutch assembly (Chapter 2).
☐ Clutch spring tension uneven. Usually caused by a sagged or broken spring. Check and replace the springs as a set (Chapter 2).
☐ Engine oil deteriorated. Old, thin, worn out oil will not provide proper lubrication for the plates, causing the clutch to drag. Replace the oil and filter (Chapter 1).
☐ Engine oil viscosity too high. Using a heavier oil than recommended in Chapter 1 can cause the plates to stick together, putting a drag on the engine. Change to the correct weight oil (Chapter 1).
☐ Clutch housing bearing seized. Lack of lubrication, severe wear or damage can cause the bearing to seize on the input shaft. Overhaul of the clutch, and perhaps transmission, may be necessary to repair the damage (Chapter 2).
☐ Loose clutch center nut. Causes housing and center misalignment putting a drag on the engine. Engagement adjustment continually varies. Overhaul the clutch assembly (Chapter 2).

6 Gearchanging problems

Doesn't go into gear or lever doesn't return

☐ Clutch not disengaging. See above.

☐ Shift fork(s) bent or seized. Often caused by dropping the machine or from lack of lubrication. Overhaul the transmission (Chapter 2).

☐ Gear(s) stuck on shaft. Most often caused by a lack of lubrication or excessive wear in transmission bearings and bushings. Overhaul the transmission (Chapter 2).

☐ Gear shift drum binding. Caused by lubrication failure or excessive wear. Replace the drum and bearing (Chapter 2).

☐ Gear shift lever pawl spring weak or broken (Chapter 2).

☐ Gear shift lever broken. Splines stripped out of lever or shaft, caused by allowing the lever to get loose or from dropping the machine. Replace necessary parts (Chapter 2).

☐ Gear shift mechanism stopper arm broken or worn. Full engagement and rotary movement of shift drum results. Replace the arm (Chapter 2).

☐ Stopper arm spring broken. Allows arm to float, causing sporadic shift operation. Replace spring (Chapter 2).

Jumps out of gear

☐ Shift fork(s) worn. Overhaul the transmission (Chapter 2).

☐ Gear groove(s) worn. Overhaul the transmission (Chapter 2).

☐ Gear dogs or dog slots worn or damaged. The gears should be inspected and replaced. No attempt should be made to service the worn parts.

Overselects

☐ Stopper arm spring weak or broken (Chapter 2).

☐ Return spring post broken or distorted (Chapter 2).

7 Abnormal engine noise

Knocking or pinging

☐ Carbon build-up in combustion chamber. Use of a fuel additive that will dissolve the adhesive bonding the carbon particles to the piston crown and chamber is the easiest way to remove the build-up. Otherwise, the cylinder head will have to be removed and decarbonized (Chapter 2).

☐ Incorrect or poor quality fuel. Old or improper fuel can cause detonation. This causes the pistons to rattle, thus the knocking or pinging sound. Drain the old fuel and always use the recommended grade fuel (Chapter 4).

☐ Spark plug heat range incorrect. Uncontrolled detonation indicates that the plug heat range is too hot. The plug in effect becomes a glow plug, raising cylinder temperatures. Install the proper heat range plug (Chapter 1).

☐ Improper air/fuel mixture. This will cause the cylinders to run hot and lead to detonation. Blocked carburetor jets or an air leak can cause this imbalance. See Chapter 4.

Piston slap or rattling

☐ Cylinder-to-piston clearance excessive. Caused by improper assembly. Inspect and overhaul top-end parts (Chapter 2).

☐ Connecting rod bent. Caused by over-revving, trying to start a badly flooded engine or from ingesting a foreign object into the combustion chamber. Replace the damaged parts (Chapter 2).

☐ Piston pin or piston pin bore worn or seized from wear or lack of lubrication. Replace damaged parts (Chapter 2).

☐ Piston ring(s) worn, broken or sticking. Overhaul the top-end (Chapter 2).

☐ Piston seizure damage. Usually from lack of lubrication or overheating. Replace the pistons and cylinders, as necessary (Chapter 2).

☐ Connecting rod bearing clearance excessive. Caused by excessive wear or lack of lubrication. Replace worn parts.

Valve noise

☐ Incorrect valve clearances. Adjust the clearances by referring to Chapter 1.

☐ Valve spring broken or weak. Check and replace weak valve springs (Chapter 2).

☐ Camshaft or cylinder head worn or damaged. Lack of lubrication at high rpm is usually the cause of damage. Insufficient oil or failure to change the oil at the recommended intervals are the chief causes. Since there are no replaceable bearings in the head, the head itself will have to be replaced if there is excessive wear or damage (Chapter 2).

Other noise

☐ Cylinder head gasket leaking.

☐ Exhaust pipe leaking at cylinder head connection. Caused by improper fit of pipe(s) or loose exhaust nuts. All exhaust fasteners should be tightened evenly and carefully. Failure to do this will lead to a leak.

☐ Crankshaft runout excessive. Caused by a bent crankshaft (from over-revving) or damage from an upper cylinder component failure. Can also be attributed to dropping the machine on either of the crankshaft ends.

☐ Engine mounting bolts loose. Tighten all engine mount bolts (Chapter 2).

☐ Crankshaft bearings worn (Chapter 2).

☐ Cam chain, tensioner or guides worn. Replace according to the procedure in Chapter 2.

8 Abnormal driveline noise

Clutch noise

- ☐ Clutch outer drum/friction plate clearance excessive (Chapter 2).
- ☐ Loose or damaged clutch pressure plate and/or bolts (Chapter 2).

Transmission noise

- ☐ Bearings worn. Also includes the possibility that the shafts are worn. Overhaul the transmission (Chapter 2).
- ☐ Gears worn or chipped (Chapter 2).
- ☐ Metal chips jammed in gear teeth. Probably pieces from a broken clutch, gear or shift mechanism that were picked up by the gears. This will cause early bearing failure (Chapter 2).

- ☐ Engine oil level too low. Causes a howl from transmission. Also affects engine power and clutch operation (Chapter 1).

Final drive noise

- ☐ Chain or drive belt not adjusted properly (Chapter 1).
- ☐ Front or rear sprocket loose. Tighten fasteners (Chapter 6).
- ☐ Sprockets worn. Replace sprockets (Chapter 6).
- ☐ Rear sprocket warped. Replace sprockets (Chapter 6).
- ☐ Differential worn or damaged. Have it repaired or replace it (Chapter 7).

9 Abnormal frame and suspension noise

Front end noise

- ☐ Low fluid level or improper viscosity oil in forks. This can sound like spurting and is usually accompanied by irregular fork action (Chapter 6).
- ☐ Spring weak or broken. Makes a clicking or scraping sound. Fork oil, when drained, will have a lot of metal particles in it (Chapter 6).
- ☐ Steering head bearings loose or damaged. Clicks when braking. Check and adjust or replace as necessary (Chapters 1 and 6).
- ☐ Triple clamps loose. Make sure all clamp bolts are tightened to the specified torque (Chapter 6).
- ☐ Fork tube bent. Good possibility if machine has been dropped. Replace tube with a new one (Chapter 6).
- ☐ Front axle bolt or axle pinch bolts loose. Tighten them to the specified torque (Chapter 7).
- ☐ Loose or worn wheel bearings. Check and replace as needed (Chapter 7).

Shock absorber noise

- ☐ Fluid level incorrect. Indicates a leak caused by defective seal. Shock will be covered with oil. Replace shock or seek advice on repair from a dealer (Chapter 6).
- ☐ Defective shock absorber with internal damage. This is in the body of the shock and can't be remedied. The shock must be replaced with a new one (Chapter 6).

- ☐ Bent or damaged shock body. Replace the shock with a new one (Chapter 6).
- ☐ Loose or worn suspension linkage components (EX250 models). Check and replace as necessary (Chapter 6).

Brake noise

- ☐ Squeal caused by dust on brake pads. Usually found in combination with glazed pads. Clean using brake cleaning solvent (Chapter 7).
- ☐ Contamination of brake pads. Oil, brake fluid or dirt causing brake to chatter or squeal. Clean or replace pads (Chapter 7).
- ☐ Pads glazed. Caused by excessive heat from prolonged use or from contamination. Do not use sandpaper/emery cloth or any other abrasive to roughen the pad surfaces as abrasives will stay in the pad material and damage the disc. A very fine flat file can be used, but pad replacement is suggested as a cure (Chapter 7).
- ☐ Disc warped. Can cause a chattering, clicking or intermittent squeal. Usually accompanied by a pulsating lever and uneven braking. Replace the disc (Chapter 7).
- ☐ Loose or worn wheel bearings. Check and replace as needed (Chapter 7).

10 Oil pressure warning light comes on

Engine lubrication system

- ☐ Engine oil pump defective, blocked oil strainer screen or failed relief valve. Carry out oil pressure check (Chapter 1).
- ☐ Engine oil level low. Inspect for leak or other problem causing low oil level and add recommended oil (Chapter 1).
- ☐ Engine oil viscosity too low. Very old, thin oil or an improper weight of oil used in the engine. Change to correct oil (Chapter 1).
- ☐ Camshaft or journals worn. Excessive wear causing drop in oil pressure. Replace cam and/or cylinder head. Abnormal wear could be caused by oil starvation at high rpm from low oil level or improper weight or type of oil (Chapter 1).

- ☐ Crankshaft and/or bearings worn. Same problems as above. Check and replace crankshaft and/or bearings (Chapter 2).

Electrical system

- ☐ Oil pressure switch defective. Check the switch according to the procedure in Chapter 9. Replace it if it is defective.
- ☐ Oil pressure indicator light circuit defective. Check for pinched, shorted, disconnected or damaged wiring (Chapter 9).

11 Excessive exhaust smoke

White smoke

☐ Piston oil ring worn. The ring may be broken or damaged, causing oil from the crankcase to be pulled past the piston into the combustion chamber. Replace the rings with new ones (Chapter 2).

☐ Cylinders worn, cracked, or scored. Caused by overheating or oil starvation. Install a new cylinder block (Chapter 2).

☐ Valve oil seal damaged or worn. Replace oil seals with new ones (Chapter 2).

☐ Valve guide worn. Perform a complete valve job (Chapter 2).

☐ Engine oil level too high, which causes the oil to be forced past the rings. Drain oil to the proper level (Chapter 1).

☐ Head gasket broken between oil return and cylinder. Causes oil to be pulled into the combustion chamber. Replace the head gasket and check the head for warpage (Chapter 2).

☐ Abnormal crankcase pressurization, which forces oil past the rings. Clogged breather is usually the cause.

Black smoke

☐ Air filter clogged. Clean or replace the element (Chapter 1).

☐ Carburetor flooding. Remove and overhaul the carburetor(s) (Chapter 4).

☐ Main jet too large. Remove and overhaul the carburetor(s) (Chapter 4).

☐ Choke cable stuck (Chapter 4).

☐ Fuel level too high. Check the fuel level (Chapter 4).

☐ Fuel injector problem (Chapter 4).

12 Poor handling or stability

Handlebar hard to turn

☐ Steering head bearing adjuster nut too tight. Check adjustment as described in Chapter 1.

☐ Bearings damaged. Roughness can be felt as the bars are turned from side-to-side. Replace bearings and races (Chapter 6).

☐ Races dented or worn. Denting results from wear in only one position (e.g., straight ahead), from a collision or hitting a pothole or from dropping the machine. Replace races and bearings (Chapter 6).

☐ Steering stem lubrication inadequate. Causes are grease getting hard from age or being washed out by high pressure car washes. Disassemble steering head and repack bearings (Chapter 6).

☐ Steering stem bent. Caused by a collision, hitting a pothole or by dropping the machine. Replace damaged part. Don't try to straighten the steering stem (Chapter 6).

☐ Front tire air pressure too low (Chapter 1).

Handlebar shakes or vibrates excessively

☐ Tires worn or out of balance (Chapter 7).

☐ Swingarm bearings worn. Replace worn bearings (Chapter 6).

☐ Wheel rim(s) warped or damaged. Inspect wheels for runout (Chapter 7).

☐ Spokes loose (see Chapter 1).

☐ Wheel bearings worn. Worn front or rear wheel bearings can cause poor tracking. Worn front bearings will cause wobble (Chapter 7).

☐ Handlebar clamp bolts loose (Chapter 6).

☐ Fork yoke bolts loose. Tighten them to the specified torque (Chapter 6).

☐ Engine mounting bolts loose. Will cause excessive vibration with increased engine rpm (Chapter 2).

Handlebar pulls to one side

☐ Frame bent. Definitely suspect this if the machine has been dropped. May or may not be accompanied by cracking near the bend. Replace the frame (Chapter 8).

☐ Wheels out of alignment. Caused by improper location of axle spacers or from bent steering stem or frame (Chapters 6 and 8).

☐ Swingarm bent or twisted. Caused by age (metal fatigue) or impact damage. Replace the arm (Chapter 6).

☐ Steering stem bent. Caused by impact damage or by dropping the motorcycle. Replace the steering stem (Chapter 6).

☐ Fork tube bent. Disassemble the forks and replace the damaged parts (Chapter 6).

☐ Fork oil level uneven. Check and add or drain as necessary (Chapter 1).

Poor shock absorbing qualities

Too hard:

   a) Fork oil level excessive (Chapter 1).

   b) Fork oil viscosity too high. Use a lighter oil (see the Specifications in Chapter 1).

   c) Fork tube bent. Causes a harsh, sticking feeling (Chapter 6).

   d) Shock shaft or body bent or damaged (Chapter 6).

   e) Fork internal damage (Chapter 6).

   f) Shock internal damage.

   g) Tire pressure too high (Chapter 1).

Too soft:

   a) Fork or shock oil insufficient and/or leaking (Chapter 1).

   b) Fork oil level too low (Chapter 6).

   c) Fork oil viscosity too light (Chapter 6).

   d) Fork springs weak or broken (Chapter 6).

   e) Shock internal damage or leakage (Chapter 6).

13 Braking problems

Brakes are spongy, don't hold

☐ Air in brake line. Caused by inattention to master cylinder fluid level or by leakage. Locate problem and bleed brakes (Chapter 7).
☐ Pad or disc worn (Chapters 1 and 7).
☐ Brake fluid leak. See paragraph 1.
☐ Contaminated pads. Caused by contamination with oil, grease, brake fluid, etc. Clean or replace pads. Clean disc thoroughly with brake cleaner (Chapter 7).
☐ Brake fluid deteriorated. Fluid is old or contaminated. Drain system, replenish with new fluid and bleed the system (Chapter 7).
☐ Master cylinder internal parts worn or damaged causing fluid to bypass (Chapter 7).
☐ Master cylinder bore scratched by foreign material or broken spring. Repair or replace master cylinder (Chapter 7).
☐ Disc warped. Replace disc (Chapter 7).

Brake lever or pedal pulsates

☐ Disc warped. Replace disc (Chapter 7).

☐ Axle bent. Replace axle (Chapter 7).
☐ Brake caliper bolts loose (Chapter 7).
☐ Wheel warped or otherwise damaged (Chapter 7).
☐ Wheel bearings damaged or worn (Chapter 7).
☐ Brake drum out of round. Replace brake drum.

Brakes drag

☐ Master cylinder piston seized. Caused by wear or damage to piston or cylinder bore (Chapter 7).
☐ Lever binding. Check pivot and lubricate (Chapter 7).
☐ Brake caliper piston seized in bore. Caused by wear or ingestion of dirt past deteriorated seal (Chapter 7).
☐ Brake caliper mounting bracket pins corroded. Clean off corrosion and lubricate (Chapter 7).
☐ Brake pad or shoe damaged. Material separated from backing plate. Usually caused by faulty manufacturing process or from contact with chemicals. Replace pads or shoes (Chapter 7).
☐ Pads improperly installed (Chapter 7).
☐ Drum brake springs weak. Replace springs

14 Electrical problems

Battery dead or weak

☐ Battery faulty. Caused by sulfated plates which are shorted through sedimentation. Also, broken battery terminal making only occasional contact (Chapter 9).
☐ Battery cables making poor contact (Chapter 9).
☐ Load excessive. Caused by addition of high wattage lights or other electrical accessories.
☐ Ignition (main) switch defective. Switch either grounds internally or fails to shut off system. Replace the switch (Chapter 9).
☐ Regulator/rectifier defective (Chapter 9).
☐ Alternator stator coil open or shorted (Chapter 9).
☐ Wiring faulty. Wiring grounded or connections loose in ignition, charging or lighting circuits (Chapter 9).

Battery overcharged

☐ Regulator/rectifier defective. Overcharging is noticed when battery gets excessively warm (Chapter 9).
☐ Battery defective. Replace battery with a new one (Chapter 9).
☐ Battery amperage too low, wrong type or size. Install manufacturer's specified amp-hour battery to handle charging load (Chapter 9).

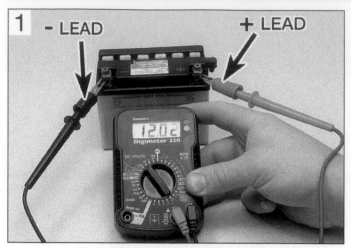

Measuring open-circuit battery voltage

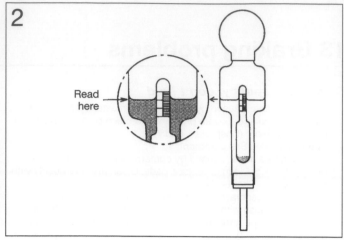

Float-type hydrometer for measuring battery specific gravity

Checking engine compression

● Low compression will result in exhaust smoke, heavy oil consumption, poor starting and poor performance. A compression test will provide useful information about an engine's condition and if performed regularly, can give warning of trouble before any other symptoms become apparent.
● A compression gauge will be required, along with an adapter to suit the spark plug hole thread size. Note that the screw-in type gauge/adapter set up is preferable to the rubber cone type.
● Before carrying out the test, first check the valve clearances as described in Chapter 1.

Checking battery open-circuit voltage

 Warning: The gases produced by the battery are explosive - never smoke or create any sparks in the vicinity of the battery. Never allow the electrolyte to contact your skin or clothing - if it does, wash it off and seek immediate medical attention.

● Before any electrical fault is investigated the battery should be checked.
● You'll need a dc voltmeter or multimeter to check battery voltage. Check that the leads are inserted in the correct terminals on the meter, red lead to positive (+), black lead to negative (-). Incorrect connections can damage the meter.
● A sound, fully-charged 12 volt battery

should produce between 12.3 and 12.6 volts across its terminals (12.8 volts for a maintenance-free battery). On machines with a 6 volt battery, voltage should be between 6.1 and 6.3 volts.
1 Set a multimeter to the 0 to 20 volts dc range and connect its probes across the battery terminals. Connect the meter's positive (+) probe, usually red, to the battery positive (+) terminal, followed by the meter's negative (-) probe, usually black, to the battery negative terminal (-) **(see illustration 1)**.
2 If battery voltage is low (below 10 volts on a 12 volt battery or below 4 volts on a six volt battery), charge the battery and test the voltage again. If the battery repeatedly goes flat, investigate the motorcycle's charging system.

Checking battery specific gravity (SG)

 Warning: The gases produced by the battery are explosive - never smoke or create any sparks in the vicinity of the battery. Never allow the electrolyte to contact your skin or clothing - if it does, wash it off and seek immediate medical attention.

● The specific gravity check gives an indication of a battery's state of charge.
● A hydrometer is used for measuring specific gravity. Make sure you purchase one which has a small enough hose to insert in the aperture of a motorcycle battery.
● Specific gravity is simply a measure of the electrolyte's density compared with that of water. Water has an SG of 1.000 and fully-charged battery electrolyte is about 26% heavier, at 1.260.

● Specific gravity checks are not possible on maintenance-free batteries. Testing the open-circuit voltage is the only means of determining their state of charge.
1 To measure SG, remove the battery from the motorcycle and remove the first cell cap. Draw some electrolyte into the hydrometer and note the reading **(see illustration 2)**. Return the electrolyte to the cell and install the cap.
2 The reading should be in the region of 1.260 to 1.280. If SG is below 1.200 the battery needs charging. Note that SG will vary with temperature; it should be measured at 20°C (68°F). Add 0.007 to the reading for every 10°C above 20°C, and subtract 0.007 from the reading for every 10°C below 20°C. Add 0.004 to the reading for every 10°F above 68°F, and subtract 0.004 from the reading for every 10°F below 68°F.
3 When the check is complete, rinse the hydrometer thoroughly with clean water.

Checking for continuity

● The term continuity describes the uninterrupted flow of electricity through an electrical circuit. A continuity check will determine whether an **open-circuit** situation exists.
● Continuity can be checked with an ohmmeter, multimeter, continuity tester or battery and bulb test circuit **(see illustrations 3, 4 and 5)**.
● All of these instruments are self-powered by a battery, therefore the checks are made with the ignition OFF.
● As a safety precaution, always disconnect the battery negative (-) lead before making checks, particularly if ignition switch checks are being made.
● If using a meter, select the appropriate

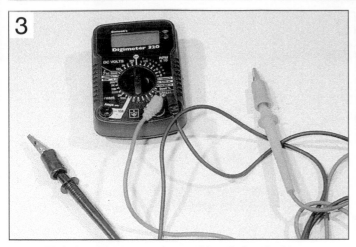

Digital multimeter can be used for all electrical tests

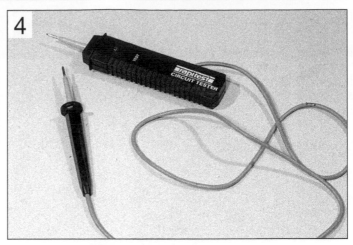

Battery-powered continuity tester

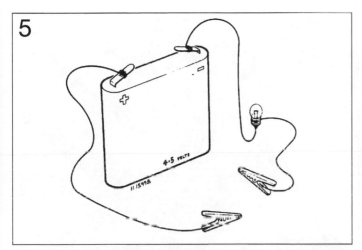

Battery and bulb test circuit

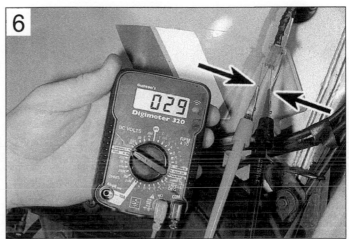

Continuity check of front brake light switch using a meter - note cotter pins used to access connector terminals

ohms scale and check that the meter reads infinity (∞). Touch the meter probes together and check that meter reads zero; where necessary adjust the meter so that it reads zero.

● After using a meter, always switch it OFF to conserve its battery.

Switch checks

1 If a switch is at fault, trace its wiring up to the wiring connectors. Separate the wire connectors and inspect them for security and condition. A build-up of dirt or corrosion here will most likely be the cause of the problem - clean up and apply a water dispersant such as WD40.

2 If using a test meter, set the meter to the ohms x 10 scale and connect its probes across the wires from the switch **(see illustration 6)**. Simple ON/OFF type switches, such as brake light switches, only have two wires whereas combination switches, like the ignition switch, have many internal links. Study the wiring diagram to ensure that you are connecting across the correct pair of wires. Continuity (low or no measurable resistance - 0 ohms) should be indicated with the switch ON and no continuity (high resistance) with it OFF.

3 Note that the polarity of the test probes doesn't matter for continuity checks, although care should be taken to follow specific test procedures if a diode or solid-state component is being checked.

4 A continuity tester or battery and bulb circuit can be used in the same way. Connect its probes as described above **(see illustration 7)**. The light should come on to indicate continuity in the ON switch position, but should extinguish in the OFF position.

Wiring checks

● Many electrical faults are caused by damaged wiring, often due to incorrect routing or chaffing on frame components.

● Loose, wet or corroded wire connectors

Continuity check of rear brake light switch using a continuity tester

can also be the cause of electrical problems, especially in exposed locations.

Continuity check of front brake light switch sub-harness

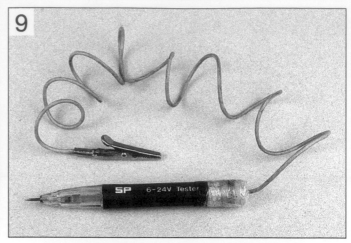

A simple test light can be used for voltage checks

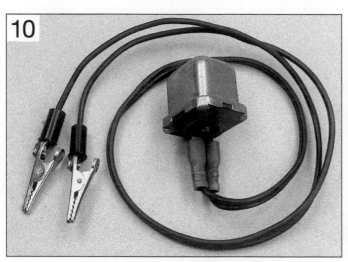

A buzzer is useful for voltage checks

Checking for voltage at the rear brake light power supply wire using a meter . . .

1 A continuity check can be made on a single length of wire by disconnecting it at each end and connecting a meter or continuity tester across both ends of the wire **(see illustration 8)**.

2 Continuity (low or no resistance - 0 ohms) should be indicated if the wire is good. If no continuity (high resistance) is shown, suspect a broken wire.

Checking for voltage

● A voltage check can determine whether current is reaching a component.

● Voltage can be checked with a dc voltmeter, multimeter set on the dc volts scale, test light or buzzer **(see illustrations 9 and 10)**. A meter has the advantage of being able to measure actual voltage.

● When using a meter, check that its leads are inserted in the correct terminals on the meter, red to positive (+), black to negative (-). Incorrect connections can damage the meter.

● A voltmeter (or multimeter set to the dc volts scale) should always be connected in parallel (across the load). Connecting it in series will destroy the meter.

● Voltage checks are made with the ignition ON.

1 First identify the relevant wiring circuit by referring to the wiring diagram at the end of this manual. If other electrical components share the same power supply (ie are fed from the same fuse), take note whether they are working correctly - this is useful information in deciding where to start checking the circuit.

2 If using a meter, check first that the meter leads are plugged into the correct terminals on the meter (see above). Set the meter to the dc volts function, at a range suitable for the battery voltage. Connect the meter red probe (+) to the power supply wire and the black probe to a good metal ground on the

motor-cycle's frame or directly to the battery negative (-) terminal **(see illustration 11)**. Battery voltage should be shown on the meter with the ignition switched ON.

3 If using a test light or buzzer, connect its positive (+) probe to the power supply terminal and its negative (-) probe to a good ground on the motorcycle's frame or directly to the battery negative (-) terminal **(see illustration 12)**. With the ignition ON, the test light should illuminate or the buzzer sound.

4 If no voltage is indicated, work back towards the fuse continuing to check for voltage. When you reach a point where there is voltage, you know the problem lies between that point and your last check point.

Checking the ground

● Ground connections are made either

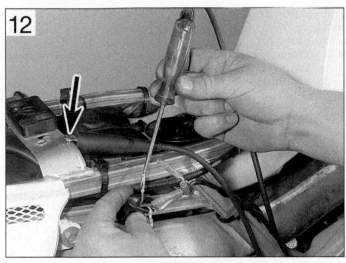

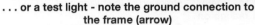

... or a test light - note the ground connection to the frame (arrow)

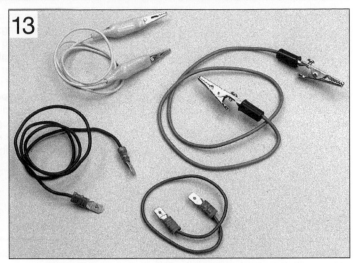

A selection of jumper wires for making ground checks

directly to the engine or frame (such as sensors, neutral switch etc. which only have a positive feed) or by a separate wire into the ground circuit of the wiring harness. Alternatively a short ground wire is sometimes run directly from the component to the motor-cycle's frame.

● Corrosion is often the cause of a poor ground connection.

● If total failure is experienced, check the security of the main ground lead from the negative (-) terminal of the battery and also the main ground point on the wiring harness. If corroded, dismantle the connection and clean all surfaces back to bare metal.

1 To check the ground on a component, use an insulated jumper wire to temporarily bypass its ground connection **(see illustration 13)**. Connect one end of the jumper wire between the ground terminal or metal body

of the component and the other end to the motorcycle's frame.

2 If the circuit works with the jumper wire installed, the original ground circuit is faulty. Check the wiring for open-circuits or poor connections. Clean up direct ground connections, removing all traces of corrosion and remake the joint. Apply petroleum jelly to the joint to prevent future corrosion.

Tracing a short-circuit

● A short-circuit occurs where current shorts to ground bypassing the circuit components. This usually results in a blown fuse.

● A short-circuit is most likely to occur where the insulation has worn through due to wiring chafing on a component, allowing a direct path to ground on the frame.

1 Remove any body panels necessary to access the circuit wiring.

2 Check that all electrical switches in the circuit are OFF, then remove the circuit fuse and connect a test light, buzzer or voltmeter (set to the dc scale) across the fuse terminals. No voltage should be shown.

3 Move the wiring from side to side while observing the test light or meter. When the test light comes on, buzzer sounds or meter shows voltage, you have found the cause of the short. It will usually shown up as damaged or burned insulation.

4 Note that the same test can be performed on each component in the circuit, even the switch.

Introduction

In less time than it takes to read this introduction, a thief could steal your motorcycle. Returning only to find your bike has gone is one of the worst feelings in the world. Even if the motorcycle is insured against theft, once you've got over the initial shock, you will have the inconvenience of dealing with the police and your insurance company.

The motorcycle is an easy target for the professional thief and the joyrider alike and the official figures on motorcycle theft make for depressing reading; on average a motorcycle is stolen every 16 minutes in the UK!

Motorcycle thefts fall into two categories, those stolen 'to order' and those taken by opportunists. The thief stealing to order will be on the look out for a specific make and model and will go to extraordinary lengths to obtain that motorcycle. The opportunist thief on the other hand will look for easy targets which can be stolen with the minimum of effort and risk.

Whilst it is never going to be possible to make your machine 100% secure, it is estimated that around half of all stolen motorcycles are taken by opportunist thieves. Remember that the opportunist thief is always on the look out for the easy option: if there are two similar motorcycles parked side-by-side, they will target the one with the lowest level of security. By taking a few precautions, you can reduce the chances of your motorcycle being stolen.

Security equipment

There are many specialised motorcycle security devices available and the following text summarises their applications and their good and bad points.

Once you have decided on the type of security equipment which best suits your needs, we recommended that you read one of the many equipment tests regularly carried out by the motorcycle press. These tests compare the products from all the major manufacturers and give impartial ratings on their effectiveness, value-for-money and ease of use.

No one item of security equipment can provide complete protection. It is highly recommended that two or more of the items described below are combined to increase the security of your motorcycle (a lock and chain plus an alarm system is just about ideal). The more security measures fitted to the bike, the less likely it is to be stolen.

Ensure the lock and chain you buy is of good quality and long enough to shackle your bike to a solid object

Lock and chain

Pros: *Very flexible to use; can be used to secure the motorcycle to almost any immovable object. On some locks and chains, the lock can be used on its own as a disc lock (see below).*

Cons: *Can be very heavy and awkward to carry on the motorcycle, although some types will be supplied with a carry bag which can be strapped to the pillion seat.*

● Heavy-duty chains and locks are an excellent security measure **(see illustration 1)**. Whenever the motorcycle is parked, use the lock and chain to secure the machine to a solid, immovable object such as a post or railings. This will prevent the machine from being ridden away or being lifted into the back of a van.

● When fitting the chain, always ensure the chain is routed around the motorcycle frame or swingarm **(see illustrations 2 and 3)**. Never merely pass the chain around one of the wheel rims; a thief may unbolt the wheel and lift the rest of the machine into a van, leaving you with just the wheel! Try to avoid having excess chain free, thus making it difficult to use cutting tools, and keep the chain and lock off the ground to prevent thieves attacking it with a cold chisel. Position the lock so that its lock barrel is facing downwards; this will make it harder for the thief to attack the lock mechanism.

Pass the chain through the bike's frame, rather than just through a wheel . . .

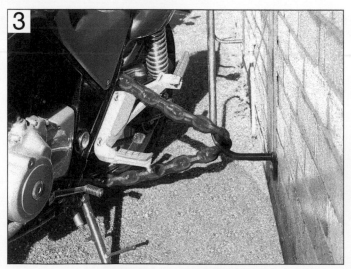

. . . and loop it around a solid object

U-locks

Pros: *Highly effective deterrent which can be used to secure the bike to a post or railings. Most U-locks come with a carrier which allows the lock to be easily carried on the bike.*

Cons: *Not as flexible to use as a lock and chain.*

● These are solid locks which are similar in use to a lock and chain. U-locks are lighter than a lock and chain but not so flexible to use. The length and shape of the lock shackle limit the objects to which the bike can be secured **(see illustration 4)**.

Disc locks

Pros: *Small, light and very easy to carry; most can be stored underneath the seat.*

Cons: *Does not prevent the motorcycle being lifted into a van. Can be very embarrassing if*

U-locks can be used to secure the bike to a solid object – ensure you purchase one which is long enough

you forget to remove the lock before attempting to ride off!

● Disc locks are designed to be attached to the front brake disc. The lock passes through one of the holes in the disc and prevents the wheel rotating by jamming against the fork/ brake caliper **(see illustration 5)**. Some are equipped with an alarm siren which sounds if the disc lock is moved; this not only acts as a theft deterrent but also as a handy reminder if you try to move the bike with the lock still fitted.

● Combining the disc lock with a length of cable which can be looped around a post or railings provides an additional measure of security **(see illustration 6)**.

Alarms and immobilisers

Pros: *Once installed it is completely hassle-free to use. If the system is 'Thatcham' or 'Sold Secure-approved', insurance companies may give you a discount.*

Cons: *Can be expensive to buy and complex to install. No system will prevent the motorcycle from being lifted into a van and taken away.*

● Electronic alarms and immobilisers are available to suit a variety of budgets. There are three different types of system available: pure alarms, pure immobilisers, and the more expensive systems which are combined alarm/immobilisers **(see illustration 7)**.
● An alarm system is designed to emit an audible warning if the motorcycle is being tampered with.
● An immobiliser prevents the motorcycle being started and ridden away by disabling its electrical systems.
● When purchasing an alarm/immobiliser system, check the cost of installing the system unless you are able to do it yourself. If the motorcycle is not used regularly, another consideration is the current drain of the system. All alarm/immobiliser systems are powered by the motorcycle's battery; purchasing a system with a very low current drain could prevent the battery losing its charge whilst the motorcycle is not being used.

A typical disc lock attached through one of the holes in the disc

A disc lock combined with a security cable provides additional protection

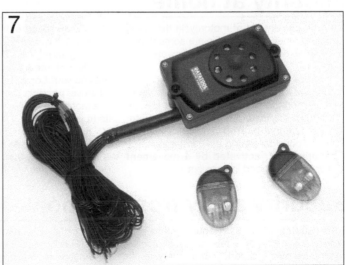

A typical alarm/immobiliser system

Indelible markings can be applied to most areas of the bike – always apply the manufacturer's sticker to warn off thieves

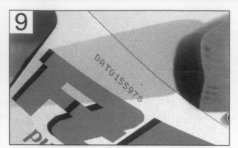

Chemically-etched code numbers can be applied to main body panels . . .

. . . again, always ensure that the kit manufacturer's sticker is applied in a prominent position

Security marking kits

Pros: *Very cheap and effective deterrent. Many insurance companies will give you a discount on your insurance premium if a recognised security marking kit is used on your motorcycle.*

Cons: *Does not prevent the motorcycle being stolen by joyriders.*

● There are many different types of security marking kits available. The idea is to mark as many parts of the motorcycle as possible with a unique security number **(see illustrations 8, 9 and 10)**. A form will be included with the kit to register your personal details and those of the motorcycle with the kit manufacturer. This register is made available to the police to help them trace the rightful owner of any motorcycle or components which they recover should all other forms of identification have been removed. Always apply the warning stickers provided with the kit to deter thieves.

Ground anchors, wheel clamps and security posts

Pros: *An excellent form of security which will deter all but the most determined of thieves.*

Cons: *Awkward to install and can be expensive.*

Permanent ground anchors provide an excellent level of security when the bike is at home

● Whilst the motorcycle is at home, it is a good idea to attach it securely to the floor or a solid wall, even if it is kept in a securely locked garage. Various types of ground anchors, security posts and wheel clamps are available for this purpose **(see illustration 11)**. These security devices are either bolted to a solid concrete or brick structure or can be cemented into the ground.

Security at home

A high percentage of motorcycle thefts are from the owner's home. Here are some things to consider whenever your motorcycle is at home:

✔ Where possible, always keep the motorcycle in a securely locked garage. Never rely solely on the standard lock on the garage door, these are usual hopelessly inadequate. Fit an additional locking mechanism to the door and consider having the garage alarmed. A security light, activated by a movement sensor, is also a good investment.

✔ Always secure the motorcycle to the ground or a wall, even if it is inside a securely locked garage.

✔ Do not regularly leave the motorcycle outside your home, try to keep it out of sight wherever possible. If a garage is not available, fit a motorcycle cover over the bike to disguise its true identity.

✔ It is not uncommon for thieves to follow a motorcyclist home to find out where the bike is kept. They will then return at a later date. Be aware of this whenever you are returning

home on your motorcycle. If you suspect you are being followed, do not return home, instead ride to a garage or shop and stop as a precaution.

✔ When selling a motorcycle, do not provide your home address or the location where the bike is normally kept. Arrange to meet the buyer at a location away from your home. Thieves have been known to pose as potential buyers to find out where motorcycles are kept and then return later to steal them.

Security away from the home

As well as fitting security equipment to your motorcycle here are a few general rules to follow whenever you park your motorcycle.

✔ Park in a busy, public place.

✔ Use car parks which incorporate security features, such as CCTV.

✔ At night, park in a well-lit area, preferably directly underneath a street light.

✔ Engage the steering lock.

✔ Secure the motorcycle to a solid, immovable object such as a post or railings with an additional lock. If this is not possible,

secure the bike to a friend's motorcycle. Some public parking places provide security loops for motorcycles.

✔ Never leave your helmet or luggage attached to the motorcycle. Take them with you at all times.

A

ABS (Anti-lock braking system) A system, usually electronically controlled, that senses incipient wheel lockup during braking and relieves hydraulic pressure at wheel which is about to skid.

Aftermarket Components suitable for the motorcycle, but not produced by the motorcycle manufacturer.

Allen key A hexagonal wrench which fits into a recessed hexagonal hole.

Alternating current (ac) Current produced by an alternator. Requires converting to direct current by a rectifier for charging purposes.

Alternator Converts mechanical energy from the engine into electrical energy to charge the battery and power the electrical system.

Ampere (amp) A unit of measurement for the flow of electrical current. Current = Volts ÷ Ohms.

Ampere-hour (Ah) Measure of battery capacity.

Angle-tightening A torque expressed in degrees. Often follows a conventional tightening torque for cylinder head or main bearing fasteners **(see illustration)**.

Angle-tightening cylinder head bolts

Antifreeze A substance (usually ethylene glycol) mixed with water, and added to the cooling system, to prevent freezing of the coolant in winter. Antifreeze also contains chemicals to inhibit corrosion and the formation of rust and other deposits that would tend to clog the radiator and coolant passages and reduce cooling efficiency.

Anti-dive System attached to the fork lower leg (slider) to prevent fork dive when braking hard.

Anti-seize compound A coating that reduces the risk of seizing on fasteners that are subjected to high temperatures, such as exhaust clamp bolts and nuts.

API American Petroleum Institute. A quality standard for 4-stroke motor oils.

Asbestos A natural fibrous mineral with great heat resistance, commonly used in the composition of brake friction materials. Asbestos is a health hazard and the dust created by brake systems should never be inhaled or ingested.

ATF Automatic Transmission Fluid. Often used in front forks.

ATU Automatic Timing Unit. Mechanical device for advancing the ignition timing on early engines.

ATV All Terrain Vehicle. Often called a Quad.

Axial play Side-to-side movement.

Axle A shaft on which a wheel revolves. Also known as a spindle.

B

Backlash The amount of movement between meshed components when one component is held still. Usually applies to gear teeth.

Ball bearing A bearing consisting of a hardened inner and outer race with hardened steel balls between the two races.

Bearings Used between two working surfaces to prevent wear of the components and a build-up of heat. Four types of bearing are commonly used on motorcycles: plain shell bearings, ball bearings, tapered roller bearings and needle roller bearings.

Bevel gears Used to turn the drive through 90°. Typical applications are shaft final drive and camshaft drive **(see illustration)**.

BHP Brake Horsepower. The British measure-ment for engine power output. Power output is now usually expressed in kilowatts (kW).

Bevel gears are used to turn the drive through 90°

Bias-belted tire Similar construction to radial tire, but with outer belt running at an angle to the wheel rim.

Big-end bearing The bearing in the end of the connecting rod that's attached to the crankshaft.

Bleeding The process of removing air from a hydraulic system via a bleed nipple or bleed screw.

Bottom-end A description of an engine's crankcase components and all components contained therein.

BTDC Before Top Dead Center in terms of piston position. Ignition timing is often expressed in terms of degrees or millimeters BTDC.

Bush A cylindrical metal or rubber component used between two moving parts.

Burr Rough edge left on a component after machining or as a result of excessive wear.

C

Cam chain The chain which takes drive from the crankshaft to the camshaft(s).

Canister The main component in an evap-orative emission control system (California market only); contains activated charcoal granules to trap vapors from the fuel system rather than allowing them to vent to the atmosphere.

Castellated Resembling the parapets along the top of a castle wall. For example, a castellated wheel axle or spindle nut.

Catalytic converter A device in the exhaust system of some machines which

Cush drive rubber segments dampen out transmission shocks

converts certain pollutants in the exhaust gases into less harmful substances.

Charging system Description of the components which charge the battery, ie the alternator, rectifer and regulator.

Clearance The amount of space between two parts. For example, between a piston and a cylinder, between a bearing and a journal, etc.

Coil spring A spiral of elastic steel found in various sizes throughout a vehicle, for example as a springing medium in the suspension and in the valve train.

Compression Reduction in volume, and increase in pressure and temperature, of a gas, caused by squeezing it into a smaller space.

Compression damping Controls the speed the suspension compresses when hitting a bump.

Compression ratio The relationship between cylinder volume when the piston is at top dead center and cylinder volume when the piston is at bottom dead center.

Continuity The uninterrupted path in the flow of electricity. Little or no measurable resistance.

Continuity tester Self-powered bleeper or test light which indicates continuity.

Cp Candlepower. Bulb rating commonly found on US motorcycles.

Crossply tire Tire plies arranged in a criss-cross pattern. Usually four or six plies used, hence 4PR or 6PR in tire size codes.

Cush drive Rubber damper segments fitted between the rear wheel and final drive sprocket to absorb transmission shocks **(see illustration)**.

D

Degree disc Calibrated disc for measuring piston position. Expressed in degrees.

Dial gauge Clock-type gauge with adapters for measuring runout and piston position. Expressed in mm or inches.

Diaphragm The rubber membrane in a master cylinder or carburetor which seals the upper chamber.

Diaphragm spring A single sprung plate often used in clutches.

Direct current (dc) Current produced by a dc generator.

Decarbonization The process of removing carbon deposits - typically from the combustion chamber, valves and exhaust port/system.

Detonation Destructive and damaging explosion of fuel/air mixture in combustion chamber instead of controlled burning.

Diode An electrical valve which only allows current to flow in one direction. Commonly used in rectifiers and starter interlock systems.

Disc valve (or rotary valve) An induction system used on some two-stroke engines.

Double-overhead camshaft (DOHC) An engine that uses two overhead camshafts, one for the intake valves and one for the exhaust valves.

Drivebelt A toothed belt used to transmit drive to the rear wheel on some motorcycles. A drivebelt has also been used to drive the camshafts. Drivebelts are usually made of Kevlar.

Driveshaft Any shaft used to transmit motion. Commonly used when referring to the final driveshaft on shaft drive motorcycles.

E

ECU (Electronic Control Unit) A computer which controls (for instance) an ignition system, or an anti-lock braking system.

EGO Exhaust Gas Oxygen sensor. Some-times called a Lambda sensor.

Electrolyte The fluid in a lead-acid battery.

EMS (Engine Management System) A computer controlled system which manages the fuel injection and the ignition systems in an integrated fashion.

Endfloat The amount of lengthways movement between two parts. As applied to a crankshaft, the distance that the crankshaft can move side-to-side in the crankcase.

Endless chain A chain having no joining link. Common use for cam chains and final drive chains.

EP (Extreme Pressure) Oil type used in locations where high loads are applied, such as between gear teeth.

Evaporative emission control system Describes a charcoal filled canister which stores fuel vapors from the tank rather than allowing them to vent to the atmosphere. Usually only fitted to California models and referred to as an EVAP system.

Expansion chamber Section of two-stroke engine exhaust system so designed to improve engine efficiency and boost power.

F

Feeler blade or gauge A thin strip or blade of hardened steel, ground to an exact thickness, used to check or measure clearances between parts.

Final drive Description of the drive from the transmission to the rear wheel. Usually by chain or shaft, but sometimes by belt.

Firing order The order in which the engine cylinders fire, or deliver their power strokes, beginning with the number one cylinder.

Flooding Term used to describe a high fuel level in the carburetor float chambers, leading to fuel overflow. Also refers to excess fuel in the combustion chamber due to incorrect starting technique.

Free length The no-load state of a component when measured. Clutch, valve and fork spring lengths are measured at rest, without any preload.

Freeplay The amount of travel before any action takes place. The looseness in a linkage, or an assembly of parts, between the initial application of force and actual movement. For example, the distance the rear brake pedal moves before the rear brake is actuated.

Fuel injection The fuel/air mixture is metered electronically and directed into the engine intake ports (indirect injection) or into the cylinders (direct injection). Sensors supply information on engine speed and conditions.

Fuel/air mixture The charge of fuel and air going into the engine. See Stoichiometric ratio.

Fuse An electrical device which protects a circuit against accidental overload. The typical fuse contains a soft piece of metal which is calibrated to melt at a predetermined current flow (expressed as amps) and break the circuit.

G

Gap The distance the spark must travel in jumping from the center electrode to the side electrode in a spark plug. Also refers to the distance between the ignition rotor and the pickup coil in an electronic ignition system.

Gasket Any thin, soft material - usually cork, cardboard, asbestos or soft metal - installed between two metal surfaces to ensure a good seal. For instance, the cylinder head gasket seals the joint between the block and the cylinder head.

Gauge An instrument panel display used to monitor engine conditions. A gauge with a movable pointer on a dial or a fixed scale is an analog gauge. A gauge with a numerical readout is called a digital gauge.

Gear ratios The drive ratio of a pair of gears in a gearbox, calculated on their number of teeth.

Glaze-busting see **Honing**

Grinding Process for renovating the valve face and valve seat contact area in the cylinder head.

Ground return The return path of an electrical circuit, utilizing the motorcycle's frame.

Gudgeon pin The shaft which connects the connecting rod small-end with the piston. Often called a piston pin or wrist pin.

H

Helical gears Gear teeth are slightly curved and produce less gear noise that straight-cut gears. Often used for primary drives.

Helicoil A thread insert repair system. Commonly used as a repair for stripped spark plug threads **(see illustration)**.

Installing a Helicoil thread insert in a cylinder head

Honing A process used to break down the glaze on a cylinder bore (also called glaze-busting). Can also be carried out to roughen a rebored cylinder to aid ring bedding-in.

HT (High Tension) Description of the electrical circuit from the secondary winding of the ignition coil to the spark plug.

Hydraulic A liquid filled system used to transmit pressure from one component to another. Common uses on motorcycles are brakes and clutches.

Hydrometer An instrument for measuring the specific gravity of a lead-acid battery.

Hygroscopic Water absorbing. In motorcycle applications, braking efficiency will be reduced if DOT 3 or 4 hydraulic fluid absorbs water from the air - care must be taken to keep new brake fluid in tightly sealed containers.

I

lbf ft Pounds-force feet. A unit of torque. Sometimes written as ft-lbs.

lbf in Pound-force inch. A unit of torque, applied to components where a very low torque is required. Sometimes written as inch-lbs.

IC Abbreviation for Integrated Circuit.

Ignition advance Means of increasing the timing of the spark at higher engine speeds. Done by mechanical means (ATU) on early engines or electronically by the ignition control unit on later engines.

Ignition timing The moment at which the spark plug fires, expressed in the number of crankshaft degrees before the piston reaches the top of its stroke, or in the number of millimeters before the piston reaches the top of its stroke.

Infinity (∞) Description of an open-circuit electrical state, where no continuity exists.

Inverted forks (upside down forks) The sliders or lower legs are held in the yokes and the fork tubes or stanchions are connected to the wheel axle (spindle). Less unsprung weight and stiffer construction than conventional forks.

J

JASO Japan Automobile Standards Organization. JASO MA is a standard for motorcycle oil equivalent to API SJ, but designed to prevent problems with wet-type motorcycle clutches.

Joule The unit of electrical energy.

Journal The bearing surface of a shaft.

K

Kickstart Mechanical means of turning the engine over for starting purposes.

Only usually fitted to mopeds, small capacity motorcycles and off-road motorcycles.

Kill switch Handebar-mounted switch for emergency ignition cut-out. Cuts the ignition circuit on all models, and additionally prevent starter motor operation on others.

km Symbol for kilometer.

kmh Abbreviation for kilometers per hour.

L

Lambda sensor A sensor fitted in the exhaust system to measure the exhaust gas oxygen content (excess air factor). Also called oxygen sensor.

Lapping see **Grinding**.

LCD Abbreviation for Liquid Crystal Display.

LED Abbreviation for Light Emitting Diode.

Liner A steel cylinder liner inserted in an aluminum alloy cylinder block.

Locknut A nut used to lock an adjustment nut, or other threaded component, in place.

Lockstops The lugs on the lower triple clamp (yoke) which abut those on the frame, preventing handlebar-to-fuel tank contact.

Lockwasher A form of washer designed to prevent an attaching nut from working loose.

LT Low Tension Description of the electrical circuit from the power supply to the primary winding of the ignition coil.

M

Main bearings The bearings between the crankshaft and crankcase.

Maintenance-free (MF) battery A sealed battery which cannot be topped up.

Manometer Mercury-filled calibrated tubes used to measure intake tract vacuum. Used to synchronize carburetors on multi-cylinder engines.

Tappet shims are measured with a micrometer

Micrometer A precision measuring instru-ment that measures component outside diameters **(see illustration)**.

MON (Motor Octane Number) A measure of a fuel's resistance to knock.

Monograde oil An oil with a single viscosity, eg SAE80W.

Monoshock A single suspension unit linking the swingarm or suspension linkage to the frame.

mph Abbreviation for miles per hour.

Multigrade oil Having a wide viscosity range (eg 10W40). The W stands for Winter, thus the viscosity ranges from SAE10 when cold to SAE40 when hot.

Multimeter An electrical test instrument with the capability to measure voltage, current and resistance. Some meters also incorporate a continuity tester and buzzer.

N

Needle roller bearing Inner race of caged needle rollers and hardened outer race. Examples of uncaged needle rollers can be found on some engines. Commonly used in rear suspension applications and in two-stroke engines.

Nm Newton meters.

NOx Oxides of Nitrogen. A common toxic pollutant emitted by gasoline engines at higher temperatures.

O

Octane The measure of a fuel's resistance to knock.

OE (Original Equipment) Relates to components fitted to a motorcycle as standard or replacement parts supplied by the motorcycle manufacturer.

Ohm The unit of electrical resistance. Ohms = Volts 4 Current.

Ohmmeter An instrument for measuring electrical resistance.

Oil cooler System for diverting engine oil outside of the engine to a radiator for cooling purposes.

Oil injection A system of two-stroke engine lubrication where oil is pump-fed to the engine in accordance with throttle position.

Open-circuit An electrical condition where there is a break in the flow of electricity - no continuity (high resistance).

O-ring A type of sealing ring made of a special rubber-like material; in use, the O-ring is compressed into a groove to provide the sealing action.

Oversize (OS) Term used for piston and ring size options fitted to a rebored cylinder.

Overhead cam (sohc) engine An engine with single camshaft located on top of the cylinder head.

Overhead valve (ohv) engine An engine with the valves located in the cylinder head, but with the camshaft located in the engine block or crankcase.

Oxygen sensor A device installed in the exhaust system which senses the oxygen content in the exhaust and converts this information into an electric current. Also called a Lambda sensor.

P

Plastigage A thin strip of plastic thread, available in different sizes, used for measuring clearances. For example, a strip of Plastigage is laid across a bearing journal. The parts are assembled and dismantled; the width of the crushed strip indicates the clearance between journal and bearing.

Polarity Either negative or positive ground, determined by which battery lead is connected to the frame (ground return). Modern motorcycles are usually negative ground.

Pre-ignition A situation where the fuel/air mixture ignites before the spark plug fires. Often due to a hot spot in the combustion chamber caused by carbon build-up. Engine has a tendency to 'run-on'.

Pre-load (suspension) The amount a spring is compressed when in the unloaded state. Preload can be applied by gas, spacer or mechanical adjuster.

Premix The method of engine lubrication on some gasoline two-stroke engines. Engine oil is mixed with the gasoline in the fuel tank in a specific ratio. The fuel/oil mix is sometimes referred to as "petrol".

Primary drive Description of the drive from the crankshaft to the clutch. Usually by gear or chain.

PS Pferdestärke - a German interpretation of BHP.

PSI Pounds-force per square inch. Imperial measurement of tire pressure and cylinder pressure measurement.

PTFE Polytetrafluroethylene. A low friction substance.

Pulse secondary air injection system A process of promoting the burning of excess fuel present in the exhaust gases by routing fresh air into the exhaust ports.

Q

Quartz halogen bulb Tungsten filament surrounded by a halogen gas. Typically used for the headlight **(see illustration)**.

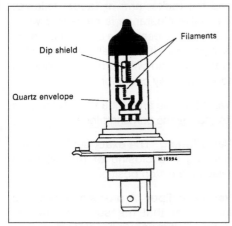

Quartz halogen headlight bulb construction

R

Rack-and-pinion A pinion gear on the end of a shaft that mates with a rack (think of a geared wheel opened up and laid flat). Sometimes used in clutch operating systems.

Radial play Up and down movement about a shaft.

Radial ply tires Tire plies run across the tire (from bead to bead) and around the circumference of the tire. Less resistant to tread distortion than other tire types.

Radiator A liquid-to-air heat transfer device designed to reduce the temperature of the coolant in a liquid cooled engine.

Rake A feature of steering geometry - the angle of the steering head in relation to the vertical **(see illustration)**.

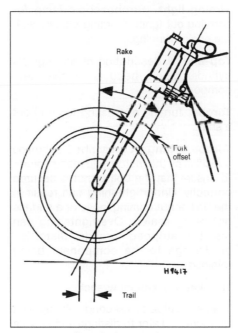

Steering geometry

Rebore Providing a new working surface to the cylinder bore by boring out the old surface. Necessitates the use of oversize piston and rings.

Rebound damping A means of controlling the oscillation of a suspension unit spring after it has been compressed. Resists the spring's natural tendency to bounce back after being compressed.

Rectifier Device for converting the ac output of an alternator into dc for battery charging.

Reed valve An induction system commonly used on two-stroke engines.

Regulator Device for maintaining the charging voltage from the generator or alternator within a specified range.

Relay A electrical device used to switch heavy current on and off by using a low current auxiliary circuit.

Resistance Measured in ohms. An electrical component's ability to pass electrical current.

RON (Research Octane Number) A measure of a fuel's resistance to knock.

rpm revolutions per minute.

Runout The amount of wobble (in-and-out movement) of a wheel or shaft as it's rotated. The amount a shaft rotates "out-of-true." The out-of-round condition of a rotating part.

S

SAE (Society of Automotive Engineers) A standard for the viscosity of a fluid.

Sealant A liquid or paste used to prevent leakage at a joint. Sometimes used in conjunction with a gasket.

Service limit Term for the point where a component is no longer useable and must be replaced.

Shaft drive A method of transmitting drive from the transmission to the rear wheel.

Shell bearings Plain bearings consisting of two shell halves. Most often used as big-end and main bearings in a four-stroke engine. Often called bearing inserts.

Shim Thin spacer, commonly used to adjust the clearance or relative positions between two parts. For example, shims inserted into or under tappets or followers to control valve clearances. Clearance is adjusted by changing the thickness of the shim.

Short-circuit An electrical condition where current shorts to ground bypassing the circuit components.

Skimming Process to correct warpage or repair a damaged surface, eg on brake discs or drums.

Slide-hammer A special puller that screws into or hooks onto a component such as a shaft or bearing; a heavy sliding handle on the shaft bottoms against the end of the shaft to knock the component free.

Small-end bearing The bearing in the upper end of the connecting rod at its joint with the gudgeon pin.

Snap-ring A ring-shaped clip used to prevent endwise movement of cylindrical parts and shafts. An internal snap-ring is installed in a groove in a housing; an external snap-ring fits into a groove on the outside of a cylindrical piece such as a shaft. Also known as a circlip.

Spalling Damage to camshaft lobes or bearing journals shown as pitting of the working surface.

Specific gravity (SG) The state of charge of the electrolyte in a lead-acid battery. A measure of the electrolyte's density compared with water.

Straight-cut gears Common type gear used on gearbox shafts and for oil pump and water pump drives.

Stanchion The inner sliding part of the front forks, held by the yokes. Often called a fork tube.

Stoichiometric ratio The optimum chemical air/fuel ratio for a gasoline engine, said to be 14.7 parts of air to 1 part of fuel.

Sulphuric acid The liquid (electrolyte) used in a lead-acid battery. Poisonous and extremely corrosive.

Surface grinding (lapping) Process to correct a warped gasket face, commonly used on cylinder heads.

T

Tapered-roller bearing Tapered inner race of caged needle rollers and separate tapered outer race. Examples of taper roller bearings can be found on steering heads.

Tappet A cylindrical component which transmits motion from the cam to the valve stem, either directly or via a pushrod and rocker arm. Also called a cam follower.

TCS Traction Control System. An electron-ically-controlled system which senses wheel spin and reduces engine speed accordingly.

TDC Top Dead Center denotes that the piston is at its highest point in the cylinder.

Thread-locking compound Solution applied to fastener threads to prevent loosening. Select type to suit application.

Thrust washer A washer positioned between two moving components on a shaft. For example, between gear pinions on gearshaft.

Timing chain See **Cam Chain**.

Timing light Stroboscopic lamp for carrying out ignition timing checks with the engine running.

Top-end A description of an engine's cylinder block, head and valve gear components.

Torque Turning or twisting force about a shaft.

Torque setting A prescribed tightness specified by the motorcycle manufacturer to ensure that the bolt or nut is secured correctly. Undertightening can result in the bolt or nut coming loose or a surface not being sealed. Overtightening can result in stripped threads, distortion or damage to the component being retained.

Torx key A six-point wrench.

Tracer A stripe of a second color applied to a wire insulator to distinguish that wire from another one with the same color insulator. For example, Br/W is often used to denote a brown insulator with a white tracer.

Trail A feature of steering geometry. Distance from the steering head axis to the tire's central contact point.

Triple clamps The cast components which extend from the steering head and support the fork stanchions or tubes. Often called fork yokes.

Turbocharger A centrifugal device, driven by exhaust gases, that pressurizes the intake air. Normally used to increase the power output from a given engine displacement.

TWI Abbreviation for Tire Wear Indicator. Indicates the location of the tread depth indicator bars on tires.

U

Universal joint or U-joint (UJ) A double-pivoted connection for transmitting power from a driving to a driven shaft through an angle. Typically found in shaft drive assemblies.

Unsprung weight Anything not supported by the bike's suspension (ie the wheel, tires, brakes, final drive and bottom (moving) part of the suspension).

V

Vacuum gauges Clock-type gauges for measuring intake tract vacuum. Used for carburetor synchronization on multi-cylinder engines.

Valve A device through which the flow of liquid, gas or vacuum may be stopped, started or regulated by a moveable part that opens, shuts or partially obstructs one or more ports or passageways. The intake and exhaust valves in the cylinder head are of the poppet type.

Valve clearance The clearance between the valve tip (the end of the valve stem) and the rocker arm or tappet/follower. The valve clearance is measured when the valve is closed. The correct clearance is important - if too small the valve won't close fully and will burn out, whereas if too large noisy operation will result.

Valve lift The amount a valve is lifted off its seat by the camshaft lobe.

Valve timing The exact setting for the opening and closing of the valves in relation to piston position.

Vernier caliper A precision measuring instrument that measures inside and outside dimensions. Not quite as accurate as a micrometer, but more convenient.

VIN Vehicle Identification Number. Term for the bike's engine and frame numbers.

Viscosity The thickness of a liquid or its resistance to flow.

Volt A unit for expressing electrical "pressure" in a circuit. Volts = current x ohms.

W

Water pump A mechanically-driven device for moving coolant around the engine.

Watt A unit for expressing electrical power. Watts = volts x current.

Wet liner arrangement

Wear limit see **Service limit**

Wet liner A liquid-cooled engine design where the pistons run in liners which are directly surrounded by coolant **(see illustration)**.

Wheelbase Distance from the center of the front wheel to the center of the rear wheel.

Wiring harness or loom Describes the electrical wires running the length of the motorcycle and enclosed in tape or plastic sheathing. Wiring coming off the main harness is usually referred to as a sub harness.

Woodruff key A key of semi-circular or square section used to locate a gear to a shaft. Often used to locate the alternator rotor on the crankshaft.

Wrist pin Another name for gudgeon or piston pin.

Notes

Note: *References throughout this index are in the form, "Chapter number"•"Page number"*

Notes